The Pricing of Crude Oil

Taki Rifaï

The Praeger Special Studies program—utilizing the most modern and efficient book production techniques and a selective worldwide distribution network—makes available to the academic, government, and business communities significant, timely research in U.S. and international economic, social, and political development.

The Pricing of Crude Oil
Economic and Strategic Guidelines for an International Energy Policy
expanded and updated edition

Praeger Publishers   New York   Washington   London

PRAEGER SPECIAL STUDIES IN INTERNATIONAL ECONOMICS AND DEVELOPMENT

Library of Congress Cataloging in Publication Data

Rifai, Taki.
   The pricing of crude oil.

   (Praeger special studies in international economics and development)
   Bibliography: p.
   1. Petroleum industry and trade. 2. Petroleum products—Prices. I. Title.
HD9560.6.R53   1975   338.2'3   75-19279
ISBN 0-275-01510-6

PRAEGER PUBLISHERS
111 Fourth Avenue, New York, N.Y. 10003, U.S.A.

Published in the United States of America in 1975
by Praeger Publishers, Inc.

All rights reserved

© 1975 by Praeger Publishers, Inc.

Printed in the United States of America

To
Lili, Kiki, and Sami

FOREWORD

In the early days of July 1969, the Organization of Petroleum Exporting Countries (OPEC) held its first seminar on petroleum economics in Vienna. In one of the discussion sessions, the chairman asked the author to comment on a paper dealing with the comparative evaluation of crude oils. These comments were made within the general framework of concepts and ideas that can be found in Chapter 11 of this book. In his conclusion, the author warned the audience against confusing relative values of crude oils with their prices and against the temptation of utilizing the former to explain or to control the latter. He emphasized the complexity and the sensitivity of the whole issue and urged that its manipulation and utilization be confined to experienced specialists.

However, in response to this last statement, a distinguished member of the audience, the governor to OPEC of a major Middle Eastern oil-exporting country, complained that policy-makers and responsible officials were most unhappy with economists who failed to provide them with a comprehensive understanding of the mechanism of price formation and evolution. This accusation was absolutely fair and justified. The confusion and complexity of the crude pricing issue has always been an important feature of the international petroleum industry, and the few contributions that have attempted to investigate it have been successful only in emphasizing the contradictions, mysteries, and obscurantism surrounding the subject.

This book is an attempt to take up the challenge made to petroleum economists. Its intention was primarily to investigate crude prices from the viewpoint of producing countries and for their own use. However, the petroleum industry is a very complex one, and there is a limit to any simplification and popularization effort intended to put its basic features within reach of the average reader from producing countries. A descriptive approach through detailed illustration can apply only to the preliminary effort of setting the over-all framework for the investigation, to defining the major problems to be tackled, and to outlining their interdependence. Thorough investigation and analysis of the mechanism of price formation and evolution can proceed only through a systematic and scientific approach, and the mechanism's built-in complexity can be penetrated and overcome only if the phenomena involved are satisfactorily represented and simulated by adequate economic and mathematical models.

This book is essentially the result of work by a single person and represents the fruits of several years of meditation and active research based on the author's direct involvement in the current world oil crisis. It was mainly written throughout 1970 and 1971, at a time when the international petroleum industry was undergoing a profound over-all mutation. It is very difficult indeed to record events while history is actually in motion. Some of the ideas and forecasts elaborated in the early months of the writing were quite judicious and justified at the time of their drafting but then appeared outmoded sometime later in the light of new facts and events. However, the basic philosophy and the early diagnosis of the fundamental pricing problems were formulated before the 1971 Teheran and Tripoli Agreements, and it was comforting to note that not only were they not contradicted by these eventful episodes for the industry but also that they help illuminate and explain the real motivations and implications of the recent metamorphosis of the industry and help in foreseeing its future evolution.

Although the author acted as the economic advisor to the Libyan Ministry of Petroleum for three years (1969-71), the views and ideas expressed in this book are exclusively and entirely personal and do not necessarily represent in any way any official opinion or commitment. However, the author is most grateful to Libyan authorities for their friendly and cooperative attitude and for having enabled him to become involved fully in the historic 1970-71 negotiations, which changed the face and future of the international petroleum industry.

It should be noted that the studies and investigations embodied in this book are exclusively based on published information and data generally available to the public. Sources are cited whenever necessary so that no controversies need arise in our investigation of the highly complex and sensitive issues of crude oil pricing.

The appellation of the world's richest oil region is a controversy in itself; the area will be referred to in this book as the "Gulf" or the "Gulf area." Special attention will be paid in order to avoid any confusion with the Gulf of Mexico (the U.S. Gulf), which is traditionally called the "Gulf" in specialized literature.

Although the author has worked at the Institut Francais du Pétrole for over eight years, during which his ideas about crude oil prices have developed and matured, the views and opinions expressed in this book are strictly personal and do not necessarily reflect in any way the formal thinking of the Institut or any of its staff. Meanwhile, the author is deeply grateful and indebted to the Institut for material assistance in drafting and editing the manuscript, without which the publication of this book would have been seriously delayed.

CONTENTS

	Page
FOREWORD	vi
LIST OF TABLES	xiii
LIST OF FIGURES	xv
ABBREVIATIONS AND SPECIAL USAGES	xvii

PART I: AN INFORMAL APPROACH TO CRUDE OIL PRICES

Chapter

1	WHY AN INFORMAL APPROACH TO CRUDE OIL PRICES?	1
2	THE NATURE AND SCOPE OF THE PROBLEM	7
	Notes	11
3	THE PSYCHOLOGICAL DIMENSIONS OF THE PRICING ISSUE	12
	Notes	19
4	WHY POSTED PRICES?	20
	Crude Oil Prices and Producing Countries	21
	Crude Oil Prices and Oil Companies	27
	Crude Oil Prices, Consuming Countries, and Consumers	33
	Notes	37

Chapter		Page
	PART II: A TECHNOCRATIC APPROACH TO CRUDE OIL PRICES	
5	THE PRICING ISSUE: BACKGROUND AND ATMOSPHERE	43
6	PRICING PATTERNS IN THE GULF AREA	47
	1944-49: The ECA Era	48
	Pricing Patterns in the Gulf Area	52
	Essentials of the Pricing Structure	62
	Notes	66
7	PRICING PATTERNS OUTSIDE THE GULF AREA	67
	The Philosophy of Over-All Pricing Patterns	68
	The Economics of Tanker Transportation	70
	The Impact of Gravity on Transportation Costs	73
	Looking for an Over-All Pricing Pattern	76
	Notes	80
8	ANALYSIS OF PRICING PRACTICES OUTSIDE THE GULF AREA	83
	The Eastern Mediterranean Postings	85
	Looking for an Over-All Pricing Pattern	87
	The Pricing of Libyan Crude Oils	89
	The Pricing of Algerian Crude Oils	99
	The Pricing of Nigerian Crude Oils	106
	Notes	109
9	SULFUR DIFFERENTIAL	111
	Sulfur Compounds in Crude Oils	112
	Sulfur Differential: Refining	115
	Sulfur Differential: Pollution	118
	Notes	123
10	FREIGHT DIFFERENTIAL	124
	Consequences of a Major Crisis	125
	Freight Differential: The Suez Allowance, 1967	129
	The Economics of Super-Tankers	134
	Notes	141

Chapter		Page
11	RELATIVE VALUES OF CRUDE OILS	143
	Prices Versus Relative Values	145
	Relative Values	150
	Crude Oil Supplies	151
	Market Structure	151
	Refining Models	156
	Refining Economics	159
	Criteria for Comparative Evaluation	160
	Notes	161

PART III: A STRATEGIC APPROACH TO CRUDE OIL PRICES

Chapter		Page
12	THE GENESIS OF PRICING PATTERNS	165
	Pricing Economics in Western Europe in the Early 1950s	165
	Background to Pricing Patterns in the Eastern Hemisphere	166
	A European Look at Crude Oil Prices	170
	Prices of Petroleum Products in Western Europe in the Early 1950s	171
	The U.S. Domestic Oil Industry	176
	The Role of Major Oil Companies	180
	Notes	187
13	A COMPREHENSIVE EXPLANATION OF PRICING GENESIS	188
	Politics and Economics of Pricing in the U.S. Domestic Market	189
	U.S. Impact on Price Formation in the Eastern Hemisphere	194
	Intercartel Conflicts: Dollar Oil Versus Sterling Oil	199
	The Majors' Strategic Objectives and Products' Pricing Patterns	205
	Notes	208

Chapter		Page
14	GRAVITY DIFFERENTIAL AND INTEGRATED ECONOMICS	210
	The Real Significance of Gravity	211
	The Economics of Integrated Oil	217
	The Significance of Gravity Differential	220
	Notes	226
15	A TENTATIVE EXPLANATION OF PRICING PATTERNS	234
	Integrated Economics at the Service of the Majors	234
	Pricing of Libyan High-Pour and Nigerian Crude Oils	238
	Notes	246

PART IV: LOOKING INTO THE FUTURE

16	CRISIS IN MOTION	249
	1970: The Libyan Year	251
	Assessment and Consequences of the Libyan Settlement	258
	February 15, 1971: The Teheran Agreement	263
	March 20, 1971: The Tripoli Agreement	270
	Further Worldwide Price Settlements	276
	The Gulf Area	276
	Nigerian Crude Oil	278
	Eastern Mediterranean Crude Oil	279
	Venezuelan Oil	283
	The Algerian Case	288
	Oil Transit Countries	294
	Assessment and Consequences of the Crisis	296
	Notes	303
17	FUTURE OUTLOOK	305
	The Ambiguities of a Most Peculiar Crisis	305
	The U.S. Energy Crisis and World Oil Economics	310
	Future Outlook	323
	Notes	335

Chapter		Page
18	POSTSCRIPT: THE UNCERTAINTIES OF THE FUTURE	338
	1973: The Vicissitudes of the Crisis	338
	Transitory Period of Mutation	338
	Immediate Consequences of the War	340
	The "Oil Weapon"	341
	First Wave of Increase, October 16, 1974	343
	OPEC's Redefinition of the Oil Price Concept	348
	New Price Structure, December 23, 1973	349
	1974: The Shaping of a New World	356
	The Evolution of Crude Oil Prices	357
	The Evolution of Supply and Demand Patterns	368
	Toward a New International Equilibrium	375
	Notes	380

Appendix

A	THE ESSO MEMORANDUM	383
B	MAIN FEATURES OF THE PARTICIPATION AGREEMENT	391

SELECTED BIBLIOGRAPHY	395
ABOUT THE AUTHOR	401

LIST OF TABLES

Table		Page
4.1	Cost and Percentage Breakdown of OPEC's "Consumer Barrel"	35
6.1	Evolution of Crude Oil Prices in Postwar Period up to 1953	50
6.2	Crude Oils in Gulf Area, Parity Prices FOB Quoin Island	54
6.3	Posted Prices of Selected Venezuelan Crude Oils, 1969-70	63
7.1	Conversion Factors Versus Gravity	74
7.2	Basic Gravity Differentials Governing CIF Prices	76
8.1	Eastern Hemisphere Crude Oils, Posted Prices Outside Gulf Area	84
8.2	Comparative Evolution of Posted Prices, 36-36.9° Light Arabian, FOB Ras Tanura and FOB Sidon	88
9.1	Maximum Sulfur Content of "Low-Sulfur" Oils	113
10.1	1967 Compromise Agreement, Suez Allowance Against Elimination of Royalty-Expensing Allowances	135
10.2	Evolution of Size Structure of World Tanker Fleet	138
10.3	Approximate Construction Cost of 250,000-Tanker	140
12.1	EEC Consumption of Primary Energy	168
12.2	Domination of U.S. Domestic Oil Industry by Majors in Late 1930s	182
13.1	Refining and Cracking Capacity, July 1, 1956	205
14.1	Eastern Hemisphere Crude Oils	215

Table		Page
15.1	Main Characteristics of Selected Crude Oils	242
15.2	Additional Costs of Disposing of High-Pour Crude Oils	244
16.1	Price Adjustments Under Teheran Conditions for Crude Oils Not Included in Teheran Agreement	277
16.2	Evolution of Iraqi Posted Prices and Oil Income	282
16.3	Selected New Venezuelan Tax Reference Prices	285
16.4	Comparative Net Earnings of U.S. Oil Companies	302
17.1	Estimated Production and Revenue	326
18.1	New Structure of Posted Prices in the Gulf Decided in Kuwait on October 16, 1973	346
18.2	Posted Prices of Main Gulf Crudes as at January 1, 1974	351
18.3	Evolution of Main Gulf Crude Postings, 1971-74	352
18.4	Evolution of the Government Unit Income per Barrel for Main Gulf Crudes, 1971-74	353
18.5	Evolution of the "Total Cost Including Taxes" of the Main Gulf Crudes, 1971-74	354
18.6	Changes in Prices and Costs of Light Arabian Crude Oil	365

LIST OF FIGURES

Figure		Page
6.1-A	Posted Prices in Gulf Area, Parity Prices FOB Quoin Island	57
6.1-B	Posted Prices in Gulf Area, Linear Price Scale FOB Quoin Island	57
6.2	Posted Prices in Gulf Area, Parity Prices FOB Quoin Island	59
6.3	Posted Prices of Venezuelan Crude Oils, Linear Price Scales	65
6.4	Linear Price Scales of Gulf and Venezuelan Crude Oils	65
7.1	Conversion Factor Versus Gravity	75
7.2	Eastern Hemisphere Crude Oils, Average Freight Differentials with Respect to Western Europe and Quoin Island	78
8.1	Crude Oil Postings Outside Gulf Area, Comparative Positions of Parity Prices FOB Quoin Island and Gulf Linear Price Scales at Different Freight Rates	90
8.2	Actual Postings in Eastern Hemisphere, Deviation of Parity Prices from Prevailing Pricing Patterns FOB Quoin Island	91
8.3	Impact on Posted Prices of Criteria for Averaging Freight Rates	91
10.1	Tanker Freight Rates, 1967-71	127
12.1	Interregional Trade in Crude Oil and Petroleum Products in 1952	169
12.2	Impact of Geography on Refining Economics	173
14.1	Comparative TBP Curves of Selected Crude Oils	213

Figure		Page
14.2	TBP Volume Yields of Major Bulk Products Versus Gravity of Crude Oils	216
14.3	Illustration of Pricing Model	222
15.1	Crude Oil Value Versus Gravity for Various Types of Crude Oils, Based on Gulf Coast Cargo Product Prices of June 23, 1954	238
16.1	Price Structure of Venezuelan Oil	287
16.2	Linear Price Scales in Gulf Area at World-scale 72 Posted Prices as Per FOB Quoin Island	299

ABBREVIATIONS AND SPECIAL USAGES

AFRA	Average Freight Rate Assessment
AIOC	Anglo-Iranian Oil Company
API	American Petroleum Institute (oil gravity measure in API degrees)
Aramco	Arabian American Oil Company
ATRS	American Tanker Rate Schedule
Bbl	Barrel
BGD	Basic gravity differential
BP	British Petroleum
CFP	Companie Française des Pétroles
CIF	Cost, insurance, and freight
dwt	Deadweight ton
ECA	Economic Cooperation Administration
EEC	European Economic Community
FOB	Freight on board
GFD	Gravity fluctuation differential
Gulf	Gulf Oil Corporation
Gulf, Gulf area	Arab/Persian Gulf
Independents	Smaller companies usually not fully integrated
Intascale	International Tanker Nominal Freight Scale
IPC	Iraq Petroleum Company
Majors	Large integrated international oil companies, mostly Standard Oil (N.J.), Shell, BP, SOCAL, Texaco, Gulf, Mobil, and CFP
MEES	Middle East Economic Survey
MER	Maximum efficient rate (of production)
OGJ	Oil and Gas Journal
OPEC	Organization of Petroleum Exporting Countries
PIW	Petroleum Intelligence Weekly
Tapline	Trans-Arabian Pipeline
TBP	True boiling point (standard distillation method)
USMC	United States Maritime Commission
Worldscale	Worldwide Tanker Nominal Freight Scale

PART

I

AN INFORMAL APPROACH
TO CRUDE OIL PRICES

CHAPTER 1

WHY AN INFORMAL APPROACH
TO CRUDE OIL PRICES?

Crude oil pricing is one of the most passionate and controversial issues of the international petroleum industry. Irrespective of the technical, economic and political aspects and implications of the industry, adhering to one pricing system or another would largely influence the ultimate pattern of cost and profit sharing for the main parties concerned, namely producers (host countries), operators (oil companies), and consumers. This is mostly due to the high degree of integration in the industry, to the complexity and diversity of taxation systems in multinational business, and to the historical fact that the largest part of the profits has been, and still is, concentrated in the production stage.

During World War I, Georges Clemenceau of France said that "one drop of oil is worth one drop of blood," and the saying has gained much veracity and strength since. Oil is the blood of modern civilization, and nobody anywhere can afford to dispense with it. Petroleum products have extensively penetrated, in one way or another, all aspects of life in industralized nations, and they now form the backbone of the welfare of individuals and the economic prosperity of nations. Economists say that demand for oil is inelastic with respect to prices and thus has to be met. Politicians and statesmen are primarily concerned with the security and continuity of supplies; for them, economic considerations should be allowed to come into play only once basic strategic interests are secure.

The international petroleum industry has been dominated by the international major oil companies, which are characterized by their very large size and high degree of integration and diversification. The study of the past history and actual performances of these "sacred monsters" is fascinating, and their fantastic achievements might raise apprehension and distrust in some outside observers and determined opposition and even hatred in others. Nevertheless, the most common feeling (often a secret one) of all observers is profound admiration for the formidable accumulation of skills and know-how that make up the real force of the majors and represent their most valuable asset.

The basic strategy governing the international petroleum industry and commanding the plans and action of the majors is a strategy of power, not of profit. It is obvious that once you have acquired power, you not only get profits but also the assurance of their permanence and continuity. The strategy of the majors is articulated upon long-term universal objectives, and within such a framework partial, short-term plans are drawn up and optimized through the multitude of affiliates making up the group. Should any conflict of interest arise between strategic objectives and occasional short-term plans or objectives, the latter are generally sacrificed. In such a case, the attitude of one company or another would appear incomprehensible on professional and economic grounds if the specific and partial framework of that company were considered individually as an isolated entity. The situation would appear more intelligible when considered within a larger scope and checked against the universal, long-term approach to the international oil industry. Unfortunately, not enough information or data are available for such an exercise, and public relations departments of oil companies are efficiently active presenting facts and events persuasively in a framework of the images and myths of the industry whose power they want to maintain.

Within the same spirit of simplification and mystification, crude oil prices are candidly presented as objective facts of life, resulting from economic laws governing supply and demand in a free competitive market. The international oil industry is essentially Anglo-American, and mainly American. Because of the overwhelming specter of antitrust laws in the United States and the haunting memory of the dismantling of the old Standard Oil in 1911, U.S. companies are very anxious to emphasize the competitive nature of prices. Such an attitude inevitably fails to explain the multitude of disparities, mysteries, and contradictions in price movements throughout the world and over the years. These shortcomings are generally disguised by active campaigns of public relations departments, by limiting the scope and minimizing the consequences of these shortcomings, and by the majors' privileged position of power and decision-making. In fact, crude oil prices do not derive from any economic or commercial concept. The key to a genuine understanding of the mysteries of the industry is to realize that crude and products prices are essentially strategic in nature and that the full control of the pricing mechanism represents one of the commanding levers of the majors' strategy of power. Prices are the most effective and convenient mechanism by which they can control and direct the international oil business, and pricing schemes—once they have been determined—provide the general framework within which professional activities are optimized. Crude prices—and products' prices—should represent the starting point, not the outcome, of any meaningful economic analysis of the industry, and this fundamental change in philosophy is the predominating feature of this survey's informal approach to crude oil prices.

Strategy cannot be separated from politics, and major oil companies cannot afford to adopt plans and objectives that do not coexist harmoniously with the political and economic interests of the big powers dominating the world, or at least the so-called free world— that is to say, the United States and occasionally some leading European countries. Tactical moves, as well as over-all strategic plans, are altered, whenever necessary, to match any significant change in the political environment or in the balance of interests of the different parties concerned. Crude oil and products' prices provide the most convenient way of implementing such a change or adapting strategic plans and objectives; and, in fact, the study of the evolution of crude oil prices during the past decades would be unintelligible if not made within a larger political scope and checked against the balance of power and the interests of the leading parties of the moment. From this particular viewpoint, it can be stated that the pricing issue is essentially and ultimately political in essence and that the basic motivations for price changes should be sought in the political environment outside the industry rather than within the industry's professional or economic setup.

In this study we will endeavor to demonstrate the accuracy and soundness of this informal approach to a full understanding of the international oil business. We will emphasize that the structure of the industry outside the United States developed into its present state in the late 1940s and early 1950s, a time when the political and economic influence of the United States was unmatched and unchallenged. The new face of the industry in the Eastern Hemisphere that emerged from this gestation period was the result of active, although indirect, cooperation between the big American majors and the U.S. State Department and the harmonious compatibility of their interests. The second phase, which lasted into the early 1960s, witnessed increasing pressure from European consuming countries, the deterioration of the balance of crude oil supplies in the United States, the appearance of marginal competition in crude oil production, and other phenomena. All these factors, combined with the insignificant political influence of producing countries, brought about the price reductions of 1959 and 1960 and the creation of the Organization of Petroleum Exporting Countries (OPEC) that immediately followed.

The third phase extended through the 1960s and was a rather confused period that saw unprecedented, fantastic expansion and development in the industry, increasing instability in the Middle East, and the progressive emergence of the front of producing countries as an organized pressure group acting through OPEC. The major crisis in crude supplies that followed the six-day war was swiftly and remarkably overcome by the industry, and the situation was progressively normalized in spite of the prolonged closure of the Suez Canal. Then there was the unexpected crisis in late 1970, which resulted in the Teheran and Tripoli Agreements of February and March 1971 and in a total victory for producing countries.

The complexity of the pricing issue is real and almost universally accepted as an insuperable barrier to a rational understanding of the problems involved and the interests of the parties concerned. Oil companies, and mainly the majors, which presumably design and to some extent manipulate the rules of the pricing game, regard their freedom to do so as one of the most vital pillars of their existence and prosperity, and they surround the game with a thick confusing cloud of secrecy and discretion. Consequently, debates and discussion of this crucial issue have been largely left to the curiosity of outsiders and academic observers of the industry and to the passionate but superficial language of financially motivated governments and politically motivated and emotional nationalists in host countries. Irritation about pricing is less sensible at the consumers' end since the relative share of the cost of crude oil in the final bill presented to the ultimate consumers is largely overshadowed by direct and indirect taxes imposed by local authorities, who are usually more preoccupied by the security and continuity of supplies and by their balance of payments than by the price of the total bill.

The world of tomorrow and the world of yesterday are increasingly different: The strategic interests and objectives of the United States have undergone profound changes, oil companies no longer exercise exclusive control of prices and no longer have total freedom of action, and producing countries will therefore play an increasingly active role. The political pressure of the latter is essentially motivated by nationalistic feelings, traditionally oriented against oil companies and charged with suspicion and distrust. This attitude is intensified by general social and economic underdevelopment, by relative lack of professional experience, and by the "colonial" background of oil concessions in the Middle East. The secretive atmosphere surrounding the pricing issue, which successfully operated in the past, would inevitably prove to be a serious obstacle to mutual understanding and trust, which are the basis of future harmonious development of the industry. The involvement of producing countries in the shaping of the destiny of the oil industry is irreversible, and the only way to guarantee their positive contribution would be to promote their professional consciousness and maturity and to help them develop into full partners so that they can appraise their real responsibilities within the international forum, not against it, and behave accordingly. This is a long and hard way, where psychological considerations are as important as political and professional ones and even more difficult to appraise and manipulate. To start with, there should be no more taboo subjects so that producing countries will no longer have the obsessive feeling that they have always been cheated and will continue to be. The most taboo issue is unquestionably the pricing issue, and this study is intended as a constructive contribution towards a demystification of the subject and thus a means to a better understanding between producing countries, oil companies, and all other parties concerned.

Simple explanations to highly complex issues such as crude oil pricing are always dangerous. Vulgarization and simplification efforts can be misleading because the relevant facts and arguments are of a very sophisticated and sensitive nature, and, if they were to be set bluntly in front of the average and unprepared observer of the industry, even though he might be a highly competent technician or a high-ranking official, a large part of their essence and significance would be missed or misunderstood. On the other hand, specialists' dissertations and discussions are often boring and unintelligible because of their crabbed style and language and their ramifications into inextricable details and calculations. Nevertheless, the debate can only remain confined to a limited forum of well-prepared and initiated people. They should have a genuine understanding of the international petroleum industry and should be able to appraise the interdependence and relative importance of its multiple aspects, both tangible and intangible. They should preferably be familiar with the historical background of the pricing issue as it has been fairly described in the specialized literature and fully aware of current and potential problems.

At such a level, one might contemplate initiating some kind of dialogue in order to pave the way to a basic discussion of the fundamental issues. Still, difficulties are abundant, and there are no clear guidelines to help in breaking through. A primary fact should be recognized and taken as a starting point: There can be no unique or universal approach to the pricing issue, but there can be a confrontation of the different concepts reflecting the specific and often conflicting interests of the parties concerned. Tentatively, one should consider the pricing issue as it may be viewed in the context of the academic approach, based on economic theories applied to the industry and on the interpretation of apparent facts; and by governments and nationalists in producing countries; governments of consuming countries; oil companies, particularly the majors; and the U.S. Government.

It seems advisable to study each of these approaches in its own context and according to its specific logic and standards. The next step would be to elaborate and develop a dynamic confrontation between two or more of them, according to some mutually agreeable standards and guidelines. The basic difficulty here is that oil companies appear to be at the center of the issue, towards which claims and accusations from all other parties involved converge. In the face of these solicitations, the attitude of the companies has always been characterized by an obvious lack of cooperation and an apparent determination to safeguard their prerogatives and their exclusive freedom in this respect. It is unfortunate to note, with J. E. Hartshorn,* that up to now, most international oil companies have left almost all serious discussion of price formation and structure in the international oil business to their critics, while decrying the incompetence of these critics and the invalidity of their findings on the grounds that an outside observer can have only a distorted view of the industry.

Our informal approach will attempt to present the case of crude oil prices as it might appear to the average observer from producing countries and to explain the underlying feelings and psychological complexes and motivations that may influence some of the nationalists' seemingly irrational attitudes or actions. The mechanism and mysteries of price formation and structure will then be investigated and analyzed at length from technical, economic, financial, and political standpoints.

---

*J. E. Hartshorn, Oil Companies and Governments (London: Faber and Faber, 1967), p. 128.

CHAPTER 2

THE NATURE AND SCOPE OF THE PROBLEM

A proper understanding of current problems concerning crude oil prices cannot be achieved without comprehensive study and analysis of the present structure of the international oil industry and of its evolution and development throughout the 1950s and 1960s, during which time the Middle Eastern and North African oil industries underwent profound changes. Such investigation will be carried out extensively in the following chapters. Special attention should be focused on the following items:

- The integrated nature of the industry and the predominant position of the major oil companies;
- The disparity of political, financial, economic, and technical conditions in oil consuming, producing, and exporting countries;
- The major role played by the United States, the laws and regulations of which apply to the world's most important oil companies, both majors and independents; and in particular, the internal structure and conflicts of the U.S. domestic market, which affect deeply, although indirectly, the international oil industry in the Eastern Hemisphere;
- The magnitude and sensitivity of the economics involved (with the 1970 average production of OPEC member countries running about 23 million barrels [Bbl] per day, an increase of 1¢/Bbl in posted prices would mean some $45-50 million per year in incremental income for host governments);
- The balance of supply and demand during the 1960s and the increasing competition in the profession under the pressure of independent producers, national oil companies, and the "buyer's market" that has long prevailed both for crude oil and petroleum products;
- The high degree of political instability in the Middle East and the gloomy prospects for the security of Arab oil—more than half of

the world's reserves—along with the dramatic deterioration in the long-standing conflict between Israel and the Arab nations.

Before proceeding further, it seems advisable to review the specialized vocabulary currently utilized and to check its common significance and understanding against the evolution of the industry and the continuous change of the industry's apparent facts and figures.

Posted prices are most relevant to our discussion since they are the real foundation of revenue of the host governments of producing countries, which are mainly interested in keeping their per barrel income as high as possible. For the industry, however, crude prices do have several different significances. Historically speaking, posted prices originated in the United States, where they have had the same significance since their inception in the early years of the century. In the United States, posted prices for crude oils are buyer's prices, not seller's prices—that is, they are announced by refiners, who thus define the prices they are willing and prepared to pay for crude oils at the well-head from any producer connected to their pipeline-gathering system. After adequate transportation costs are added, posted prices are determined at export terminals on the U.S. Gulf Coast. In the early decades of the century, these prices gradually emerged as the keystone of the international pricing structure at a time when the U.S. Gulf was the main source for crude oil export in the open market.

Outside the United States, where most of the world's production is in the hands of a few highly integrated and diversified large-scale producers, posted prices are set by the seller of the crude oil, freight on board (FOB) at export terminals. The evolution of posted prices in the Middle East as related to the U.S. Gulf postings has been widely discussed in specialized literature[1] and popularized articles and essays. This story will be discussed and analyzed later in this study. The simplified popular image is that the Gulf-plus system of price determination prevailed throughout the world until the Middle East emerged as a major crude oil supplier at the end of World War II and was formally recognized as a second price-basing point. The system was then replaced by a "net-back formula" calling for the equalization of CIF prices on an imaginary line that was initially set at Italy and gradually moved westward to Northwestern Europe, and then further still, to the Eastern Coast of the United States and later to the midcontinent, with a corresponding decrease in Middle Eastern posted prices as compared to the U.S. Gulf postings. This relative degradation of Eastern Hemisphere oil was operant even when prices were increased in both price-basing areas. For example, crude prices were increased in early 1957 shortly after the first Suez crisis by an average of about 35¢/Bbl in the United States, but the increase was limited to only 13-15¢/Bbl in the Gulf area, less than half the average increase in the Western Hemisphere.[2]

Further unilateral reductions of Middle Eastern postings in 1959 and 1960 prompted an immediate reaction by the host governments of producing countries leading to the creation of the Organization of Petroleum Exporting Countries (OPEC), which thus manifested the emergence of a political grouping that had been largely ignored in strategy and policy-making up to that date. By its very existence, OPEC succeeded in freezing posted prices at their 1960 level for the duration of the 1960s in spite of the continuation of apparently adverse economic conditions in the industry.

Posted prices have direct and clear significance for host governments, whose income is directly related to them in terms of fixed percentages. Therefore, since the interest of these governments is obvious, they concentrate on maintaining prices at the highest possible level while pushing production upwards and striving to improve the income percentage. It is understandable that any price reduction would readily develop into a fundamental clash with oil companies since the largest part of the income of producing countries (up to 99 percent for some) is based on oil revenue.

The matter is more subtle and complex for oil companies. With the weakening of crude markets throughout the 1950s, posted prices gradually lost their significance as an indicator of the commercial value of crude oils. Companies' profits are related to effective (realized) sales' prices in nonintegrated transactions, while the matter is much more complex for large integrated companies.

Although they remain restricted to a small share of world crude oil movements, nonintegrated transactions have great political and economic significance because they indirectly reflect the evolution of the structure and built-in forces and interests of the international oil business. In the late 1950s, realized prices gradually departed from posted prices by an increasing margin, reflecting the abundant oversupply capacity and severe competition from new producers with no downstream outlets of their own. Discounts on posted prices became a common practice for independents and for majors' sales to nonaffiliates. Nevertheless, comparison of published figures can be misleading when one scrutinizes the level of realized prices since effective sales prices were very frequently a kind of "hidden" reality disguised by a variety of nonprice discounts. In other words, in order to study the so-called open market prices (in Japan, Argentina, Brazil, and Uruguay, and with independent refiners in Europe and Asia), the compiling of price statistics is meaningless unless quoted figures are adjusted so that hidden advantages are reflected in the price level in terms of cents a barrel. In fact, this is highly difficult to calculate due to lack of information on the specific conditions of each deal and the large variety of indirect discounts such as long-term financing of crude oil purchases, tied loans for refining investment, acceptance of soft currencies as partial payment, extra concessions for quality differences, free technical assistance, buy-back of excess products, and freight allowance on delivered prices and recharters of vessels at rates favorable to FOB buyers. There have been several instances where some independent refiners had their plants built virtually for free against medium- or long-term crude purchase commitments.

In integrated operations, which make up the major bulk of the market, crude oil is "transferred" from one affiliate to another at a "transfer price" that is, in most cases, a mere indicator of the internal accountings of the group. The level of these transfer prices is generally determined within the framework of an over-all optimization of companies operations and a maximization of consolidated net profits, taking into consideration the different taxation systems and fiscal regulations in the various countries where the many segments of their integrated activities are scattered. In this respect, U.S. tax regulations play a major role because of the predominance of American companies in the international oil business.

Nevertheless, with the standardization of the financial terms of most concessions in producing countries (the so-called OPEC formula), posted prices have acquired a specific significance to oil companies, both majors and independents, since they determine quasi-automatically the total effective cost to the companies of each barrel of crude produced and exported. This total cost is made up of the production cost, which varies from one field to another, and the tax cost resulting from the application of simple arithmetical rules to posted prices. Under these conditions, Middle Eastern crude oils are not as cheap as they may appear, since total tax costs generally exceed \$1/Bbl. as compared to production costs commonly announced as about 10-20¢/Bbl. The tax-paid cost was largely increased after the 1971 Teheran and Tripoli Agreements.

In the light of the foregoing, it would be logical to put the debate on economic and financial grounds and to define further the extent of the distortions and centrifugal forces that continue to reinforce the disparity between posted prices and market realities. However, logic has never been a characteristic virtue of the oil industry, and to ignore the psychological and political dimensions of the most important issue would be the culmination of the most unrealistic views of the industry. Such intangible factors have currently become the predominant and most powerful underlying forces governing actual facts and contradictions in the industry, especially in the Middle East. Political instability in this area is a real danger and an overwhelming threat to the international oil industry, and particularly to the security and continuity of supplies. This instability is maintained at its paroxysm by the Palestinian tragedy, which has commanded Middle Eastern politics for several decades. As a result, most Arab nations and nationalists have been carried so far into "socialism" and "antiimperialism" that the approaches to the most vital problems of the area, and particularly to the oil and pricing issues, cannot be made and studied without using "leftist" language and terminology. An understanding of the profound currents agitating the area cannot be achieved through the old-fashioned "colonial" or "neoimperialist" viewpoints that have long governed the thinking and actions of most oil executives and policy-makers of "imperialist" Western nations with respect to the "exploitation" of the natural resources of developing countries.

Producing countries will play an active role in shaping the future of the oil industry, and their political and bargaining weight can no longer be ignored or minimized. The key to future stability and harmonious development lies in the desire and ability of developed nations to forge their relations with developing countries on the new and original basis of true justice and mutual interests, with special emphasis on the developed nations' "humanitarian debt," which requires them to help the poor out of their misery and underdevelopment. How this general statement might apply to the specific question of crude oil prices should stimulate the energies and imagination of those interested in, or in charge of, predicting and influencing the future of the petroleum industry. This study is intended to help them in their difficult and thankless task.

NOTES

1. H. J. Frank, Crude Oil Prices in the Middle East (New York: Praeger Publishers, 1966); and W. A. Leeman, The Price of Middle East Oil (Ithaca, N.Y.: Cornell University Press, 1962).
2. Frank, op. cit., p. 86.

CHAPTER
3

THE PSYCHOLOGICAL DIMENSIONS OF THE PRICING ISSUE

Oil-rich Middle Eastern countries are underdeveloped to varying degrees; until very recently, the living conditions of some of them have been the same as those of their medieval ancestors. But even the most advanced of them are handicapped by serious shortages of elite and skilled labor, by over-all economic and social underdevelopment, and by lack of professional experience. Most of them were politically dependent when concession agreements were concluded and the oil industry underwent its fantastic development as a foreign entity insulated from the daily life of native citizens and immunized against national aspirations and nationalist policies. The contrast between the healthy and sophisticated industry and the awkward living conditions of the natives was flagrant. It is unquestionable that the oil industry has greatly contributed to the material welfare and social development of the area and to the improvement of standards of living, and it should be credited for that. Nevertheless, the very fact that "physical" misery has been largely overcome in most producing countries has actually generated a more harmful, hostile psychological attitude towards the industry. Because of popular agitation, mostly in oil-lacking countries, there often has been a distinct disparity between official and popular attitudes towards oil matters. The latter played a very active, although inefficient, role in shaping oil destiny in the Middle East in the 1950s and 1960s.

Physical welfare, the generalization of education, and the emergence of ardent nationalism have actually contributed to an awakening and crystallization of feelings of frustration and irrational attitudes towards oil companies. Nationalism and anti-Western positions are permanent underlying factors that determine popular attitudes and behavior; they help explain and clarify the rather confused view of big oil companies as "colonial" or "imperialist" that is deeply rooted in public opinion in the Middle East; this posture is further obscured and aggravated by the general emotional repulsion against Western support of Israel currently prevailing in Arab countries.[1]

It should be emphasized that the nationalist forces that have attacked and threatened oil companies and Western interests have been generally located in oil-lacking countries. Subversion originated outside, but its scope and force have steadily gained larger audiences because of the generalization and popularization of information media, the highly passionate and emotional atmosphere throughout Arab countries in response to the Palestinian tragedy, and the specific features of the Arab masses, which are characterized by sentimentality and emotionalism. It was a rather easy task for some nationalist leaders to mobilize public opinion against oil companies and to exploit the feelings of frustration that generally flourish in conjunction with nationalism. At times, a psychological approach is more relevant to an explanation of some popular attitudes than political or economic analysis. Nevertheless, although these wild forces had no direct access to the levers of command in producing countries, they imposed political limitations on the actions of the latter and on the nature and extent of the countries' relations with oil companies. Public opinion in the Middle East, and particularly in Arab countries, has become a highly powerful, and feared, censor of the oil industry, and cannot be neglected when considering any past or future action.

Public opinion is a versatile and elusive concept; its force and action are materialized through leading nationalists, pushed and supported by an overheated atmosphere of passion and emotion. The oil industry has been the victim of the anti-Western feelings of the Arab masses, but it has also greatly contributed to the creation of this mood. Apart from the political considerations that will be evoked shortly, its distant and unpopular public relations image and its intrigues and maneuverings have caused serious damage to its image and contributed to making oil companies appear as champions of exploitation and imperialism. Such developments in public opinion were facilitated by the public's general ignorance and tendency towards xenophobia and by the masses' skillful manipulation by traditional and occasional leaders, some of them more inclined to rely on easy demagogy than on sound economic or political programs. One of the easiest ways to inflame the public imagination was to state that oil companies were constantly cheating, that they are presently cheating, and will continue to cheat and that they can never be trusted. According to David Hirst,

> It is impossible to understand the image the oil companies have earned for themselves in the Middle East without taking into account the historical and political setting of

which they are very much a part. Coming after four hundred years of Turkish rule, the recent political experiences of the Arab people have been disillusioning, to say the least.... All things considered, it is not surprising that the word "imperialism" has become one of the most highly charged words of the Arab language; that this word, and all that it stands for, has entered into the very culture of the people, and colours their thinking on all the more particular issues, such as the oil industry, on which they may have an opinion.[2]

Indeed, the historical background of Western presence and interests in the Middle East was characterized by direct and indirect colonization of Arab countries by the British and the French and by the emergence and development of Arab nationalism to combat these European powers. Arab nationalism was developed during World War I, under the push and paternalism of the British and their famous Lawrence, and was designed to mobilize the Arabs against the Turkish presence with a formal promise of independence after the war. But the earliest experiences of the Arabs with the British were bitter and disillusioning. Instead of acquiring autonomy, or independence, as they had been promised, Arab countries, unified under the Ottoman Empire, were parceled out to the British and the French: Iraq, Jordan, Palestine, and Egypt to the British; Syria and Lebanon to the French. Popular and nationalist feelings have been continuously mobilized since then in an effort to recover liberty. Political independence was achieved in the years after World War II.

Individual domestic nationalism in Arab countries, upon their independence, was ephemeral. In 1948, the area was confronted with the Palestinian tragedy and the creation of Israel, directly supported by Western countries, to the detriment of legitimate Arab natives. The moderate bourgeoisie, which had led the struggle for independence and could have ensured relative political stability in the area and relatively sympathetic attitudes towards Western powers, was progressively eliminated from most Arab countries by revolutionary regimes determinedly hostile to the West. Nationalistic feelings developed throughout the area, ignoring borders and independent of actual political regimes. Progressively and imperceptibly, popular hostility crystallized against "imperialism" and found a propitious field of action in concerted campaigns against oil companies. Arab hostility to Western imperialism culminated during the first Middle East crisis of 1956, when the Suez Canal was closed and the Iraq Petroleum Company (IPC) line was sabotaged in Syria. Anti-Western feelings reached unprecedented levels when the tragedy was repeated in June 1967, and the very existence of oil interests in Arab countries was threatened.

The Palestinian tragedy lies at the heart of Arab nationalism and should be regarded as the ultimate key to an understanding of popular attitudes and actions in Arab countries. We are not concerned here with official stands and the policies of governments, but rather with the popular imagination and emotions. To many observers, these feelings and some of the attitudes (and actions) they generate might seem irrational and demagogic, and indeed, some of the underlying motivations are far from noble. But moral standards have never played a significant role in shaping the history of humanity, and Arab nationalism, apparently opposed to "Western imperialism," should be considered as a basic fact of life. Its motivations embody a large array of frustration, humiliation, and inextricable psychological and social complexes.

These considerations are of the utmost importance to an understanding of popular attitudes towards the oil industry and the oil companies. In their helpless disarray and confusion, due to continuous humiliation in the face of successive defeats that emphasized their weakness, incapacities, and underdevelopment, the Arab masses naturally turned to oil as their only and ultimate weapon of delivery from humiliation and misery. Any serious analysis of oil crises in the Middle East cannot be complete if such human and psychological considerations and background are not given due attention. The forces of passion that have always characterized the petroleum industry were brought to paroxysm in the Arab world. Taking all the above into consideration, and keeping in mind the inevitable multiple conflicts that naturally develop with the confrontation of different civilizations, we might come to the conclusion that efficient and realistic solutions to some of the more basic problems of the industry can germinate only on irrational grounds. We should attempt to understand the basic features of this irrationality so that efforts to clarify present and potential problems will be fruitful in shaping the future to the mutual advantage of all.

In fact, early anti-Western campaigns were the responsibility of a few leaders who relied more on the emotive imagination of the popular masses and on their ready response to slogans and occasional demagogy than on consistent and coherent political programs. Arab nationalism was a most versatile and efficient weapon, even though it may appear incomprehensible and rather contradictory in some respects. Words have always had a fascinating effect on Arab masses, and the images and dreams they evoke enjoy a magical power irrespective of their actual significance. In some cases, this situation may make responsible officials prisoners of slogans, even when these slogans become outdated and void of any significance.

David Hirst outlines the four schools of thinking that have played an active role in shaping public opinion in Arab countries during the 1950s and 1960s: Three of them have some intellectual support, but

As for the fourth classification—the popular demonology—
this operates at the lowest level of opinion, as purveyed
by the yellow press and certain Arab leaders in their
more demagogic moments. On this level, imperialism
tends to acquire a sort of monstrous transcendental quality like the force of evil itself. Fitted into this picture
of allegoric simplicity are the oil companies, which not
only possess the aura of imperialism, but sometimes
have influence over its workings. But imperialism, as
it lurks among the people has an adversary—Arab nationalism. Like the forces of good and evil, these two are
locked in mortal combat. Oil is directly associated with
the struggle.[3]

It is the fourth component of Arab public opinion that is the most threatening to oil companies because its motivations are basically irrational and emotional and because it embodies the tremendous potential of easily manipulated human forces; the danger lies in the fact that, when these forces are inflamed, they can easily get out of control, and collective demonstrations and eventually unwarranted riots often develop. This "savage" public opinion acts as a vigilant censor of the policies and actions of governments in crisis periods. The best illustration of its impact might be found in the stoppage of all Libyan oil exports in the aftermath of the six-day war. This decision was taken on the spot and implemented by the oil workers despite the hostility and the repression of the Libyan Government, then headed by King Idris.

The recent history of Arab oil nationalism has been characterized by the emergence of a new technocratic elite with solid academic background and good professional experience in the numerous sectors of the oil industry. These technocrats were generally in responsible positions within the governments and helped to lend a minimum of reality and common sense to state policies as well as popular moves.

The technocratic elite in Arab countries has contributed to and succeeded in demystifying many of the mysterious and taboo issues of the industry and demonstrated that the ability to acquire techno-logical and managerial skills is not solely a privilege of Westerners. This resulted, indirectly, in an adverse popular reaction towards oil companies, which were regarded as enjoying exorbitant privileges and as striving continuously to keep nationals out of key positions so as to perpetuate their domination. In fact, public relations depart-ments as well as senior executives of oil companies are to be blamed for their distant and paternalist attitude when dealing with employment and national affairs and for not having attempted to improve their public image, which was a queer mixture of paternalism and disregard.

The highly active role of the technocrats led to the creation and development of a unified Arab approach to oil affairs. Since the technocrats had no political commitment, they inevitably found themselves moving closer and closer to their counterparts in other Arab countries, whose problems and aims were essentially the same. Thus, the unification of Arab oil policies developed into an integral part of the technocrats' ambitions. In this respect, technocrats were very active in the late 1950s, when crude pricing was the hot issue in the area. The mobilization of public opinion was accomplished through press campaigns and such official forums as the Arab League and Arab Petroleum Congresses. They helped create an atmosphere in which OPEC could come into being and, isolated from everyday Arab politics, pursue effective policies of its own making.

Under these conditions, the price reductions that took place in the Middle East in 1959 and 1960 were felt as an intolerable challenge, which technocrats and nationalists were nevertheless prepared and determined to accept. For the first time in the history of developing nations, a common front was formally organized, and the organization thus created represented an effective political force that oil companies could not ignore. By its very creation, OPEC materialized the nationalist aspirations for dignity and sovereignty and readily achieved substantial success in inducing price stabilization from that time on.

It is obvious, then, that the crude oil pricing issue in the Middle East developed essentially within a political context. In fact, the masses as well as informed technocrats were very sensitive and responsive to this highly emotional and controversial issue. First of all, national income in all producing countries was essentially based on oil revenues, which are directly related to posted prices. Therefore, officials and nationalists were naturally very eager to stabilize revenues on a steadily increasing basis by keeping prices at the highest possible level or at least by avoiding their deterioration. Generally speaking, oil companies were suspected, and occasionally accused, of price manipulation designed to increase their profits at the expense of producing countries. Leftist propaganda in this respect insisted on the monopolistic structure of big oil companies and on their cartel collusion in exploiting helpless producing countries. The companies themselves contributed to a further exasperation of public suspicion by their general attitude towards crude prices, primarily characterized by secretiveness and paternalism. Their position was generally confined to the statement that complete freedom in price setting was an unquestionable privilege that they were not prepared to debate. To the popular imagination, such an attitude could have only one explanation: that the companies were striving to disguise their ill-intentioned practices. Frustration resulting from such a negative attitude was reinforced by early studies by technocrats of the historical background and evolution of prices, which proved that the issue was not as inaccessible to public investigation as the companies pretended. Although these studies and investigations were only preliminary, they succeeded in focusing public attention on crude prices as one of the main areas of struggle against monopolistic companies and in crystalizing nationalist confrontation with Western imperialism on this highly complex issue.

That is to say, the issue of crude oil prices emerged and developed in the Middle East essentially within a political context. Technicalities behind pricing were as mysterious as oil companies wanted them to be, and in any case, they were out of the reach of the common understanding of the average nationalist observer or antagonist. They were not even debated because the principle of participation in price formulation was denied to producing countries. Crude oil prices became a symbol of the over-all confrontation with imperialism and monopolistic oil companies, and any common understanding or solution to standing and underlying problems was necessarily influenced by psychological and political considerations extending beyond the specific limits of the oil industry.

In the past, oil companies could afford to act without taking political stands and professional claims of nationalists into consideration when drawing up their pricing strategies. The 1970-71 crisis and the Teheran and Tripoli settlements that followed prove that producing countries must be full partners in shaping the future of the oil industry and that oil companies must adapt to this new fact of life. The elaboration of the future on peaceful grounds, to the common benefit of all, should be undertaken in a trustful atmosphere of cooperation and good will, and thus all possible efforts should be made to dissipate the thick cloud of suspicion that has always characterized the relations between the two future partners. In particular, one of the important factors for the future is the dissipation of the mysteriousness and dissimulation that have always overshadowed the pricing issue.

Producing countries should be enabled to reach a degree of genuine understanding of the industry and to develop high-level professional and managerial skills compatible with the new responsibilities they will inevitably enjoy in the coming years. It would be in the interest of oil companies, and consuming countries as well, to abide by the Arab saying "A wise, qualified opponent is preferable to an ignorant ally." This study is intended to help establish such a mood of good will and cooperation by demystifying the main object of conflict and suspicion. It is designed for those who may have the ambition of imagining a new, peaceful picture of the future of the oil industry, beyond present conflicts and passions.

This effort to make the debate more reasonable and to minimize the impact of irrational psychological and emotional considerations will be particularly appreciated by those who know, or who will learn through this study, that the fundamental interests of producing countries and major oil companies are far from being divergent or contradictory over the pricing issue, at least in the coming decades. Bright prospects for the future are still beclouded by a thick, heavy veil of past conflicts. This study, it is hoped, will initiate further efforts and contributions to ensure that the chance of a new era of peaceful international cooperation between developing and developed nations will not be ruined. The passionate divorce from the past can help the concerned parties achieve a marriage de raison with the future. The task is not incommensurate with the potential capabilities and resources of the partners, should they be motivated by a sincere desire for cooperation and good will. Participation is the name of the new game.

NOTES

1. The importance of the political and psychological considerations discussed in this chapter is emphatically outlined by David Hirst in Oil and Public Opinion in the Middle East (London: Faber and Faber, 1966).
2. Ibid., p. 13.
3. Ibid., pp. 23-25.

CHAPTER

4

WHY POSTED PRICES?

The general attitude of oil companies vis-à-vis the pricing issue has always been a very cautious one. When pressed to give an opinion, they usually try to avoid discussion or to water down the issue with such statements as "prices are determined by competitive market forces." J. E. Hartshorn best described this attitude as follows:

> Even inside the American oil industry, which has the most elaborate price reporting and statistical series of any oil business in the world, it is not easy to get rational discussion of the way in which these prices are in fact formed. In the United States, the spectre of antitrust proceedings broods over any such discussion, and its shadow extends wherever in the world American companies operate. Even so, it is notable and unfortunate that up to now, most international oil companies, in particular, have left almost all the serious discussion of price formation in the international oil business to their critics—usually contenting themselves with the claim that oil pricing is the result of "competition," naively defined.[1]

In light of this attitude, there is a large array of opinions and arguments. Posted prices for crude oil are said to be buyers' prices in the United States, whereas they are sellers' prices outside the United States. Everybody seems to be convinced that crude prices in the United States should be kept high and even increased. Throughout the 1960s some observers came to the conclusion that crude prices outside the United States were unrealistically high and would be inevitably reduced. M. A. Adelman and P. H. Frankel seem to have adopted such a competitive approach to crude prices; in their opinion the marginal cost of production, which is very low indeed in the Gulf area, should ultimately control the trend of future price movement.[2]

Meanwhile, producing countries claim that crude prices are too low, while outside nationalists, like Abdullah Tariki, have greatly contributed to a popularization of the debate in the highly passionate and emotional public opinion media of the Middle East. On the other side, officials in consuming countries have generally refrained from making public statements because they were, in reality, in a quite uncomfortable position. They would have certainly liked to see crude prices decrease in order to minimize their over-all disbursements in hard currency, but no further than a certain limit beyond which it would become very difficult to coordinate and harmonize the economics of alternative sources of energy and to help national coal industries survive.

Posted prices are occasionally presented as effective reflections of the real level of commercial transactions. More often, it is claimed that they bear no relation to market realities, which are described in terms of discounts, realized prices, transfer prices, and so on. They are instead considered as providing a bench mark for taxation in host countries, even though the picture is far from being simple because of differences in taxation mechanisms from one country to another: taxation based on posted, realized, or tax reference prices, rates, and the use or nonuse of royalty expensing, and so on. Oversimplified conclusions are generally drawn in terms of the per barrel revenue of host countries and tax-paid costs to oil companies.

In light of the above, the concept of crude oil prices might seem veiled with ambiguities and contradictions. In fact, almost all these opinions are valid, but not simultaneously. The international oil business is very complex and can easily accommodate the coexistence of genuine contradictions that change in nature and scope according to the time and place. A situation or opinion cannot be understood outside of its own professional context and political environment and motivations. This chapter will attempt to provide appropriate explanations for the multiple questions suggested by this introduction.

CRUDE OIL PRICES AND PRODUCING COUNTRIES

To producing countries, the significance of crude oil prices was primarily related to their revenue from oil exports. In almost all cases, oil revenue accounted for the largest part of total government income, amounting to 99 percent and more in some cases. Oil revenues were composed of two factors: the total volume of exports and unitary income per barrel. Increasing production was never a serious problem, either for producing countries or for oil companies, since demand was steadily increasing at high rates and oil reservoirs in the Middle East were very great indeed, while production costs remained low.

Every producing country would have preferred to see its own production go up preferentially, and some of them undertook strong campaigns to that end (for example, Iran in the late 1960s). However, uplift programs were decided by the major oil companies, which thus found themselves in the position of arbitrator. Since production economics was quite similar in all the producing countries of the Middle East, the majors' decisions were dictated mainly by political considerations, such as maintaining balance between friendly nations and the need to retaliate against hostile governments by restricting development investment and production rates. This was particularly the case in Iraq during the 1960s after the dramatic confrontation with General Abdul Karim Kassem in 1961.

Producing countries themselves had no direct way to increase production. Furthermore, their interests in this respect were conflicting so that they could not engage in collective actions against oil companies, and all attempts to regulate production under common proration programs proved abortive. By contrast, they were almost unanimous in deciding to press for higher prices and better financial performances. Developments during the 1950s and 1960s were characterized by profound changes in the political framework and in the psychological reactions that governed the attitude of host governments towards oil companies and crude prices.

During the 1950s, which witnessed the rapid expansion of crude oil production in the Middle East, producing countries were generally considered minor and negligible entities in the forum of forces and interests that directed and controlled the majors' decisions with respect to the development of Middle Eastern oil and its pricing. Because of a lack of political influence and professional maturity, host governments were forced into a passive attitude; furthermore, they had no genuine reasons for complaining since their total income was rapidly increasing due to constant increases in production.

The seeds for radical change germinated in emotional, nonprofessional soil, outside official circles. The omnipresent conflict in the Middle East greatly contributed to the activation of popular feelings in the area against Western interests. Oil companies were presented to the popular imagination as monopolistic ogres that appropriated, to their exclusive profit, the only wealth the Arab nations had, a wealth unquestionably of vital strategic interest to the West. With the 1956 crisis, oil dramatically emerged as the most powerful political weapon in the over-all battle of Arab nationalism. Fortunately for the oil companies, this popular agitation took place in oil-lacking countries, and its fury was limited to the sabotage of the IPC line across Syria. Upon the creation of OPEC in 1960, the main preoccupation of its member countries centered on crude oil prices, and all changes that took place thereafter in that respect were instituted within the framework of the organization. The priority of crude oil prices was clearly formulated in the first OPEC resolution adopted in Baghdad.[3] It is commonly accepted that OPEC succeeded in stabilizing crude prices at their August 1960 level and preventing further cuts. The organization has always been very satisfied with this accomplishment because it rapidly realized it was almost impossible to achieve its primary objective of increasing prices and restoring them to pre-August 1960 or pre-February 1959 levels. Without formally rejecting this primary objective and while awaiting better circumstances for its attainment (which occurred unexpectedly in the 1971 Teheran and Tripoli Agreements), OPEC recognized that an increase in governments' income could be achieved by the improvement of profit-sharing agreements with oil companies.

Generally speaking, under the 50/50 profit-sharing pattern prevailing in the Middle East, host governments were entitled to a 12.5 percent royalty of crude exports, with the option of lifting it in kind, an option that was never used until very recently. Net profits, obtained by deducting production costs from sales proceeds, were to be equally shared. Both royalty and proceeds were to be evaluated on the basis of posted prices, but royalty payment was credited against total payments to governments, which thus could not exceed 50 percent, and different allowances and discounts could be deducted from posted prices for the purpose of assessing tax liability. In particular, marketing expenses and discounts to third parties were often deductible, so that profit sharing was actually based on what is commonly known as realized prices, a situation that was particularly true of Libya and was in accordance with the stipulations of the Libyan Petroleum Law. The situation was different for Indonesia and Venezuela in terms of royalty percentage, tax rate, allowances, discounts, and so on.

The first manifestation of a common philosophy and adequate policy guidelines was publicly formulated in the resolutions of the Fourth Conference held in Geneva in April-June 1962.[4] These resolutions called for the standardization of fiscal and taxation regulations, royalty expensing, and the elimination of allowances in all member countries. After episodic negotiations, an over-all agreement was reached between the Middle East governments and the international oil companies. Its main terms were as follows:

1. The royalty on crude oil will be expensed to the extent of 12.5 percent of posted prices at export ports (border value for Iraqi oil moving by pipelines to the Mediterranean). Where higher royalties exist, as in Saudi Arabia and Iraq, the amounts above 12.5 percent will not be "expensed" but will continue to be credited fully against the 50 percent income tax.

2. The companies will be specifically allowed only a 0.5¢/Bbl marketing allowance off posted prices.

3. The companies will also be allowed a basic 8.5 percent discount off posted prices in computing taxable income in the first year of the agreement (1964).

4. In 1965 and 1966, the companies will increase their payments by progressively reducing the basic 8.5 percent discount. This will be done through a complicated formula designed to recognize (in the companies' opinion) the stronger market demand for heavier crudes than for the light crudes:
● for heavy crudes of 27° API and below, the 8.5 allowance will be reduced to 7.5 in 1965 and to 6.5 in 1966;
● for lighter crude oil, 0.13235¢/Bbl for each degree of gravity above 27° API will be added to the 7.5 percent. In 1966, 0.26470¢/Bbl for each degree of gravity above 27° API will be added to the 6.5 percent discount.

5. No specific reductions in the discounts were provided for in the agreement beyond 1966. However, it was stated that oil companies will "from time to time determine whether the level of such allowance should be changed for a year or years after 1966 in the light of the competitive, economic and market situation of each quality and gravity of crude oil exported which is expected at the time to prevail during the reasonably foreseeable future as compared with that situation in 1964."[5]

The agreement was ratified by all Middle Eastern OPEC members, except Iraq because of its political confusion and pending problems with IPC resulting from the unresolved conflicts created by Law 80 of General Kassem, which expropriated more than 99 percent of the company's existing concession area. In Libya, the enforcement of the agreement was an occasion to amend the Petroleum Law so as to shift tax liability from realized to posted prices. Shortly after, on September 30, 1966, an agreement was signed between the Arabian American Oil Company (Aramco) and Saudi Arabia for the elimination of discounts off posted prices with respect to oil sales to nonaffiliated third parties.

This agreement was considered by the concerned OPEC members as a temporary half-way achievement since they were determined to attain full elimination of discounts and allowances off posted prices. However, attention was diverted from this issue by the stoppage of oil flow through the IPC pipeline across Syria in late 1966 and early 1967 and by the major Middle East crisis of June 1967. The matter, however, regained momentum precisely because of the crisis and the closure of the Suez Canal, and, in fact, oil flow from Saudi Arabia through the Tapline was stopped, and Aramco was not allowed to reopen it until it recognized that oil exported from Sidon should enjoy an extra premium because of the closure of the Canal. In recognition of this temporary Suez premium, Aramco agreed to eliminate the 6.5 percent basic discount. Shortly thereafter, OPEC members held several meetings to discuss the matter (September 15-17 in Rome, October 5 in Taif, November 27-29 in Vienna). The Saudi Oil minister and the Iranian Finance minister were jointly asked to conduct a final round of negotiations with the oil companies concerned on behalf of other member countries. After somewhat animated negotiations (not only with oil companies but also between member countries themselves) a final agreement was reached in early January 1968 for the gradual total elimination of discounts and allowances. Libya did not take part in this agreement and planned separate negotiations of its own, but this attitude was only a question of form since discounts and allowances were already "temporarily" eliminated to compensate for the Suez premium. Iraq was also in an exceptional position since it had not yet ratified the original royalty expensing agreement.

Under this new agreement, the basic percentage discount was to be eliminated gradually over five years through 1971, and gravity allowances would be abolished over eight years through 1974.[6] The marketing allowance was to be maintained at 0.5¢ a barrel. The main feature of this compromise agreement was to maintain and emphasize gravity allowances, thus penalizing lighter crude oils. The expressed motive of the companies for this provision was to modify the effect of the traditional 2¢ a barrel gravity differential, which, they argued, was no longer justified by market conditions.[7] This agreement came to a premature end with the 1971 Teheran and Tripoli Agreements, which provided for the immediate and total elimination of all discounts and allowances.

Although this study is primarily concerned with Eastern Hemisphere oil, it would be most interesting to indicate how Venezuelan oil economics compares with the above-mentioned fiscal and taxation mechanism, since Venezuelan oil has played a direct role in shaping international oil economics in both the Eastern and Western Hemispheres. Compared to the Middle East, Venezuela has much smaller reserves and a higher production cost, averaging 55-65¢/Bbl against 10-20¢/Bbl in the Gulf area. Furthermore, about half of the crude production is actually exported as refined products, and Venezuelan oil has always enjoyed a privileged situation in the U.S. market.

Venezuelan income from crude oil and products' exports was composed of a fully expensed royalty of 16-and-2/3 percent based on posted prices plus an income tax of 47.5 percent of net profits based on actual sales at realized prices. This resulted in a relatively higher average per barrel income than in the Middle East. Now, in the early 1960s, when Middle Eastern OPEC member countries were striving to obtain royalty expensing—a formula already enjoyed by Venezuela at an even higher rate—it was reported that Venezuela was not concerned with this debate but supported its OPEC sister countries against the oil companies. In fact, Venezuela was conducting its own campaign to improve the terms of its deals with oil companies. The Venezuelan move was not publicized and was never put on the agenda of OPEC, but it was, nevertheless, a swift and efficient one. By the time Middle Eastern countries obtained trimmed-off royalty expensing under the conditions described above, Venezuela had attained its own set of improvements.[8]

Venezuelan tactics were based on retroactive tax claims going back to 1958, based on the charge that the companies had been selling their oil at prices that were too low, thus reducing "artificially and unduly" the revenue paid to the country in taxes based on actual realized prices. In 1965, Venezuela presented the big producing companies, like Esso and Shell, with back tax bills for the period 1958-60 and indicated in early 1966 that unless an over-all settlement could be reached on the issue, back tax claims for the period 1961-65 would also be handed to the companies. Total back claims on the entire industry for the nine-year period have been estimated at about $750 million.

The new pattern agreed upon was to cover a period of five years from 1967 through 1971. Its main features were as follows:

- Expensed royalty is maintained at 16 and 2/3 percent based on posted prices;
- Basic tax rate will be increased from 47.5 percent to 52 percent, with various credits or rebates allowed under certain conditions;
- "Tax-reference prices" for each grade of crude and product will be established starting January 1, 1967. These reference prices will become the basis for computing minimum tax payments;
- There will be a quit clause by which this arrangement represents the settlement of existing Venezuela tax claims covering the nine-year period from 1958 through 1966.

Taxes will be computed at the new higher rate, based on tax reference price or realized price, whichever is higher. Tax reference prices for crude oils were set according to a mathematical formula articulated on a basic gravity differential (BGD) of 0.364¢/Bbl for each one-tenth of a degree (°API) of gravity and on a theoretical crude oil of zero gravity priced at $1.04/Bbl. The formula is expressed as follows:

$$X \ \$/Bbl = 1.04 + 3.64G$$
$$G, \ °API: \text{ crude gravity}$$

Tax reference prices for products were not published, but they were generally set part way between actual market prices and posted prices. In the case of heavy fuel oil, for instance, the new reference price of $1.65-1.68/Bbl is slightly above the real market value of around $1.60/Bbl but far below the posted price of $2/Bbl.

To complete the picture, it would be interesting to give some indication of the Indonesian situation and to outline its specific differences in relation to other OPEC sister countries. The oil industry of Indonesia is one of the world's oldest, petroleum having been produced and exported for more than 75 years. The history of Indonesian oil, however, has been a spotty one. Originally an important early contributor to world crude supply under Dutch occupation, the industry suffered a series of crises and disasters—resulting from the country's invasion, foreign occupation, severe damage in two world wars followed by postwar dislocations—plus continuing political disruptions, the changeover to independence from former Dutch rule, and the nationalization of oil beginning in 1960 when President Sukarno issued a decree, the Oil Mining Regulation of 1960, whereby future oil production for export was to be reserved to the state. After three years of intense crisis, an agreement was reached in late 1963 with the three major groups (Caltex, Shell, and Standard Vacuum) accounting for 90 percent of Indonesian production.

According to the new agreement, oil companies shifted from conventional concessionary statutes to joint-venture agreements by which they became "contractors" to national oil companies. The fiscal modes provided that the government would receive taxes and royalties equal to 60 percent of the company's taxable income, leaving the company 40 percent. Taxable income, in turn, was determined according to an elaborate formula that, in essence, based taxes on the weighted average value of a company's sales to third parties. Taxes were payable on actual realized prices in deals with nonaffiliated third parties; taxable income on sales to affiliates were based on the weighted average of the value of third-party sales.[9]

Under these conditions, it is obvious that Indonesia could not be concerned with the fiscal negotiations initiated by Middle Eastern countries over royalty expensing and related allowance elimination. Posted prices as such did not even exist in the main producer of the Far East, which, nevertheless had to compete with Middle Eastern production in its own natural markets, principally in Japan.

CRUDE OIL PRICES AND OIL COMPANIES

The largest part of crude oil production in the Eastern Hemisphere is controlled (except in Libya and Algeria) by the big international majors, and the largest part of this production is disposed of through integrated channels. It is currently admitted by most observers that an average minimum of about 85 percent of the crude moved in international trade never changes hands but passes from one affiliate to another. Prices "shown" at such levels of transfer cannot be explained or justified by conventional economic or commercial considerations since they are insulated from the competitive forces of free market transactions. They cannot be considered outside the framework of the integrated operations of the majors.

The remaining 15 percent or so of crude production is sold in so-called arm's-length transactions to third parties under the competitive conditions of supply and demand in a free market. Prices effectively represent the commercial level at which crude ownership changes hands. These are usually called realized prices, but their real significance and level can be blurred and confused by a multitude of nonprice discounts that make their direct comparison rather meaningless. Such nonintegrated oil is produced and sold by both independent producers and major oil companies, and the relative importance of the former tends to stabilize and diminish because some of them strive to integrate their downstream operations so as to dispose of their production in a secure and captive market, thus protecting themselves from the irresistible pressure of the integrated majors trying to force them out of business.

We will investigate below both aspects of crude oil pricing. It should be noted, however, that integrated oil pricing is highly significant to the international oil business although it is often left in the shadows. A great deal of publicity is generated around realized prices, and the uninitiated observer tends to believe that they represent the oil industry as a whole.

Let us begin with the really significant "prices" of integrated oil. To major oil companies, the posted prices of integrated oil have the same significance as to producing countries. They are no more than bench marks or indicators designed to determine their tax liability at the production stage. Producing countries are interested only in their per barrel income and do not really care about the level of accounting prices at which crude oil is transferred from one affiliate to another. The net "profit" shown at the production stage of integrated operations—transfer prices minus tax-paid costs—has no real significance for the oil companies concerned. The only significant figure corresponds to consolidated net profits from their multinational, worldwide, diversified, and integrated operations—that is to say, proceeds from sales to ultimate consumers minus all costs incurred in exploration, production, transportation, refining, and distribution minus all taxes paid at these different stages of the industry, under tax regulations that vary from one country to another. Integrated oil companies, as clearly demonstrated by Mrs. E. T. Penrose, "allocate overheads costs among their foreign branches, subsidiaries and affiliates, and adjust their transfer prices, in order to reduce their total tax outlays."[10]

The minimization of tax disbursements obviously demands discrimination between the different countries in which the different branches of the industry are traditionally located. For historical, political, and strategic reasons, the largest part of profits are "shown" at the production stage, whereas other downstream operations are occasionally reported as profitless, if not losing, ventures. Nevertheless, the largest part of the investment made by the majors is located precisely in these downstream operations.[11]

A barrel of crude oil has no value as such, although production operations in the Middle East are known to yield high rates of profitability. Its value materializes only after it has passed through the different downstream operations showing only a few percent profitability if that. The consolidated over-all performance of the oil industry is currently reported to be 11 to 12 percent, that is to say of comparable magnitude to the other main industrial sectors in the United States. In this complex setup, none of the conventional significances of crude oil "prices" can be taken as representative of the decision-making mechanism of integrated oil companies, which is largely based on minimization of total tax outlays.

More precisely, according to M. Lipton,

> transfer prices of crude oil—prices at which crude oil is invoiced from producing affiliates to refining-marketing affiliates—are largely devoid of any economic content as a price. Throughout the fifties, the corporate transfer or invoice prices was posted price; then perhaps it carried a slight discount; later it occasionally has been the tax-reference price. In most instances, the price that is established as the basis of tax corporation in the Middle East tends also to be the price at which crude is invoiced from producing to refining affiliates—but this has solely to do with tax position of the oil companies. It is a way in which they try to minimize their aggregate tax liability. Strange thing. The function of the producing department is to find the oil. The function of the marketing department is to give the oil away. And the function of the tax attorney is to make the profit for the company.[12]

In a very real sense, however, the transfer price, which is relevant to decision-making within the integrated oil company, is not the invoice price but the economic transfer price, which can be defined as the value that should be credited to crude oil in the over-all balance of the consolidated economics of integrated operations so as to obtain an acceptable over-all rate of return on all investments engaged in the whole spectrum of integrated operations. In a numerical exercise, stated in the literature, the true economic transfer price in integrated channels is estimated at $1.20/Bbl for 34° API crude oil from the Middle East.[13]

The posted price structure of international oil was originally as closely linked to "competitive market reality" as the pricing departments of the integrated oil companies could make it. However, by the mid-1950s, "market reality" started to change in the Eastern Hemisphere in both producing and consuming countries. Products' realization in Western Europe started to weaken under the increasing pressure of governments, the development of excess refining capacity, and wild competition from independent operators as well as from emerging European national oil corporations. However, these political and professional considerations would not have been sufficient to induce continuous erosion of products prices if there had not been an uncontrolled oversupply of crude oil. Middle Eastern reserves are huge and prolific, with very low production costs yielding substantial profit margins. As long as supply was exclusively under the oligopolistic control of the majors, market prices both for crude oil and products could be controlled and competitive market forces could be easily manipulated according to the majors' strategic plans.

The increasing availability of low-cost crude oils from independent producers without integrated outlets progressively led to an oversupply situation since the majors had to enter into a price-cutting competition in order to keep the market under control. Producing countries were largely responsible for the encouragement of independents by granting concessions or joint-venture agreements under which taxation was assessed on the basis of realized prices. This was particularly true for Libya, where the fulgurant development of production in the early 1960s under the 1955 Petroleum Law largely contributed to making arm's-length transactions, or sales to third parties at realized prices much lower than posted prices, a permanent and dominant feature of the international oil industry throughout the 1960s.

"Posted prices are unrealistic figures that no longer have any connection with market reality; they are too high and should be adjusted downward, sooner or later." Statements such as this have been very fashionable in recent years. The fact that arm's-length transactions represented no more than some 15 percent of oil movements was seldom mentioned, and the type of judgment referred to above was presented as a general one and was widely publicized by consuming countries and oil companies alike. The latter used this argument to resist government claims for higher per barrel income. Furthermore, the political and strategic significance of the price cutting associated with commercial competition between independents and majors in the so-called open market was much greater than the specific share of such transactions in the overall market for crude oil and petroleum products. To allow competition to operate effectively in a significant part of the market would have been a kind of cancer in the consolidated oligopolistic structure of the international oil business; it represented a potential threat to the power strategy of the majors and could have led, in the long run, to the collapse of the entire system. The attitude of the majors during the confused period of the 1960s was dictated by the necessity of ensuring the continuation of their operations, with due regard to the actual considerations of the moment, and by their determination to squeeze independents out of crude oil production or at least to bring them under control in some way.

The literature is abundant with data and information about realized prices, discounts off posted prices, arm's-length transactions, sales to third parties, and so forth. A tradition has been established for following and reporting market evolution: bids for crude supply to some Latin American countries (such as Argentina, Brazil, and Uruguay), the official publication of the Ministry of International Trade and Industry (MITI) in Japan, spot deals with independent refiners in Europe and Asia, and so forth. Several studies have been devoted to the subject, and one can hardly find a publication dealing with crude oil prices that ignores this issue. Special mention should be made of the studies of M. A. Adelman and Ching Chin Chen.[14]

The crude oil market in the Eastern Hemisphere in the late 1950s and in the 1960s in generally described as buyers' market characterized by such oversupply and competition between sellers that the preponderant bargaining position of buyers dictated the level of sales prices to third parties. Such transactions were limited to no more than 10 percent of oil movement, as Adelman concluded after an extensive survey of the period 1957-67. These transactions were generally made between independent refiners with no crude oil production of their own, independent producers with no downstream outlets, and/or major oil companies. The situation of independent producers is quite clear. They had to dispose of their production in spite of severe control of the industry by the majors. Independent refiners served as providential loopholes, and the only way to break through was by the price competition afforded by low production costs and large profit margins. The motivations of integrated majors are less clear. Profit considerations are not negligible, although they are very difficult to assess. The picture changes according to whether crude oil for sales to third parties is considered at very low marginal production costs or evaluated at its economic transfer price at parity with the bulk of integrated production. However, it seems that the majors' motivations were essentially strategic in nature. Since competition in sales to third parties is unavoidable, it is better to be in control and hope ultimately to squeeze independent producers out of the market. In fact, major oil companies have played an increasingly important role in international bids for crude supply, while independent producers have concentrated on securing term supply contracts.

Realized prices are, no doubt, lower than posted prices, but what is their actual level? This question is highly difficult to answer since market realities are much more complex than the oversimplified picture often presented to support the case of still lower prices. To begin with, realized prices are alleged to reflect the market value of crude oil as governed by supply and demand under open competition. The so-called open market often has practically no knowledge of real competition because of lack of information. In an economic sense, competition assumes that every buyer or seller in the market is well-informed of the price behavior of other competitors. Competition is far from being the general rule in crude sales to third parties, which sales are mainly based on astute price camouflages and a dramatic shortage of information. C. C. Chen has demonstrated how each seller would attempt to increase his total profits precisely by relying on the ignorance of competitors in regard to the details of his transactions.[15] Sales prices are widely reported in trade journals, but they are rarely comparable, even when one compares bids offered for the same tender. Moreover, they are generally unrepresentative of the real value at which crude oil changes hands because of the multitude of nonprice discounts that have been widely utilized to obscure and camouflage actual "market realities." Unfortunately, the subject is too large to be covered extensively within the limited scope of this chapter. Only general indications will be presented and discussed in what follows.

The first difficulty lies in the nature and reliability of sources of information, which include trade publications, international bids, customs statistics, and so forth. The figures given are seldom accompanied by specific details about their scope and actual significance. The same set of prices reported about the same transaction might lead to different interpretations according to the various sources of information, and it is quite difficult to distinguish between showy information and significant figures. The methods of nonprice discounts are numerous. One of the most common methods operates as a freight concession whereby the crude price is quoted as an FOB price plus a freight rate much lower than the open-market, competitive freight rate. The picture can be further confused by the seller's directly quoting CIF prices or by offering, free of charge, one or more tankers to the purchasers. Another way would be to grant tied loans at favorable interest rates, a method very difficult to assess in terms of cash equivalents because it depends on specific financial and credit conditions and possibilities of the parties involved and on the alternatives offered to them, which vary with time. Other possibilities of nonprice discounts would deal with such things as quality and specifications, delivery and terms of payment, buy-back commitments, technical assistance, and the currency to be paid.

To illustrate the impact of such non-price competition, open-market prices in the mid-1960s for 34° Arabian or Iranian crude oils could vary from about $1.40 for least-favored clients to nearly $1.10-1.20 for well informed large sized European refiners, as compared to posted prices of $1.80 (Arabian) and $1.79 (Iranian). These figures represent the best guessing of market performances by competent observers of the industry. This tends to confirm, but not precisely, our estimates that the net-back at the Gulf on such a crude for an integrated producer-refiner is approximately $1.10 even. But what of these considerable differences? The most simple explanation is undoubtedly part of the truth: imprecise sets of observations made on the same underlying phenomenon cannot coincide. Hence the difference is to be ascribed to chance, to our imperfect observational scheme, and to the haphazard accumulation of errors.

In light of the above, competition in the open market seems almost mythical. Alternative tactics governing the market appear to the outside observer of the industry almost as industrial spying techniques: The better informed you are about others, the better deals you can get. The intrusion of Soviet oil in the late 1950s contributed to and aggravated the general confusion due to the Soviet Union's alleged political motivations.

CRUDE OIL PRICES, CONSUMING COUNTRIES, AND CONSUMERS

This aspect of international oil business is one of the most obscure and confused although it is, apparently, highly publicized and popularized. The parties and people directly involved have always been curiously silent, and their interests or opinions are generally presented and discussed, directly or indirectly, by third parties.

Let us begin with the consumers, who ultimately pay, in one way or another, for all the cash flow generated by the oil industry. In a general sense, they are the millions and millions of people who pay to run their cars, to heat their houses, and to cook their meals. These arc thc rctail customers. Then come middle-sized groups such as small and medium-sized enterprises and other groups that are primarily consumers of distillates for transportation, heating, and power generation. These we will call wholesale customers. Finally, we have big industries and power generation plants, which are very high consumers of heavy fuel. They will be referred to as bulk customers. This classification is very approximate and arbitrary in some respects, but it is significant and very convenient for the purposes of our discussion.

Retail customers are the eternal victims, not of producing countries or of oil companies, but of their own governments, since most of the price they pay consists of levies and taxes imposed by their governments. The records in this area are set in France and Italy, where taxes make up more than 75 percent of the sale price of gasoline. Customers are helpless since their demand is inelastic to prices and their bargaining position as individuals is limited not only because of the omnipresent political and administrative power of governments but also because of increasing costs of living and the large disparity between initial investment (the price of a car) and operating costs (price of gasoline, service, and so on). Moreover, car ownership has taken on mythical significance in Western Europe, so that local administrations are tempted to exploit this situation in order to increase income through indirect taxation.

Wholesale customers enjoy a relatively stronger bargaining position and can occasionally implement alternative technical solutions. Their tax situation is generally a moderately acceptable one, and the government usually adopts a flexible taxation policy in order to control and direct the important economic sectors making up this category of consumers.

Bulk customers are generally the pillars of economic life and prosperity in industrialized countries. Taxes on heavy fuels are normally kept very low in order to hold the basic cost of energy at a minimum so as to support and protect the competitive position of the nation. On the other hand, since the "natural" price of fuel oil is very low, it can, and must, support a few dollars in taxes per ton in order to bring it within a minimum parity of alternative sources of energy such as nuclear power and, far more important, the subsidized national coal industries. As fuel oil prices decrease, subsidies for the shaky coal industry tend to increase.

OPEC has estimated that European consumers in 1967 paid a weighted average price of about $10,739 per barrel of composite products.[16] The detailed costs and percentage breakdown of OPEC's so-called consumer barrel are shown in Table 4.1. These figures are obviously approximate and open to challenge in several respects. Nevertheless, their relative magnitude is accurate: Taxes paid in consuming countries are six times higher than those paid in producing

TABLE 4.1

Cost and Percentage Breakdown of
OPEC's "Consumer Barrel"

	Cost ($)	Percentage
Cost of production	0.285	2.7
Cost of refining	0.350	3.3
Tanker freight	0.680	6.3
Storage, handling, distribution, and dealer's margin	2.790	26.0
Oil company net profits	0.681	6.3
Indirect and turnover oil taxes in consuming countries	5.100	47.5
Revenue of producing countries	0.853	7.9
Total	10.739	100.0

Source: OPEC bulletin, September/October 1969.

countries, and there is no hope of seeing the tax burden lightened in the consuming countries.

The governments of consuming countries are, in fact, the only groups concerned with crude oil prices. The main factor that could induce them to limit the increase of products' sales prices to individuals and groups is related to the present policies regarding inflation, a very acute problem in Western Europe in recent years. The primary issue in setting the over-all philosophy and policy of consuming countries regarding oil, and more generally the energy issue, is the continuity and security of supplies. None of the industrialized nations of today can afford to have even a few hours' shortage of energy; the security premium is therefore a permanent component of energy cost and policy, although it is not always clearly formulated. However, because the major consuming countries do not have significant crude oil reserves of their own, they are almost totally dependent on big international oil companies for the security and regularity of supplies. The latter have always lived up to their responsibilities and proved to be very efficient in crisis periods (1956, 1967, and so on). The "security premium" conceded to them by the governments of consuming countries is represented, in fact, by the governments' "political" acceptance that their "national" or, better still, domestic oil industry be largely controlled by the same big oil companies. This political concession is significant when one considers the strong nationalistic feelings in Western Europe and the drive for a European unity independent of the economic and political domination of the United States. In this respect, the attempts of some governments to achieve relative energy independence and to set up their own national oil companies have not been convincing (ENI in Italy, Elf-Erap in France, and others). Furthermore, all attempts to establish a common European energy policy, within or outside the framework of the Common Market—something like a counter-OPEC—have also proved unsuccessful precisely because this sector is not under the full political control of the governments concerned.

The real area of interest for governments of consuming countries in regard to crude oil prices is actually related to their balance of payments. They are eager to reduce to a minimum their total bill in foreign currency for crude purchases. Since imports have been steadily increasing over the years, the only way to achieve this would be to bring prices down. The only "relative" exceptions in this respect are the British and Dutch governments since the crude production of the British Petroleum Company (BP) and of Royal Dutch/Shell is much higher than the crude consumption of these two countries, so that the net balance of price increase would be in their favor; the higher profits of the two companies are larger than the incremental costs of domestic supplies.

An important remark should be made at this point. The net balance of trade between producing and consuming countries is almost at equilibrium, according to OPEC's chief economist, F. Hussein.[17] He has noted that

> payments incurred by major industrialized countries for importing oil have been mostly recovered by means of substantial exports of a variety of goods to OPEC member countries, coupled with receiving comparable amounts generated mainly by oil investments and to a lesser extent by other related investments and services. The extent of this recovery in 1967 was about 87 percent, [meaning] that every dollar paid by the OECD [Organization for Economic Cooperation and Development] countries to OPEC countries had generated a reciprocal payment of 87 cents. The remaining 13 percent which is the ultimate oil import cost could hardly be described as a substantial deficit incumbering the economics of the industrialized nations with any significant burden.

It should be added that the above-mentioned figures do not include the incremental liquidities retained by individuals and governments in some producing countries and reinvested in industrialized countries;

the sums concerned are estimated to be several hundred million dollars a year.

For all the above considerations, it is obvious that the case of consuming countries throughout the 1960s calling for lower crude oil prices was very weak indeed and that the arguments could probably be used against them. The benefit of the ultimate consumers is largely overshadowed by indirect taxes imposed by their own governments. Balance of payments is not a convincing argument especially since industrialized countries have regularly exported the inflation built into the prices of their goods whereas crude prices have been stable if not decreasing. And, finally, the cost of energy derived from oil was not too expensive. In a commercial and economic sense, the cost of energy in the marketplace—as that of any other commodity—is determined by the cost of alternative sources such as subsidized coal and uncompetitive nuclear energy, and it is quite clear that fuel oil was much cheaper than such alternatives. This fact was openly admitted by the consuming countries themselves in order to explain and justify the price rises decided at Teheran and Tripoli; they recognized that oil has been too cheap for many years and that this was only a normal readjustment of the situation. In fact, European countries have benefited from the joint effect of several political considerations that helped to keep crude prices down and even put these nations in a position to exert direct and indirect pressure for further reductions.

In light of the fundamental changes that have affected the industry in the recent crises, it is clear that this debate is irremediably outdated and that a new era has started within which producing countries will play a determining political role in parity with consuming countries and oil companies.

NOTES

1. J. E. Hartshorn, Oil Companies and Governments (London: Faber and Faber, 1967), p. 128.
2. E. T. Penrose, The Large International Firm in Developing Countries (London: George Allen and Unwin, 1968), p. 65.
3. The relevant OPEC resolution reads as follows:

> That members can no longer remain indifferent to the attitude heretofore adopted by the Oil Companies in effecting price modification;
>
> That members shall demand that Oil Companies maintain their prices steady and free from all unnecessary fluctuation; that members shall endeavour, by all means available to them, to restore present prices to the

level prevailing before the reductions; that they shall ensure that if any new circumstances arise which in the estimation of the Oil Companies will necessitate price modifications, the said companies shall enter into consultation with the member or members affected in order to fully explain the circumstances;

That members shall study and formulate a system to ensure the stabilization of prices by, among other means, the regulation of production, with due regard to the interests of the producing and consuming nations; and to the necessity of securing a steady income to the producing countries, an efficient, economic and regular supply of this source of energy to consuming nations, and a fair return on their capital to those investing in the petroleum industry.

4. The relevant OPEC resolutions read as follows:

Resolution IV.32: ". . . to ensure that oil produced in member countries shall be paid for on the basis of posted prices not lower than those which applied prior to August 1960 . . . An important element of the price structure to be devised will be the linking of crude oil prices to an index of prices of goods which the Member Countries need to import."

Resolution IV.33: ". . . to work out a formula where under royalty payments shall be fixed at a uniform rate which Members consider equitable, and shall not be treated as a credit against tax liability."

Resolution IV.34: ". . . Member countries affected should take measures to eliminate any contribution to the marketing expenses of the Companies concerned."

5. Technical details of the royalty expensing agreement and an assessment of its impact on per barrel income can be found in Petroleum Intelligence Weekly (PIW), January 18, 1965, p. 6; and Middle East Economic Survey (MEES), November 11, 1966.

6. MEES, January 12, 1968.

7. The impact of these agreements on the revenues of Middle Eastern exporters was relatively limited; average per barrel income was about 79.9 cents in 1965, 83.1 in 1967, and 85 in 1968. By contrast the increase in Libya was spectacular: 62.9 cents a barrel in 1964, 83.8 in 1966, and 101.6 in 1967. For further details see MEES, September 10, 1971, or Petroleum Press Service, September 1972, p. 321.

8. "Higher Prices and Taxes Agreed in Venezuela," PIW, October 10, 1966, pp. 6-7; and "Here is How Venezuelan Tax Prices Will Work," PIW, January 23, 1967, pp. 3-4.

9. PIW, October 7, 1963, and October 24, 1966; A. R. Martinez, Our Gift, Our Oil (Vienna, 1966), p. 138.

10. Penrose, op. cit., p. 43.

11. M. Lipton, "Government and Future of World Oil," 2d Management Conference on Economics of Petroleum Distribution, Northwestern University, March 27, 1966, pp. 16-20.

12. Ibid., p. 20.

13. Ibid., p. 21.

14. M. A. Adelman, The World Oil Outlook (Baltimore: Johns Hopkins Press, 1964), and "Oil Prices in the Long Run (1963-75)," Journal of Business of the University of Chicago, April 1964; and C. C. Chen, "Crude Prices and the Postwar Japanese Oil Industry" (Ph.D. dissertation, Massachusetts Institute of Technology, 1967).

15. C. C. Chen, op. cit., p. 16.

16. OPEC Bulletin, September/October 1969; and MEES, November 14, 1969, p. 6.

17. F. Hussein, "Some Aspects of Trade Between OPEC and OECD Groups of Countries," OPEC Seminar, Vienna, June 1969.

PART II
A TECHNOCRATIC APPROACH TO CRUDE OIL PRICES

CHAPTER

5

THE PRICING ISSUE:
BACKGROUND AND ATMOSPHERE

Although the political, psychological, and emotional considerations outlined in the preceding chapters should not be taken as a justification for continuous claims from producing countries for higher crude oil prices, they do represent an effective and important factor in the present debate. They are mainly intended to clarify and explain some of the irrational attitudes of host countries towards pricing problems.

Nevertheless, problems do exist, emanating from professional facts and realities and evolving along with the continuous changes and developments in the industry and its environment. The ever changing context of crude oil prices is a most complex one, and its outcome could be directed or forced through different channels in order to serve specific interests. This forum of conflicting influences and pressures provides the background to what we shall call, in general terms, the political dimensions of prices. A brief survey of how these dimensions have actively shaped the present structure of prices in the Eastern Hemisphere will help in establishing the over-all framework for our approach to crude oil prices.

The history of Middle Eastern oil began only recently, although the area has been known to be rich in oil-bearing sites since ancient times and exploration has been pursued since the early years of this century. The strategic importance of Eastern Hemisphere supplies was emphasized during World War II and was enhanced by the relative decline of crude oil reserves and the deterioration of the energy balance in the United States. Extensive development and production of Middle Eastern fields started in the late 1940s and the fabulous crude reserves have supported the subsequent relentless expansion of production.

The pricing issue developed in the Middle East during the 1950s, in an emotional and political context where the most active role was played by nationalists and public opinion outside governmental circles and independent of official action. Governmental involvement began with the creation of OPEC in 1960 and was accompanied by a marked evolution in the philosophy of the approach to the pricing issue. The political and emotional context that was the dominant feature of popular debates over crude prices in the late 1950s was progressively replaced by a systematic and technocratic approach to pricing economics. On political and psychological grounds, the creation of OPEC was a salutary measure for the whole industry since the debate could now be conducted more rationally and could be transposed from uncontrollable popular forums to professional confrontation and brainstorming in a peaceful atmosphere in Geneva and then in Vienna, far from the troubled Middle East.

Within this new mood of insulation from popular emotional contamination, crude prices were mainly regarded as a bench mark for governmental revenue; and from this viewpoint, the terms of profit sharing were as important as the level of the price itself. The performance and achievements of OPEC during the 1960s were chiefly in fiscal and financial areas. The revenues of host governments were increased by improving their per barrel income through the royalty expensing agreement of 1964 and through the elimination of allowance for Mediterranean exporters in 1967.

The creation of OPEC added a new dimension to the pricing issue whether it is regarded from the emotional popular viewpoint or through the technocratic, financial approach of responsible officials. Once producing countries form a collective front to confront oil companies, they need to coordinate their plans and to harmonize their claims. Consequently, crude prices should be checked against each other to determine how they would fit into the collective demands fromulated by the organization and to ensure that member countries receive comparable, if not identical, benefits and advantages from them. Furthermore, this exercise is indispensable for averting possible conflicts between the specific interests and objectives of individual countries and for elaborating the compromises and accommodations that are necessary to establish a common group stand.

However, the harmonization and standardization of taxation procedures on the occasion of the royalty expensing agreement did not provide a satisfactory answer to the subjective question of whether member countries would receive equal benefits from the application of the new uniform rules to existing prices, and the question remained unanswered for the simple reason that nobody knew how these existing prices would compare with each other. In other words, if the existing pattern of postings were to embody a specific built-in advantage or penalty in the price level of one crude oil in comparison with others, then the same disparity would be reflected in the governmental per barrel income through the application of identical taxation formulas. The complexity and ambiguity of this aspect of pricing economics result from the fact that there is no evidence that posted prices do fit into a consistent, over-all pattern, although it is currently believed that they generally vary in terms of 2 cents a barrel for each degree API of gravity and that they are more or less related by net-back formula equalizing their delivered values in the marketplace. These general remarks underline the two most important factors in price disparities, namely gravity and geographical location, but they fail to clarify how posting levels would compare with each other or to explain the mechanisms of price formulation and evolution.

Crude oil prices, and especially posted prices in the Middle East, are generally described as being "historical" in nature, and the disparities that can be occasionally discerned are accepted as necessary "facts of life." Notwithstanding, no serious study of pricing economics and mechanisms can be undertaken without prior investigation of any consistent pattern that might govern crude prices. Once the existence of such an over-all structure has been demonstrated, it would be possible to investigate the mechanism of its formulation and evolution.

The first steps towards professional maturity of producing countries materialized through their efforts to penetrate the mysteries of the taboo issue of crude oil pricing. They were motivated by the naive desire to understand its basic economics and then to demonstrate the soundness of their claims. Because the producing countries started from almost complete ignorance of the realities of the industry, such a technocratic approach was highly stimulating since it served to establish a kind of dialogue on professional grounds with the industry and on intellectual grounds with outside observers and critics. However, it proved to lead nowhere in the confrontation over the necessity or desirability of price readjustments, although it enabled the debates to be put on increasingly higher levels of elaboration and sophistication. Indeed, the technocratic approach enabled the actual situation to be understood in a given place at a given moment and the possible mechanisms of evolution to be elaborated through certain models of simulation. However, too many factors and parameters had to be taken into consideration, and their number and complexity increased with the sophistication and refinement of the models considered, but none of the latter were able to reflect the apparently simple and universal structure of posted prices, which appear insulated from market evolution and fluctuation. Economic analysis would make it possible to understand operational plans and professional decisions within the industry, as well as individual commercial transactions dealing with crude oils and petroleum products, but it fails to explain the past history of prices, and its predictions about prospective evolution are often contradicted by reality.

The shortcomings of any technocratic approach to crude oil prices through economic analysis will be outlined throughout this study. Roughly speaking, they derive from a fundamental conflict in philosophy and concept in two respects. Firstly, almost all models proposed to simulate the pricing process are based on the assumption that price formulation and evolution are subject to economic laws in competitive markets; this assumption is in contradiction to the integrated and oligopolistic structure of the oil industry, and checking the pricing issue against integrated economics would show that crude oils "cooperate" to meet market demands instead of competing. Secondly, the evolution of crude oil prices is largely influenced by extraprofessional considerations, so that it cannot be explained by any objective analysis of market realities. Quite often, price changes are policy-induced and are intended to force the market into a predetermined evolution instead of being influenced by its recent history and performances.

In other words, crude oil prices do not seem to derive from an economic concept relating them to the economics of production or from a commercial concept governed by the dynamics of supply and demand but rather from a strategic concept that aims to insulate prices from the continuous fluctuation and evolution of the industry. This fundamental conclusion is an innovation within the framework of conventional approaches to the economics of the international oil industry. The technocratic approach could enable one to elaborate specific pricing patterns that satisfactorily reflect market realities. However, further investigation and deeper economic analysis of the mechanisms of formulation and evolution of these patterns would fail to provide any satisfactory answer, even though they might enable one to have a highly sophisticated and detailed vision of the industry. These dead-end proceedings have progressively led to the conclusion that the basic concept that aimed to explain prices through economics was unsatisfactory and that the only way out of the dilemma was to proceed through the strategic and political approach outlined above.

Nevertheless, it is essential to have a full understanding of the technocratic approach to crude oil prices before proceeding with the analysis of the underlying political forces and strategic considerations so that one can better visualize the problems outlined by the conventional economic analysis. The structure of crude oil pricing patterns will be briefly investigated in the following chapters to provide the necessary technical and economic background and to outline the nature and extent of conflicts and problems that result from the conventional concept of crude oil prices.

CHAPTER

6

**PRICING PATTERNS
IN THE GULF AREA**

Crude oil postings in the Eastern Hemisphere were formally introduced to the Gulf area with the establishment of the 50/50 profit-sharing pattern by Aramco in 1950. In the postwar period, the price of 36° Arabian light crude oil, FOB Ras Tanura, was considered as representative for all crude oils produced in the area, irrespective of their gravity, quality, or geographical origin. In fact, there was very little of these crude oils, which originated in Iran, Iraq, Kuwait, and Saudi Arabia, and in the absence of postings, their prices were not known since no significant quantities were traded outside integrated channels. In other words, there was no significant pricing structure in the Gulf area in the period from 1945 to 1950 during which the collective level of crude oil prices, represented by 36° Arabian light, was successively reduced with respect to prices prevailing in the U.S. Gulf as a consequence of the shift from the Gulf-plus system to the dual-basing system, according to which the CIF prices of all crude oils were equalized at a given marketplace.

It should be recalled that crude oil prices outside the United States were determined by the Gulf-plus system until 1944 when the Gulf area was recognized as a basic pricing point in conjunction with the U.S. Gulf. The predominance of the Gulf-plus system was due to the historical fact that the U.S. oil industry was by far the largest in the world and that the U.S. Gulf was practically the only place where importers could get supplies and spot cargoes on the open market to cover any likely requirements. The contribution of the Middle East to world production was limited to some 5-6 percent of the world total during the war period against about 10 percent from Venezuela and 64 percent from the United States. The necessity of elaborating a specific pricing pattern in the Gulf area in the postwar period was evidenced by the fact that the deterioration of the energy balance of the United States was aggravated to such a point that the still largest crude oil producer in the world became a net oil importer as of 1948, so that the highly expanding markets in Western Europe and Japan had to be supplied from the only area where reserves could stand the fantastic increase of demand, namely the Middle East.[1]

The establishment of a specific pricing pattern in the Gulf area was achieved in two successive stages. During the period from 1944 to 1949, the general level of prices, as symbolized by the price of a reference crude oil, was adjusted in comparison with the general level of prices in the U.S. Gulf to determine their relative positions through the net-back formula. Once the relative position of Middle Eastern oil was settled in the new over-all pricing pattern of the international oil industry, a specific pricing pattern was elaborated thereafter for the whole spectrum of crude oils available in the Gulf, articulated on the price of the reference crude oil. The elaboration of this specific pattern was achieved in 1953; almost all posted prices published since then fitted into that pattern with satisfactory approximation, and all price variations that took place affected prevailing posted prices collectively without altering their relative positions.

1944-49: THE ECA ERA

The Gulf-plus system remained in force as a general rule for crude oil pricing until 1944, when the Arabian Gulf area was recognized as a separate price basing point independent from the U.S. Gulf-Caribbean area. This was achieved under pressure from the British admiralty to secure low-cost fuel oil for the British fleet and to stop paying the phantom freight rates from the U.S. Gulf associated with the Gulf-plus pricing system. The main negotiations were conducted with Aramco, and the British negotiator agreed to a price of $1.05 a barrel for 36° Arabian crude oil FOB Ras Tanura, the same as for similar crude oils in the U.S. Gulf. At this level, the net-back formula, based on prevailing freight rates, would equalize CIF prices in Southern Europe somewhere around Italy.

Up to 1944, the concept of crude oil prices outside the United States did not even exist since all crude oils were priced as if they originated in the U.S. Gulf. The real significance of the 1944 agreement is that it introduced the concept of crude oil FOB prices to the international oil industry. However, this new concept had little significance to Middle Eastern oil in general until posted prices were formally introduced in the area in 1950 with the introduction of the 50/50 profit-sharing pattern by Aramco. In the meantime most crude oil production was disposed of through integrated channels at unpublished transfer prices. Few arm's-length sales were made to non-affiliated companies, but very little information is available about these transactions in which Aramco seems to have played a significant role. Most of the data reported refer to 36° Arabian light crude oil, which thus emerges as a pricing leader.

By the end of World War II, prices in the Gulf area and in the U.S. Gulf were equalized at $1.05 a barrel for 36° crude oil. (See Table 6.1.) Demand for petroleum in the United States and abroad rose sharply during the immediate postwar period, and the general price level also shot up after the lifting of wartime controls gave free rein to pent-up inflationary forces. U.S. crude oil prices underwent several increases totaling 45 cents a barrel in 1946 and 95 cents in 1947. In addition, pipeline charges and other costs rose, so that the delivered cost of Texas sour crude at Gulf of Mexico ports reached $2.75 a barrel by December 1947.

Changes in the United States were generally followed by changes in the Middle East, though often with a time lag and occasionally by smaller amounts. In no case did Middle East price changes preceed those in the United States, nor did Middle East prices change entirely independently of United States prices. By April 1948, when the Economic Cooperation Administration (ECA) began to operate, the prevailing price of 36° Arabian crude oil was $2.22 a barrel. The total postwar price increase in the Gulf area thus amounted to $1.17 a barrel as compared to a total advance of $1.70 a barrel in the U.S. Gulf.

The $2.22 a barrel price set the equalization line with U.S. Gulf postings at $2.75 a barrel in Northwestern Europe (Southampton-U.K.) at prevailing U.S. Maritime Commission (USMC) freight rates. The net-back formula relating these two prices was applicable to Iranian oil FOB Abadan and was therefore called the Iranian formula. However, it remained in force only for a few months and was replaced by an Arabian formula, which led to a further price reduction for Arabian crude oil down to $2.03 a barrel FOB Ras Tanura in May-June 1948. The change of formulas was motivated by the fact that, in 1948, the United States had become a net oil importer and the Caribbean area emerged as the effective alternative source of supply to Europe. Under the Arabian formula, CIF prices of Venezuelan and U.S. Gulf crude oils were equalized at New York, whereas CIF prices of Venezuelan and Arabian crude oils were equalized at Southampton at USMC freight rates.

In July 1949, the price of 36° Arabian crude oil was further reduced to $1.75 a barrel, without any similar move in other producing areas. At this level, CIF prices of 36° crude oils from the U.S. Gulf, the Caribbean, and the Gulf area were equalized at New York under a North American formula based on freight rates at USMC minus 35 percent. The price of 36° Arabian light, considered as a reference crude oil for the Gulf, remained stable at $1.75 a barrel until the worldwide post-Korean price increase in mid-1953.[2]

TABLE 6.1

Evolution of Crude Oil Prices
in Postwar Period up to 1953
(dollars a barrel)

Crude Oil Gravity (°API) Export Terminal	Arabian Light 36° R. Tanura	Kuwait 31° M. Ahmadi	Oficina 36° P. La Cruz	West Texas Sour 36° Gulf Coast
1944-45	1.05	—	—	1.05
1946 April	—	—	—	1.15[a]
July	—	—	—	1.40[a]
November	—	—	—	1.50[a]
December	1.20[a]	—	—	—
1947 March	1.59[a]	—	—	1.80[a]
October				2.10[a]
December	2.22[b]	—	—	2.75
Difference 1945-48	+1.17	—	—	+1.70
1948 March	2.22	2.15	2.65	2.75
July	2.03	1.97	2.65	2.75
1949 April	1.88	1.82	2.65	2.75
July	1.75	1.75	2.65	2.75
October	1.75	1.65	2.65	2.75
1950, 1951, 1952	1.75	1.65	2.65	2.75
1953 April	1.75	1.50	2.65	2.75
June	1.75	1.50	2.90	3.00
July	1.97	1.72	2.90	3.00
Difference 1948-52	-0.47	-0.65	—	—
Difference 1948-53	-0.25	-0.43	+0.25	+0.25

[a]Estimated.
[b]Price representative of all crude oils in the Gulf area.

Source: Compiled from H. J. Frank, Crude Oil Prices in the Middle East (New York: Praeger Publishers, 1966); W. Leeman, The Price of Middle East Oil (Ithaca, N. Y.: Cornell University Press, 1962); and C. Issawi and M. Yeganeh, The Economics of Middle Eastern Oil (New York: Praeger Publishers, 1962).

It is obvious that the genesis of crude oil prices in the Middle East, as described above, did not derive from any economic concept by which the price of crude oil, like any other commodity, would be governed by the economics of its production, the cost of which was quite limited indeed. (Roughly speaking, a price of about 50-60 cents a barrel would have corresponded to almost 100 percent return on investment, taking into consideration production costs and fixed royalty payments.) The original concept of crude oil prices in the Gulf area seems to derive from an established historical fact, the evolution of which was closely linked to the evolution of petroleum economics in the United States. One might be tempted to check the above-mentioned price movements against the evolution of market demand and to explain them in commercial terms. Such an exercise has been conducted in the literature, mainly by Helmut J. Frank and W. A. Leeman.[3] It appears however, that price movements in the Eastern Hemisphere were not always consistent with the requirements of sound commercial policies. In fact, the relative reduction of crude oil prices in the Gulf with respect to the U.S. Gulf postings was achieved under sustained pressure from the U.S. Government through its ECA.[4]

A large part of the U.S. aid under the European Recovery Program, the Marshall Plan to help reconstruct devastated Europe, was devoted to the purchase of crude oil and petroleum products. The ECA and its successor, the Mutual Security Agency (MSA), were responsible for financing at a time when dollars were scarce. ECA was consequently very much concerned to see that its funds were well-spent and that prices paid for oil were lowered to a minimum. These objectives also matched the desire of securing cheaper crude oils for U.S. domestic markets, which began to import increasing supplies in 1948.

Consequently, ECA pushed towards extensive development of Middle Eastern reserves and started a large-scale action to induce price cuts of these crude oils. As a government agency holding the cash, ECA could mobilize a wide range of direct and indirect means of pressure, which proved to be very effective. Middle Eastern oil was designed to help Europe's dollar shortage by the shifting from imports of high-cost finished products from the Western Hemisphere to cheaper crude oils. To achieve this, Europe's war-damaged refining facilities were restored and expanded as rapidly as possible.

It is obvious then that the evolution of Middle Eastern prices in the postwar period was directly governed by political pressure from the U.S. Government to comply with the necessities of its economic policy and European commitments. Oil companies, namely the majors, aligned their prices with the demands of the ECA because (1) this was a commanding condition for their access to the rapidly expanding markets in Europe; (2) the huge reserves and very low production costs of Middle Eastern crude oils compensated for the relative loss in prices as compared with alternative sources, so that profit margins remained healthy and substantial; and (3) there was no significant counterpressure to their move from producing countries.

Under these conditions, one may wonder why the ECA deemed the price level of $1.75 a barrel satisfactory and did not attempt to push for further price reduction. This question will be investigated later in Chapter 13, where it will be demonstrated that the limitation of the downward price movement was due to the necessity of complying with U.S. strategic objectives concerning the stability of domestic prices.

PRICING PATTERNS IN THE GULF AREA

In the postwar period the general practice of oil companies was to use the price of Arabian crude oil FOB Ras Tanura as a uniform price reference for the Eastern Hemisphere. By June 1948, all FOB prices in the Gulf were computed from Ras Tanura.[5] In other words, there was no specific pricing pattern in the Gulf area in the period from 1944 to 1949, during which the general level of prices was settled. A significant price differentiation took place on October 1949, when the price of 31° Kuwait crude oil was unilaterally reduced by 10 cents down to $1.65 a barrel shortly after the final settlement of the general price level in the Gulf area at $1.75 a barrel. Compared to the reference price of 36° Arabian light, this reduction displayed a gravity differential of 2 cents per degree °API that seems to have governed the price structure in the Gulf area since then. It is worthwhile to mention that the gravity of the reference crude oil, namely 36° Arabian light, was changed in February 1959, when it began to be priced as a 34° crude oil.

In the late 1960s there were some 25 postings in the Gulf for crude oils with gravities extending from 17° (Eocene) to 41° (Qatar). The main question that mobilized substantial effort within and outside OPEC was whether there was a rational and consistent pattern governing the postings. The general attitude of oil companies has often been to escape raising the question, to obscure the matter, and to avoid any direct reference to it by evoking the "historical" nature and origin of posted prices and their limited significance with respect to market realities. Producing countries and the general public were under the impression that crude oil prices were related in terms of a 2 cents gravity differential per each full degree of gravity difference, but no direct relationships could be established accordingly between actual postings in the area. In particular, Kuwait and Iraq long claimed that their crude oils were underpriced but were unable to demonstrate their case strongly enough to induce an appropriate adjustment.

It is quite significant that the existence of a pricing pattern governed by the rule of thumb of a 2 cent price variation per each full degree gravity is not supported by any professional evidence. Posted prices are published FOB export terminals for a specific value of gravity for each crude oil, and the price is reported to vary by 2 cents per degree, should the gravity of the same crude oil shift from the announced value. In other words, this adjustment is to be applied to crude oils individually and separately without any indication about its validity when gravity variation from one crude oil to another is considered. This formal adjustment as announced in specialized trade journals[6] in association with posted prices will be called the gravity fluctuation differential (GFD).

The other absolute fact is that posted prices are given for a full range of one degree gravity (°API), irrespective of the exact value of actual crude oil gravity in that range. For example, the price of 36° Arabian crude oil is valid over the range 36-36.9°. Consequently, any pricing scheme would have to tolerate an inherent margin of imprecision of 2 cents a barrel and to allow for a gravity fluctuation of one degree.

Posted prices are quoted FOB export terminals, which have different geographical locations in the Gulf area, thus introducing small freight disparities that are significant enough to shadow the existence of any presumed over-all pricing pattern. These disparities should be eliminated by calculating parity prices under the assumption that all crude oils are available at the same reference point representing the Gulf, namely Quoin Island (the reference point for the Gulf area in the International Tanker Nominal Freight Scale [Intascale] and the Worldwide Tanker Nominal Freight Scale [Worldscale] freight rates schedules). This exercise is presented in Table 6.2, in which parity prices FOB Quoin Island are obtained by adding to the actual postings the proper freight cost from the corresponding export terminal. The choice of a proper freight rate is somewhat arbitrary at this stage, and the impact of its fluctuation might be illustrated by considering different values of discounts off basic flat rates (Intascale minus 50, minus 35, and flat). Intascale rates are considered (and not Worldscale rates) because of the historical scope of the present exercise.

The freight issue will be investigated in some detail in the next chapter. In particular, the fluctuating nature of freight rates and its impact on pricing patterns will be outlined. In the absence of any rational criteria for the selection of an average constant freight rate for the purpose of pricing, there could be several pricing patterns corresponding to the different values of freight rates that could be considered. In the present exercise, the freight rate will be taken at the historical reference value of Intascale minus 35, and it may be easily demonstrated that the distortion induced by freight fluctuation

TABLE 6.2

Crude Oils in Gulf Area,
Parity Prices FOB Quoin Island
(dollars a barrel)

Key Number	Crude Oil	Terminal	Gravity (°API)	Sulfur (weight percent)	Intascale ($/long tons)
1[a]	Qatar	Umm Saīd	41-41.9	1.07	0.48
2[a]	Qatar	Umm Saīd	40-40.9	1.07	0.48
3	Murban	Jebel Dhanna	39-39.9	0.80	0.55
4	Murban	Jebel Dhanna	38-38.9	0.80	0.55
5	Umm Shaif	Das Island	37-37.9 34-34.9	1.10	0.25 0.25
6	Qatar	Halul Island	36-36.9	1.50	0.41[b]
7[a]	Basrah	Fao	35-35.9	1.90	0.90 + 0.37
8	Basrah	Khor el Amaya	35-35.9	1.90	0.81
9	Darius	Darius Terminal	34-34.9	2.40	0.61[b]
10[a]	Iranian Light	Abadan	34-34.9	1.32	0.77 + 0.37
11	Iranian Light	Bandar Mahshar	34-34.9	1.32	0.67
12	Iranian Light	Kharg Island	34-34.9	1.32	0.46
13[a]	Arabian Light	Ras Tanura	34-34.9	1.63	0.36
14	Oman	Mena Fahal	33-33.9	1.15	-(0.03)[b]
15	Dubai	Fateh	32-32.9	—	0.44[b]
16[a]	Iranian Heavy	Abadan	31-31.9	1.57	0.77 + 0.37
17	Iranian Heavy	Kharg Island	31-31.9	1.57	0.46
18[a]	Kuwait	Mena el Ahmadi	31-31.9	2.50	0.50
19[a]	Arabian Med.	Ras Tanura	31-31.9	3.05	0.36
20[a]	Arabian Heavy	Ras Tanura	27-27.9	3.20	0.36
21	Khafji	Ras el Khafji	26-26.9	2.85	0.43
22	Burgan	Mena Saud	23.5-24.4	1.70	0.48
23	Ratawi	Mana Saud	23.5-24.4	4.30	0.48
24	Cyrus	Cyrus Terminal	19-19.9	2.90	0.61[b]
25	Eocene	Mena Saud	16.5-17.4	4.75	0.48

TABLE 6.2 (continued)

Con-version Factor	Intascale $/Bbl	Posted Prices $/Bbl	Parity Prices FOB Quoin Island, $/Bbl		
			Case I	Case II	Case III
0.12770[a]	0.061	1.950	1.980	1.990	2.011
0.12840	0.061	1.930	1.960	1.970	1.991
0.12925	0.071	1.880	1.915	1.926	1.951
0.12995	0.071	1.860	1.895	1.906	1.931
0.13075	0.033	1.860	1.877	1.881	1.893
0.1331	33	1.780	1.797	1.801	1.813
0.13155	0.054 [b]	1.830	1.857	1.866	1.884
0.13230	0.119 + 0.049	1.720	1.829	1.847	1.888
0.13230	0.107	1.720 (1.780)[c]	1.774 (1.834)	1.790 (1.850)	1.827 (1.887)
0.13310	0.081	1.630	1.670	1.683	1.711
0.13310	0.102/+0.049	1.730	1.830	1.845	1.881
0.13310	0.089	1.780	1.825	1.838	1.869
0.13310	0.061	1.790	1.820	1.830	1.851
0.13310	0.048	1.800	1.824	1.830	1.848
0.13390	-(0.004)[b]	1.820	1.818	1.817	1.616
0.13475	0.060[b]	1.690	1.720	1.729	1.750
0.13555	0.104 + 0.050	1.580	1.682	1.698	1.734
0.13555	0.062	1.630	1.661	1.670	1.692
0.13555	0.068	1.590	1.624	1.634	1.658
0.13555	0.049	1.590	1.615	1.622	1.639
0.13900	0.050	1.470	1.495	1.502	1.520
0.13986	0.060	1.420	1.450	1.459	1.480
0.14210	0.068	1.480	1.514	1.524	1.548
0.14210	0.068	1.410	1.444	1.454	1.478
0.14630	0.089[b]	1.340	1.385	1.398	1.429
0.14880	0.071	1.280	1.315	1.326	1.351

[a]Estimated.
[b]Posted price FOB Khor el Amaya by Compagnie Française des Pétroles.
[c]Crude oils produced and posted prior to 1960.

Case I: Intascale minus 50 percent (i = -0.50)
Case II: Intascale minus 35 percent (i = -0.35)
Case III: Intascale flat (i = 0)

Source: Compiled by the author.

would not significantly affect the validity of the pricing pattern, which will be outlined below, mainly because of the short distances between export terminals and Quoin Island in the tiny Gulf area.

The existence of any pricing pattern governing parity prices FOB Quoin Island thus obtained would be indicated if they could be derived from the price of the reference crude oil (34° Arabian light) by appropriate adjustments in terms of gravity and/or quality differentials. To start with, let us check whether the general rule of thumb of 2¢/Bbl/degree gravity differential can be satisfactorily representative of the presumed pricing pattern. The best way to represent this investigation is to plot parity prices against crude oil gravities and to check the existence of any linear correlation between them. An examination of the array of parity prices as represented in Figure 6.1-A would show that there is no such relationship governing the whole spectrum of crude oil postings in the Gulf area. In other words, gravity differential alone cannot account for the presumed pricing pattern.

However, a linear correlation seems to be satisfactorily workable for crude oils of 33° gravity and above, and it might be easily checked that a gravity differential of 2¢/Bbl/degree represents a good approximation for such a statistical correlation. The linear relationship as illustrated in Figure 6.1-B will be called the linear price scale, the slope of which corresponds to the value of the gravity differential relating crude oil prices. The scheme thus outlined is determined within the range of the acceptable approximation of 2 cents a barrel.

(Parity prices of the following crude oils display significant deviations exceeding the tolerated approximation of 2 cents a barrel:

• 40° and 41° Qatar crude oils seem to be overpriced by about 2 cents, which is at the fringe of the tolerated approximation. This marginal deviation is suspected to be a premium for the relatively low sulfur content of Qatar crude oils, but this has not been confirmed for other crude oils of equally low sulfur content such as Abu Dhabi.

• A 6 cent deviation is displayed by the posted price of 35° Basrah crude oil FOB Khor el Amaya by all IPC mother companies except the Compagnie Française des Pétroles (CFP).

• The 14-15 cent deviation displayed by 34° Darius crude oil is substantial but not significant. This crude oil was originally priced while the fields were still being developed and even before it was declared as commercial.[7] Crude oil gravity at that early stage was about 27-28°, which would account for the above-mentioned deviation in combination with the relatively high sulfur content (2.4 percent) of the crude oil. The real question is why the price was not adjusted upwards when gravity finally settled at 34°. This was probably due to the absence of integrated markets to dispose of this crude oil, which was the first to be produced under a joint venture agreement.)

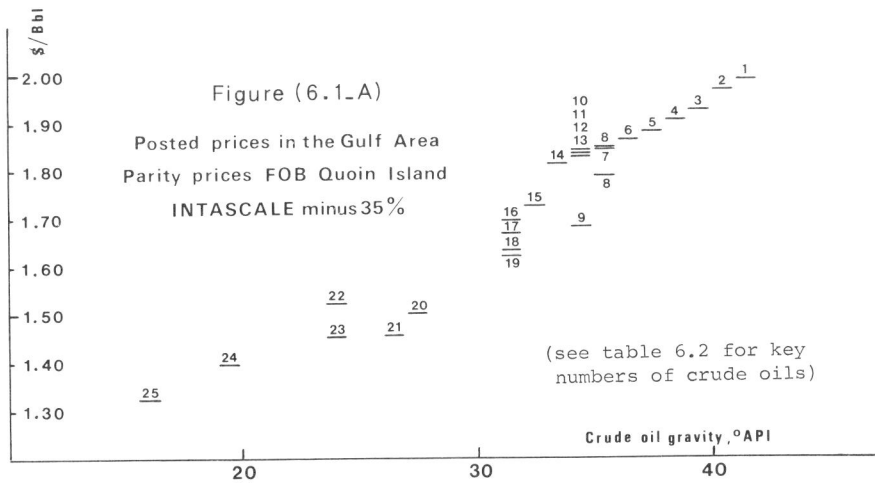

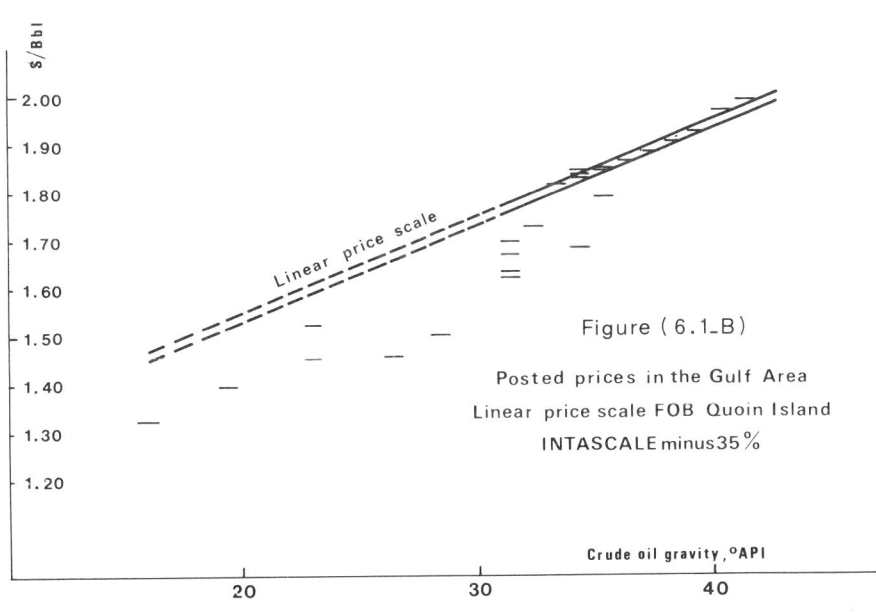

The basic pattern outlined above may be extended to crude oils of 32° gravity and below if the deviations displayed by their parity prices can be accounted for in terms of specific quality differentials over and above the gravity differential. An examination of crude oil characteristics would suggest that disparities in sulfur content might be responsible for the price deviations mentioned above. In order to check the validity of this assumption, price deviations are plotted in Figure 6.2 against corresponding differences in the sulfur contents of the crude oils considered and the reference crude oil (1.63 percent sulfur content of 34° Arabian light). Price deviations are allocated 1 cent fluctuation above and below the exact value obtained by calculation to account for the over-all approximation of the pricing pattern. Price deviations are significant only if corresponding segments do not cut into a range of 1 cent above or below the zero line.

Sulfur differentials should also be allowed some range of fluctuation since sulfur content does not seem to be absolutely constant. The available data often give different sulfur contents for the same crude oil (depending on the timing and sampling conditions of the crude assay), and it is not unusual for the sulfur content of a given crude oil to change and fluctuate over the years with the development of the field and with the relative contribution of one horizon or another of the same reservoir to over-all production. It is very difficult, indeed, to know what was the actual sulfur content of a given crude oil by the time its price was posted, and the prevailing pricing pattern does not allow for price adjustment in connection with sulfur fluctuation (as for gravity).

The range of sulfur fluctuation that can be tolerated for the purpose of working out a statistical correlation between price and sulfur differentials cannot be easily assessed on purely technical grounds. However, it should be pointed out that the parity prices of 1.32 percent Iranian light and 1.9 percent Basrah crude oils do not display any significant deviation from the reference price of 1.63 percent Arabian light crude oil. Therefore, sulfur differentials will be allocated a 0.3 percent fluctuation, as shown in Figure 6.2.

It is not unreasonable to expect sulfur to play a significant role in affecting the basic pattern of posted prices since the latter developed at a time when it was intended to reflect market realities and economics. It is a fact that higher-sulfur crude oils are more costly to handle and to refine while lower-sulfur oil would lead to a savings. It is significant that the parity prices of crude oils of 33° gravity and above do not display any deviation although they cover a wide range of sulfur content. Therefore, it may be assumed that the upper limit of sulfur content tolerated by the industry would have been around 2 percent, so that the sulfur differential might be expected to apply to crude oils of only more that 2 percent sulfur.

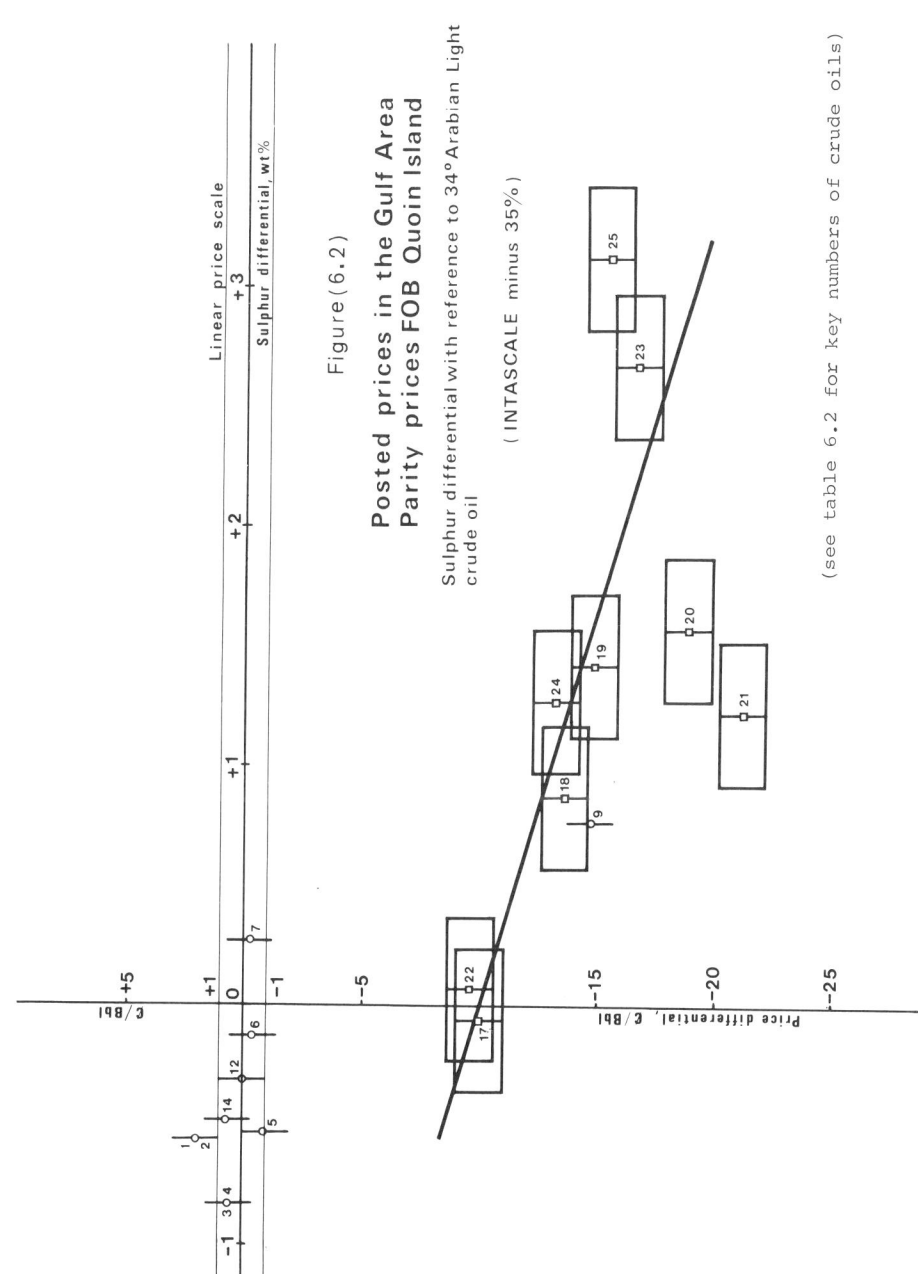

An examination of Figure 6.2 suggests that a linear correlation might be reasonably considered to account for price deviations in terms of a sulfur differential of about 3 cents a barrel for each 1 percent sulfur over and in excess of 1.6 percent for crude oils with sulfur contents exceeding 2 percent. The precision of this statistical relationship is fairly satisfactory although there are a few exceptions that display residual price deviations that may be tentatively explained and justified.*

If we eliminate price deviations attributable to sulfur differentials, then the parity prices of all crude oils of 31° gravity and below would be left with a flat difference of about 9-11 cents a barrel without any obvious connection to the factors already considered in the elaboration of the above-mentioned correlations. This price penalty seems to affect all crude oils collectively irrespective of any difference in their gravity and/or quality. This penalty, which was first introduced by the Anglo-Iranian Oil Company (AIOC) in 1953, will be called the investment discontinuity penalty for reasons that will be indicated later in this study.

To sum up the above, it may be fairly stated that posted prices in the Gulf are governed by the following over-all pattern:

1. The pattern is shadowed by freight disparities due to the different geographical locations of export terminals. They can be eliminated by considering parity prices FOB Quoin Island taken as a representative reference point. The validity of the pattern is not affected by freight fluctuation.

2. Parity prices are related to the price of a reference crude oil (36° and later 34° Arabian light) in terms of a basic gravity differential of about 2 cents for each full degree gravity difference.

3. The parity prices of all crude oils of 31° gravity and less are affected by a flat investment discontinuity penalty of about 9-11 cents a barrel.

4. Parity prices of crude oils with a sulfur content higher than 2 percent are affected by a sulfur penalty in terms of about 3 cents for each 1 percent sulfur in excess of the reference value of 1.6 percent.

---

*It seems that 17° Eocene crude oil is overpriced by about two cents—which is at the fringe of tolerated approximation—probably because this very heavy crude oil is used mainly in the manufacture of bitumen, the value of which is not affected by high sulfur content. The 6-8 cent deviation displayed by 26° Khafji crude oil and the 3-5 cent deviation by 27° Arabian heavy probably are due to the fact that these heavy crude oils are primarily destined to supply the Japanese market, where there is great effort to cut down the cost of oil supplies.

5. The whole pricing pattern is determined with an approximation of 2 cents a barrel corresponding to a gravity fluctuation of one degree.

Under these conditions, posted prices in the Gulf may be represented by the price of the reference crude oil except for a few crude oils that display significant residual price deviations, which may be tentatively explained and justified on professional grounds. However, it is worth pointing out the specific case of 35° Basrah crude oil posted FOB Khor el Amaya at $1.78 a barrel by Compagnie Française des Pétroles (CFP) in conformity with the oil's parity value and at $1.72 a barrel by BP, Esso, Mobil, and Shell, thus displaying a 6 cent penalty. This deviation is not due to a failure of the pricing pattern but seems to have been policy induced as retaliation by the majors, not including CFP, against the expropriation of about 99.5 percent of the IPC concession under Law 80.*

Only CFP took the initiative to set the new posted price at its parity value of $1.80/Bbl. All other member companies of the IPC group maintained the same postings of $1.74/Bbl FOB both Fao and Khor el Amaya, thus penalizing crude exported via the new terminal by 6¢/Bbl. This penalty is essentially and exclusively political in nature and may be easily understood within the framework and atmosphere of conflict and hostility with the Iraqi Government. Z. Mikdashi reports that the motives of CFP to set a higher price were also exclusively political in nature, "among which were: (a) the desire to obviate any possible political objection to French commercial interests in Iraq (especially at the time of the Algerian war) by offering the Iraqi Government a higher yield on Iraqi Exports, and (b) the hope that the Iraqi authorities would refrain from exacting heavy port dues on oil loaded from Khor el Amaya."[8]

---

*Relations between the IPC and the Iraqi Government have always been characterized by suspicion and lack of confidence. Disputes were constant, and most of them are still unsettled. The situation worsened dramatically from mid-1958 during the Kassem regime and culminated with the proclamation of Law 80 on December 12, 1961, depriving the three operating companies of the IPC group of 99.5 percent of their concessionary area and leaving them with only 1,938 square kilometers of producing fields. After the law took effect, 36° Basrah crude oil was exported from the shallow terminal at Fao with high port charges, which were taken into consideration to set the posted price at $1.74 a barrel. Early in 1962, a new deepwater terminal was commissioned at Khor el Amaya, where the parity price of $1.80 a barrel would not be affected by the same port charges.

ESSENTIALS OF THE PRICING STRUCTURE

The pricing pattern of FOB postings in the Gulf, after appropriate adjustments for freight disparities, investment discontinuity and sulfur penalties, and others, whenever necessary, can be represented by the following general relation:

$$X = a \cdot G + b \quad (6.1)$$

where

X = \$/Bbl—parity posted price (FOB Quoin Island);
G = °API—crude oil gravity covering the full range of one degree;
a = \$/Bbl/degree—basic gravity differential;
b = \$/Bbl—pricing factor.

An investigation of pricing patterns, FOB export terminals outside the Gulf (Eastern Mediterranean, Libya, Algeria, Nigeria, and others) would show that posted prices in each area are governed by the same basic pattern of linear relationships with respect to gravity in terms of 2¢/Bbl/degree. However, an investigation of the pricing pattern of Venezuelan crude oils, as illustrated in Table 6.3 and Figure 6.3, would demonstrate that posted prices do fit into linear correlations with respect to gravity, but that the basic gravity differential governing them—that is, the slopes of the corresponding linear price scales—are of a magnitude of 4-5¢/Bbl/degree as against 2 cents in the Eastern Hemisphere, despite the fact that the posted price of each crude oil, considered separately, does vary in terms of 2 cents for each degree of gravity fluctuation in most cases. These particular findings clearly show that the conventional concept of gravity differential might have more than one meaning. The picture can be clarified by the introduction of the following new concepts.

As already outlined, the only official fact published by the industry is that posted prices for each crude oil considered individually are to be adjusted by 2 cents a barrel for each degree of gravity fluctuation above or below the specific value associated with the prices considered. This conventional concept will be called the gravity fluctuation differential.

Now, let us consider a group of crude oils each with a given and constant gravity so that there is no need to consider any gravity fluctuation. In fact, it would be more appropriate to consider gravity variation from one crude oil to another. Should it be established that actual prices do fit into a linear correlation with respect to gravity, then it would be possible to consider that such a pricing pattern is governed by a basic gravity differential, relating prices of different

TABLE 6.3

Posted Prices of Selected Venezuelan Crude Oils, 1969-70

No.	Crude Oil	Gravity (°API)	Sulfur (weight percent)	Terminal	Posted Price ($/Bbl)	Gr. Diff. (¢/Bbl/ Degree)	Conversion Factor (φ)
1	Condensate	50	0.03	La Cruz	3.150	flat	0.12170
2	San Joaquin	41-41.9	0.13	La Cruz	3.100	2	0.12770
3	Mulata	37-37.9	0.80	Caripito	2.840	2	0.13075
4	Lagocinco	36-36.9	—	Cardon	2.760	2	0.13155
5	Officina	35-35.9	0.80	La Cruz	2.800	2	0.13230
6	Officina	34-34.9	0.80	La Cruz	2.760	2	0.13310
7	Paconsib	33-33.9	1.30	Cardon	2.700	2	0.13390
8	Jusepin	32-32.9	1.05	Caripito	2.740	2	0.13475
9	Lagomar	31-31.9	1.40	Cardon	2.550	2	0.13555
10	Guanipa	30-30.9	0.73	La Cruz	2.530	2	0.13640
11	Mara	29-29.9	2.10	Cardon	2.400	2	0.13720
12	Lagotreco	28-28.9	1.55	Miranda	2.310	2	0.13810
13	Oscurote	26-26.9	1.25	La Cruz	2.300	2	0.13985
14	Leona	24-24.9	1.24	La Cruz	2.210	2	0.14160
15	Cabimas	22-22.4	2.06	Miranda	2.070	2.5*	0.14380
16	Temblador	20-20.9	0.81	Caripito	2.150	2.5*	0.14540
17	Tamare	18	—	Miranda	1.870	flat	0.14779
18	Merey	17	2.50	La Cruz	1.880	flat	0.14878
19	Lagunilas	15-16	2.43	Miranda	2.070	flat	0.15030
20	Pilon	14-15	1.70	Caripito	1.700	flat	0.15140
21	Bachaquero	13-14	2.70	Miranda	1.700	flat	0.15240
22	Tia Juana	12-13	2.66	Miranda	1.670	flat	0.15350
23	Laguna	11	2.90	Miranda	1.620	flat	0.15506
24	Boscan	10	5.50	Miranda	1.570	flat	0.15615

*2.5¢/Bbl per full half-degree.

Source: Compiled by the author.

crude oils to gravity variations and which may have no direct connection with the gravity fluctuation differential affecting the price of each individual crude oil of the group.

Statistical correlations of posted prices in the Gulf area have shown that retaining 2 cents a barrel per degree was a satisfactory approximation to the basic gravity differential. However, it should be stated that any other value in the 1.8-2.2 range could also have been retained as a satisfactory approximation. The concept of basic gravity differential is an empirical one, and its value is determined within a degree of approximation consistent with that by which the linear pattern of prices is established. By contrast, the concept of gravity fluctuation differential is an official public fact with a precise value published by the industry. It so happens that the values of the two concepts are of the same magnitude in the Gulf, but this might be due to mere coincidence, and there is no evidence that they should be officially related to each other. In fact, their values are quite different for Venezuelan crude oils, as outlined above.

Having established that posted prices in each exporting area are governed by linear patterns with respect to gravity, we might conveniently represent posted prices by the price of a reference crude oil upon which the pricing pattern of the area considered would be articulated. In particular, 36° Arabian light played the role of such a reference crude oil in the elaboration of the pricing pattern in the Gulf area as outlined earlier. The general level of postings in this area underwent successive changes in the postwar period and finally settled at a value determined through a net-back formula by which the delivered prices of 36° crude oils from the Gulf area, the Caribbean, and the U.S. Gulf were equalized CIF the East Coast at New York.

The conventional approach outlined above is a simplified approximation of the mechanism of price formulation by which the equalization of CIF prices is made for the specific gravity of the reference crude oil, historically set at 36° API. This will be called the reference crude approach. Now, if we consider the whole range of gravity of all available crude oils in each export area, the equalization of CIF prices related to the reference crude oils from each area does not necessarily lead to the equalization of CIF prices related to other gravity values. Indeed, such an equalization over a large range of gravity is theoretically impossible, as will be demonstrated in the next chapter, since the value of the basic gravity differential is affected differently by the freight cost from one area to another. However, the impact of this phenomenon is not very significant in the Eastern Hemisphere, where pricing patterns are governed by the same (BGD) value of 2¢/Bbl/degree. The situation is fundamentally different when one compares pricing patterns for the Gulf area and Venezuela because

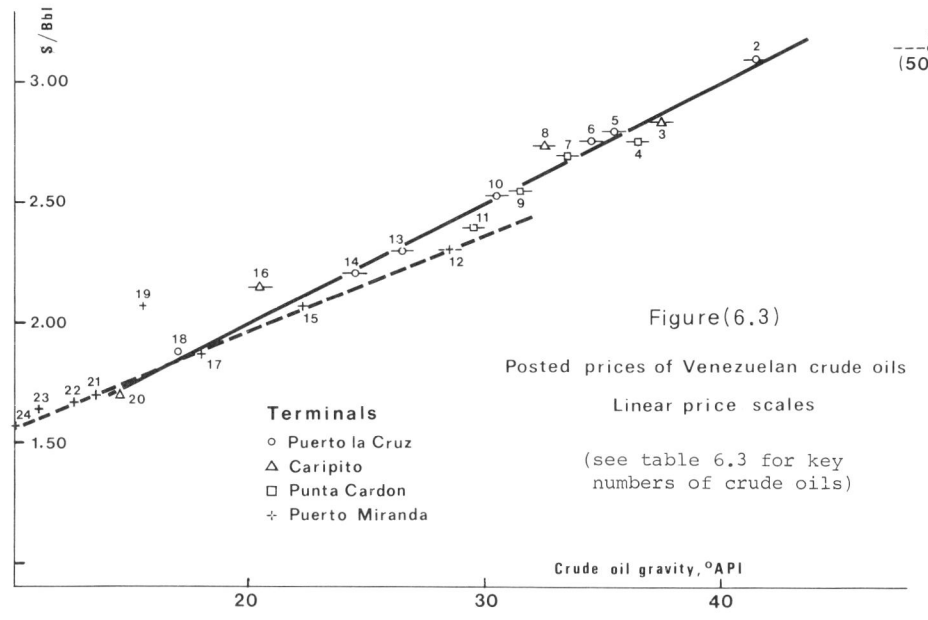

Figure (6.3)

Posted prices of Venezuelan crude oils

Linear price scales

(see table 6.3 for key numbers of crude oils)

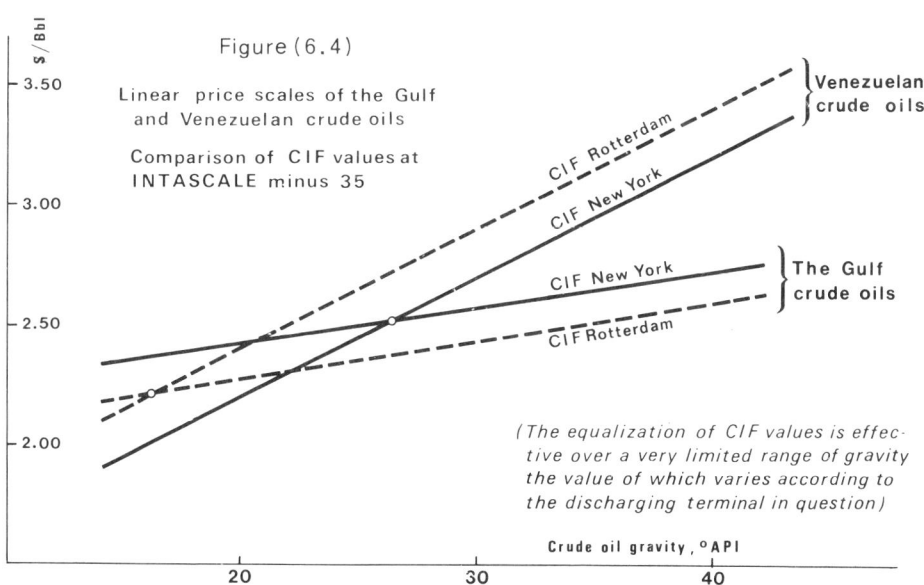

Figure (6.4)

Linear price scales of the Gulf and Venezuelan crude oils

Comparison of CIF values at INTASCALE minus 35

(The equalization of CIF values is effective over a very limited range of gravity the value of which varies according to the discharging terminal in question)

of the large difference between the values of their basic gravity differentials. The delivered values of all crude oils from these two areas, CIF New York and Rotterdam, are plotted in Figure 6.4 over the whole range of gravity—that is, linear price scales corresponding to a freight rate taken at the historical value of Intascale minus 35. It is obvious that the relationship between the price levels in the two areas, as it was determined in the late 1940s through the net-back formula applied to the reference crude approach, was valid only for a very limited range of gravity, the value of which varies from one discharging port to another. The disparity between the two pricing patterns beyond this limited range is so great that it would be quite improper to compare their economics in general terms. These findings remain valid for any other value of freight rates that would correspond to different values of the gravity ranges of restricted equalization of CIF values. These findings remain equally valid when considering Venezuelan tax reference prices, which are officially governed by a basic gravity differential of 3.64¢/Bbl/degree, as already indicated in Chapter 4.

NOTES

1. Helmut K. Frank, Crude Oil Prices in the Middle East (New York: Praeger Publishers, 1966), pp. 37-38. The evolution of the interhemisphere crude oil net flow was as follows (thousands of barrels daily):

	1947	1948	1949	1950
West to East	96	21	—	—
East to West	—	—	83	65

2. More details can be found in ibid.
3. Ibid., and Wayne A. Leeman, The Price of Middle East Oil (Ithaca, N.Y.: Cornell University Press, 1962).
4. Leeman, op. cit., has made a deep and exhaustive analysis of ECA action, and the reader is referred to Chapter 5 of his valuable book for further details.
5. Frank, op. cit., p. 32.
6. In particular, Platt's Oilgram Price Service.
7. Stephen H. Longrigg, Oil in the Middle East, 3d ed. (London: Royal Institute of International Affairs, Oxford University Press, 1968), pp. 380-381.
8. Zuhayr Kikdashi, A Financial Analysis of Middle Eastern Oil Concessions, 1901-65 (New York: Praeger Publishers, 1966), p. 199.

CHAPTER 7

PRICING PATTERNS OUTSIDE THE GULF AREA

After World War II, crude oil exports from the Gulf area developed so extensively as to provide the major bulk of supply in the Eastern Hemisphere. In the early 1950s, the Gulf area emerged as a principle price basing point that governed crude oil prices in the international oil business through a net-back formula in conjunction with Western Hemisphere prices.

However, other sources of supply have contributed increasingly to the bulk of crude oil outflow in the Eastern Hemisphere. Export terminals at the Eastern Mediterranean, such as Banias, Sidon, and Tripoli, evacuate Iraqi and Saudi Arabian crude oils transported by the Trans-Arabian Pipeline (Tapline) and IPC pipelines. They were actively operating during the 1950s, and corresponding posted prices have fully participated in the up-and-down movements of the Gulf postings to which they are related.

In 1958 Algeria, then under French control, was the first newcomer, and exports rapidly developed at Bougie, at la Skhira terminals, and later at Arzew. Libya and Nigeria soon followed in the early 1960s, and production developed and rapidly expanded. Proximity to major consuming markets in Western Europe, better qualities of crude oils, and specific political conditions in the Middle East were largely responsible for the unusual rapidity of this expansion.

Posted prices were published for crude oils exported from these areas although their context and implications were not exactly the same. Algerian postings were initially made by French authorities in light of the "French National Oil Policy" dealing with national crude oils rather than in the context of conflicting interests between host countries and major oil companies prevailing in the Middle East. Later evolution of Algerian postings was influenced by the general over-all framework of relations between independent Algeria and

France, which makes objective assessment of these postings difficult. Libyan postings were initially merely formal since the fiscal obligations of oil companies were based on realized prices until 1965, when conditions in Libya were aligned with those currently prevailing in the Middle East. This alignment directly motivated the Nigerian Government to request that posted prices be published in Nigeria for the purpose of income tax calculation as a result of the application of the "most favored African nation clause" with reference to the Libyan precedent. Consequently, Nigerian postings were published in 1967 with retroactive effect to January 1, 1966. Only posted prices at the Eastern Mediterranean developed within the same context as those prevailing in the Gulf, and they underwent the same series of fluctuation throughout the 1950s and were also frozen as of 1960.

Without our questioning for the moment the nature and significance of these posted prices, the major question is whether the prices fit consistently into an over-all pricing pattern governing posted prices in the Eastern Hemisphere or whether they are determined specifically and independently in each case. The subject is quite complex and should be thoroughly investigated on the grounds of both actual facts and philosophy.

THE PHILOSOPHY OF OVER-ALL PRICING PATTERNS

We have noted that the pricing issue has been a permanent source of conflict between producing countries and oil companies in the Eastern Hemisphere. It developed within an emotional and political context that helped make it a favored field of confrontation. However, it was mainly appraised by producing countries through its fiscal consequences—that is, government revenue from oil exports. Host governments are interested not only in increasing their per barrel income but also in seeing that they are not getting relatively less than other producing countries. This "revenue" approach is very important since it reflects the deep psychological motivations of producing countries that govern indirectly their relations with oil companies over the pricing issue.

Since production costs have fixed values for each field, the first objective of increasing the per barrel income could be achieved by increasing the general level of reference prices. However, producing countries came to realize that reference prices are governed by strategic considerations beyond their reach. The other alternative would be to improve the fiscal terms of profit sharing, which was effectively achieved by OPEC (through royalty expensing and elimination of allowances, as noted in Chapter 4). The second objective of getting relatively equal treatment was partially achieved by the above-mentioned harmonization of fiscal terms. Hence, its full achievement is related to subjective convictions of the fairness and the equity of prevailing patterns, which are presumed to relate prices from one area to another in terms of freight, gravity, and quality differentials.

However, these considerations were not significant in the Gulf area since production costs are of the same magnitude, crude oil qualities are comparable, and freight disparities are quite small. Furthermore, the pricing pattern developed at a time when producing countries were not fully aware of these problems. The situation is completely different for producing countries outside the Gulf area, where production costs are of a different magnitude, crude oils are of different qualities, and freight disparities are substantial. Furthermore, most problems associated with these questions were raised during the 1960s, when producing countries were politically and psychologically very sensitive to the pricing issue. This was particularly true for Libya, which was the only place where the pricing game actually materialized: Eastern Mediterranean postings were aligned with Gulf postings as of the early 1950s, Algerian oil was imprisoned in the special context of Algerian-French relations, and there were no posted prices in Nigeria until 1967.

In light of the foregoing, the definition of a basic concept for the elaboration of an over-all pricing pattern in the Eastern Hemisphere seems to be quite confused and confusing from a governmental viewpoint. Producing countries were not involved in the elaboration of prevailing prices, and their attitudes and actions were mainly concerned with the appraisal and the challenging of presumed pricing mechanisms instituted by oil companies. The approach of producing countries is mainly based on two underlying assumptions:

1. They adhere to the historically established rule of equalizing CIF values in a given marketplace, which was traditionally located in Western Europe.

2. They are very sensitive to the subjective concept of relative equal treatment and expect oil companies to comply automatically with the most favored nation rule. The strong impact of this psychological motivation is illustrated by the fact that Libyan governments requested appropriate price adjustments to align with Eastern Mediterranean postings in the early 1960s, when income was related to realized prices and would not have been affected by any adjustment of posted prices.

In the absence of any objective criteria to relate posted prices in the different export areas, it is quite impossible to visualize, a priori, on which grounds an eventual over-all pattern should be considered. Therefore, we will approach this question from the viewpoint of producing countries by investigating, a posteriori, whether actual posted prices fit into an eventual over-all pattern consistent with the net-back formula.

Our investigation will proceed by checking whether disparities displayed by freight, gravity, and quality can be correlated. It should be noted, however, that the freight component is the most controversial issue, since gravity adjustment in terms of 2 cents per degree was not challenged or criticized and the claim for a low sulfur premium was an innovation calling for an extra advantage over and above Middle Eastern prices. On the contrary, the application of the net-back formula, starting from a commonly accepted value of CIF prices, would lead to different values of FOB prices when one is considering different values of freight costs. Freight rates are essentially fluctuating in nature and vary continuously in time and space, whereas crude oil prices are intended to be stable and universal. Therefore, a constant average freight rate should be considered for the purpose of pricing through the net-back formula, irrespective of the actual performances of the tanker market. The selection of such an average freight rate is essentially a policy-motivated action, and retaining one value or another would lead to different values of FOB prices. Under these conditions, the "equal treatment" motivation of producing countries would reflect on their understanding of any over-all pricing pattern in that any such pattern should be based on the same value of the policy-induced average freight rate retained for the application of the different net-back formulas.

Consequently, investigation of pricing behavior in the Eastern Hemisphere depends on analysis of the freight element, which will be undertaken in this chapter. The investigation of gravity and quality differentials will be made later in this study.

THE ECONOMICS OF TANKER TRANSPORTATION

Tanker transportation economics is a most complex issue, and freight rates are largely determined by supply and demand forces in an open and very competitive market, which, to a large extent, does not come under the control of the integrated scheme of the international oil business. Freight rate fluctuation is the permanent and predominant feature of the market.

For a given transportation route, freight rate fluctuation may be influenced by the type of contract (for example, spot, consecutive voyages, time charter, and so on); the timing of the contract according to the balance between supply and demand with special reference to crisis periods; the route considered; the size of the tanker; and basic nominal rates, which are subject to unpredictable modifications and corrections to keep them continuously up to date following changes in ports and the discharging of dues and costs.

Numerous contracts each day are signed, and their terms and conditions are assessed and published by specialized brokers. The multitude of rates and conditions would make the actual values of freight rates a very poor and impractical criterion for assessing the market situation and its evolution. Instead, freight rates are measured against permanent theoretical reference scales and expressed in terms of percentages of corresponding basic rates given for round trips for each couple of loading and discharging ports. There exist several reference scales, namely USMC, American Tanker Rate Schedule (ATRS), and Intascale. The last two were replaced by Worldscale on September 1, 1969. The Worldscale schedule is revised and updated annually. (The first scales to be used in the postwar period were the USMC and the British MOT [Ministry of Transport] scales. The former was replaced by the ATRS in 1956; the latter was replaced by a scale schedule introduced in 1952, revised in 1954 and 1958 [Scale I, II, and III], and replaced by Intrascale in 1962. ATRS and Intrascale were replaced by Worldscale in September 1969. The rates of these different scales are determined under well-defined theoretical conditions [tanker size, speed, operating costs, and so on] for each couple of loading and discharging ports, the dues of which are incorporated into corresponding rates.)

The fluctuation of freight rates, given in terms of discounts off basic rates, is most unpredictable and helps create a major difficulty in the choosing of the proper type of contract at a given period—that is, predetermined cost chartering against spot quotations that could be higher or lower for the given period. In fact the oil companies settled on an average scheme by which the largest part of transportation requirements would be covered by self-owned tankers and time charter contracts. Fluctuations in both requirements and rates are balanced and adjusted through spot transactions, which account for some 15 percent of the market under normal conditions. Changes in market conditions are promptly reflected in spot quotations, whereas charter terms are affected more moderately and with some delay.

In the face of the complexity and uncertainty of the market, the main problem is to get, in time, full and proper information and data. Actually this is done for Eastern Hemisphere transactions by specialized brokers traditionally based in London. Possible discrepancies in data from different sources could be another disturbing, although minor, factor.

Information and data are readily available from specialized brokers upon request or subscription, and they are currently reported in some professional periodicals for the benefit of the average observer of the industry.

Harley, Mullion and Co. (Shipping) Limited of London publishes a <u>Weekly Freight Index</u> averaging values of weekly rates for single-voyage (Dirty) contracts, thus giving a general overview of the spot market. Petroleum Press Service publishes monthly averages of these indexes. Similar information is published in Platt's Oilgram, Oil and Gas Journal, and elsewhere.

The most commonly used criteria to assess the over-all average market situation where all factors and parameters are integrated is the well-known AFRA (Average Freight Rate Assessment), regularly published by the London Tanker Brokers' Panel since spring 1954 and currently reported in Petroleum Press Service. AFRA is a weighted average covering all tankers trading, over a given period, for spot and consecutive voyages and time charter transactions as well as an estimate of that portion of the market that represents integrated transportation operations by tankers owned and operated by oil companies.

In fact, the necessity of an objective assessment of average transportation costs was first felt and appraised by the major oil companies themselves in order to use reliable data for their long-term projections. In 1949, the Anglo-Iranian Oil Company (AIOC and later BP) and Royal Dutch/Shell requested a commission of London-based brokers to expand such an average index. This evaluation was based on the assessment of freight costs, under given technical conditions, for two-year charter contracts.[1] The index, known as the London Brokers' Award, was first published and utilized formally in pricing calculation in 1949.[2]

The London Brokers' Award was a first attempt that proved insufficient in reflecting actual long-term transportation costs. Royal Dutch/Shell requested the elaboration of a more representative index, and the Brokers' Commission finally agreed on the formulas and procedure for calculating AFRA rates,[3] in 1954, to the satisfaction of oil companies. These rates were felt to reflect long-term average costs effectively and were largely adopted and used by major oil companies for the purpose of assessing freight rates in their integrated operations.

AFRA was published as percentage discounts off basic Intascale rates. All factors governing the tanker transportation markets are integrated into weighted average figures, which vary with tanker size. Up to 1960, the size of tankers was relatively small and the space limited so that a single AFRA value was published for a size range of 15,000-25,000 deadweight tons (dwt), known as "general purpose."

With the ever increasing size of tankers, AFRA rates were differentiated by the introduction of a specific rate for medium-size tankers in the 25,000-45,000 dwt range. The appearance of huge tankers induced a further differentiation by the introduction of a new specific rate, in July 1, 1964, for large tankers in the range of 45,000-75,000 dwt, medium rates being restricted to 25,000-44,999 dwt, and general purpose to 15,000-24,999 dwt.

These revisions were necessary to reflect the increase in tanker size and its impact on transportation economics—in terms of dollars per long ton (LT)—and to preserve the essential characteristics of AFRA as an indicator of the level of freight rates per ton of crude oil carried throughout the world during the period covered by each assessment.

Up to June 1967, AFRA rates were published quarterly and semiannually. Since the 1967 Middle East crisis, AFRA rates are published monthly to cope with the necessity of following and clarifying deep perturbations occasioned by the Suez Canal closure and other hazards such as occasional pipeline stoppage (Tapline) and production cuts in Libya.

Further differentiation was introduced in November 1968 to take into consideration the introduction of super-tankers, and on that occasion, the ranges of different AFRA rates were changed to the following: general purpose, 16,500-24,999 dwt; medium, 25,000-44,999 dwt; large I, 45,000-79,999 dwt; and large II 80,000-159,999 dwt.

With the replacement of Intascale in September 1969, AFRA rates have been quoted in terms of percentages of Worldscale rates since October 1969.

To sum up, the transportation cost (F) for a round trip from a given export terminal to a given discharging port is made up of two components:

1. The corresponding basic rate (R) of a given schedule (Intascale, Worldscale, and so on), which is a fixed sum accounting for the distance and which varies with the geographical location of the ports considered.

2. The average freight rate, which reflects the market fluctuation by retaining an average value that might vary in time but that is applicable to all ports at a given moment. It will be represented as a percentage discount off Intascale rates $(1 + i)R$ and as a percentage of Worldscale rates $W \cdot R$, so that

$$F = (1 + i)R \quad R: \text{Intascale rate}$$
or
$$F = W \cdot R \quad R: \text{Worldscale rate}$$

THE IMPACT OF GRAVITY ON TRANSPORTATION COSTS

The impact of gravity on transportation costs results from the fact that freight rates are given in terms of dollars per long ton ($/LT), whereas prices are quoted in dollars per barrel ($/Bbl), and that higher-gravity crude oils (lighter ones) would represent more barrels per ton to be transported for the same cost. Consequently, the transportation cost per volume unit decreases for higher-gravity crude oils.

The magnitude of this impact is reflected by the conversion factor (φ) from $/LT into $/Bbl so that f ($/Bbl) = $\varphi \cdot$ F($/LT).

Values of the conversion factor (φ) are listed in Table 7.1 and plotted in Figure 7.1, which shows that the conversion factor is not a linear function of gravity. Therefore, the freight component of CIF prices introduces a nonlinear distortion even though FOB prices are governed by a linear pattern with respect to gravity. Nevertheless, Figure 7.1 equally shows that a linear relationship between gravity and the conversion factor could be considered over the limited range of gravity of most crude oils, say from 20° to 45° API, bearing in mind that pricing patterns are based on and tolerate an approximation of one degree gravity.

Now, if we assume that FOB prices (X) are governed by a linear pattern with respect to gravity (X = a G + b), it follows that CIF prices (Y) are equally governed by a linear pattern of their own (Y = X + f = a' G + b'). However, the value of the basic gravity differential governing CIF prices (a') is affected by the impact of gravity on the freight component, which results in diminishing the original value (a) governing FOB prices.[4] The magnitude of the decrease (a - a') is proportionate to the transportation cost (F), so that it varies with the

TABLE 7.1

Conversion Factors Versus Gravity
($/Bbl to $/LT)

°API	φ	°API	φ	°API	φ	°API	φ
10	0.15615	20	0.14584	30	0.13680	40	0.12881
11	0.15506	21	0.14488	31	0.13596	41	0.12806
12	0.15397	22	0.14392	32	0.13511	42	0.12731
13	0.15291	23	0.14301	33	0.13431	43	0.12660
14	0.15186	24	0.14209	34	0.13348	44	0.12587
15	0.15083	25	0.14117	35	0.13268	45	0.12516
16	0.14979	26	0.14027	36	0.13189	46	0.12444
17	0.14878	27	0.13939	37	0.13112	47	0.12375
18	0.14779	28	0.13851	38	0.13033	48	0.12306
19	0.14681	29	0.13764	39	0.12956	49	0.12236
						50	0.12169

Source: OPEC, Selected Conversion Factors (Vienna, October 1968).

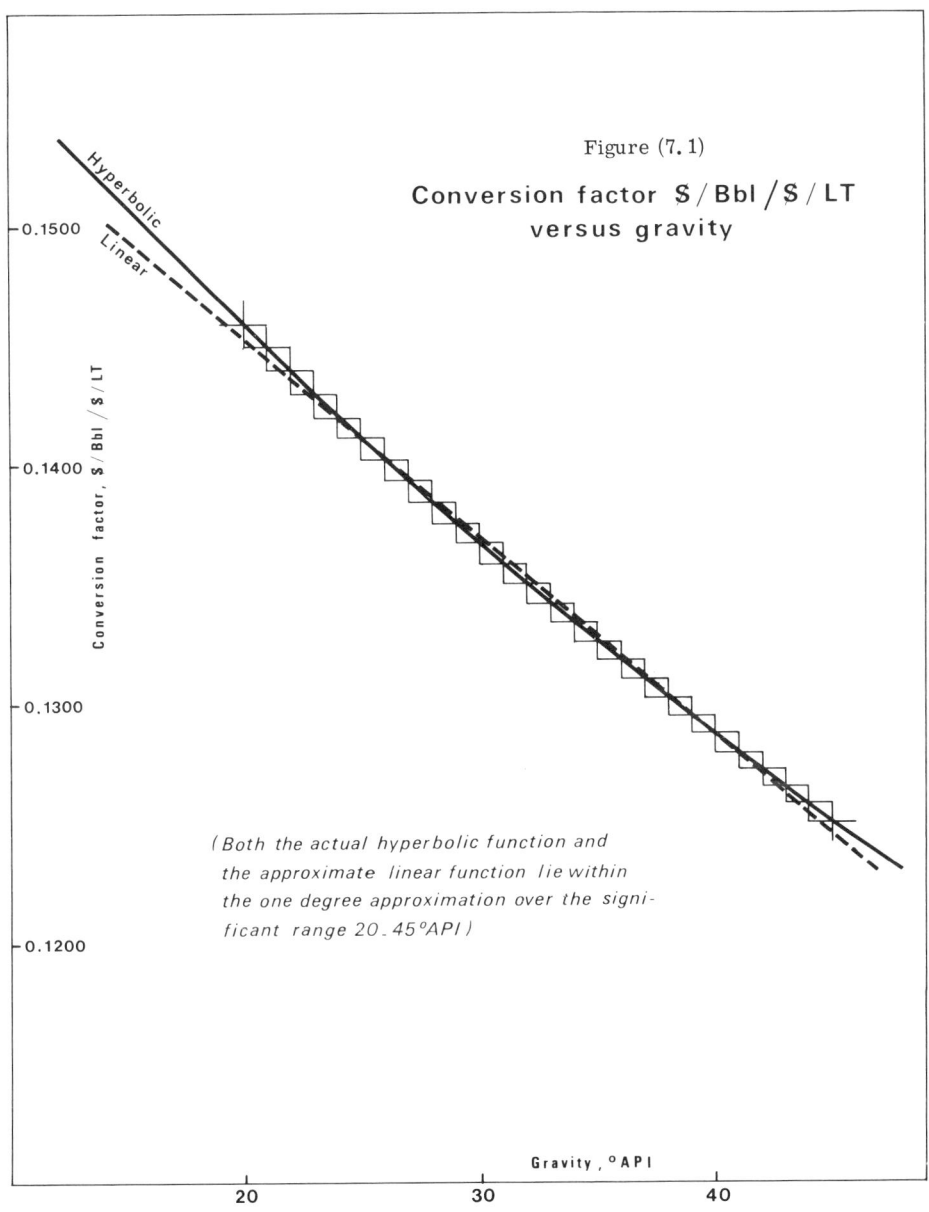

distance between loading and discharging ports corresponding to different values of Intascale rates (R) and with market fluctuations represented by different values of average discounts off Intascale rates (i).

This impact is not significant for the elaboration of pricing patterns in the Gulf area, where the basic gravity differential governing prices FOB Quoin Island has a constant value of 2 cents irrespective of freight fluctuation. This is due to the very small values of (R) from export terminals to Quoin Island.

By contrast, the impact is very significant indeed when one considers how the linear pattern of FOB prices would be affected if transposed to CIF values over long distances such as to Western Europe or the United States, which values were retained in the late 1940s to determine the level of the reference price in the Gulf area. With (a) being equal to 2 cents a degree, corresponding (a^1) values are listed in Table 7.2 at three different locations and for three different freight rates.

LOOKING FOR AN OVER-ALL PRICING PATTERN

Let us examine the mechanism of price formulation in the different export areas in the Eastern Hemisphere under the net-back formula concept by which the CIF prices of all crude oils of similar gravity and quality are equalized in a given place irrespective of their geographical origin. Since FOB and CIF prices are governed by linear

TABLE 7.2

Basic Gravity Differentials Governing CIF Prices

FOB Value of 2¢/Bbl/degree in Gulf Area Would Become:

	Discharging Port		
	Marseilles	Rotterdam	New York
Nominal rate from Quoin Is. (R), $/LT	4.77	6.38	7.92
Suez Canal toll $/LT	0.88	0.88	0.88
Intascale minus 50 (i = -0.50)	1.73	1.66	1.60
Intascale minus 35 (i = -0.35)	1.67	1.59	1.50
Intascale flat (i = 0)	1.54	1.40	1.28

Source: Compiled by the author.

relations with respect to gravity, each pricing pattern might be determined in two successive phases: (1) set the relative pricing levels of the different areas with respect to each other by determining the values of the reference price (Xo) corresponding to the reference gravity (Go) in each area, and (2) determine the basic gravity differential (a), which articulates crude oil prices in each area on the corresponding reference price (Xo).

The Gulf area was recognized as a principal price-basing point characterized by the reference terminal at Quoin Island and by Arabian light (36° and later 34° API) as the basic reference crude oil. The level of the reference price in any given export area outside the Gulf (X) corresponding to a crude oil of the same reference gravity can be derived from the basic reference price (Xo) at Quoin Island by adding the value of the freight differential (fo - f) with respect to a given marketplace where CIF values are equalized, since we have

$$Y = X_0 + f_o = X + f$$

or

$$X = X_0 + (f_o - f).$$

The value of the freight differential varies with three different factors:[5] (1) the gravity of the crude oil considered through its impact on the conversion factor; (2) the geographical location of export and discharging terminals through the nominal freight rates (Intascale flat); and (3) the market fluctuation through the average freight rate (i).

In the present exercise, the two first factors have fixed values, whereas the average freight rate (i) is normally subject to continuous fluctuation. However, since posted prices are intended to be stable and universal in scope, a fixed average freight rate has to be retained for the purpose of price determination over a given period at a policy-induced value. Consequently, the value of the reference price in each export area (X) depends directly on the policy-induced value of freight rate (i) retained by oil companies for the purpose of pricing. Since the selection of such an average value is somewhat arbitrary, it follows that the reference level in a given area could be set at different values, at the exclusive discretion of oil companies, by retaining appropriate average rates (i). If we bear in mind the pricing philosophy based on the subjective concept of equal treatment as outlined earlier in this chapter, the only way to consider a consistent over-all pricing pattern would be to retain the same policy-induced average freight (i), at a given moment, for all export areas, irrespective of its actual value, which might bear no direct relation to actual market performances.

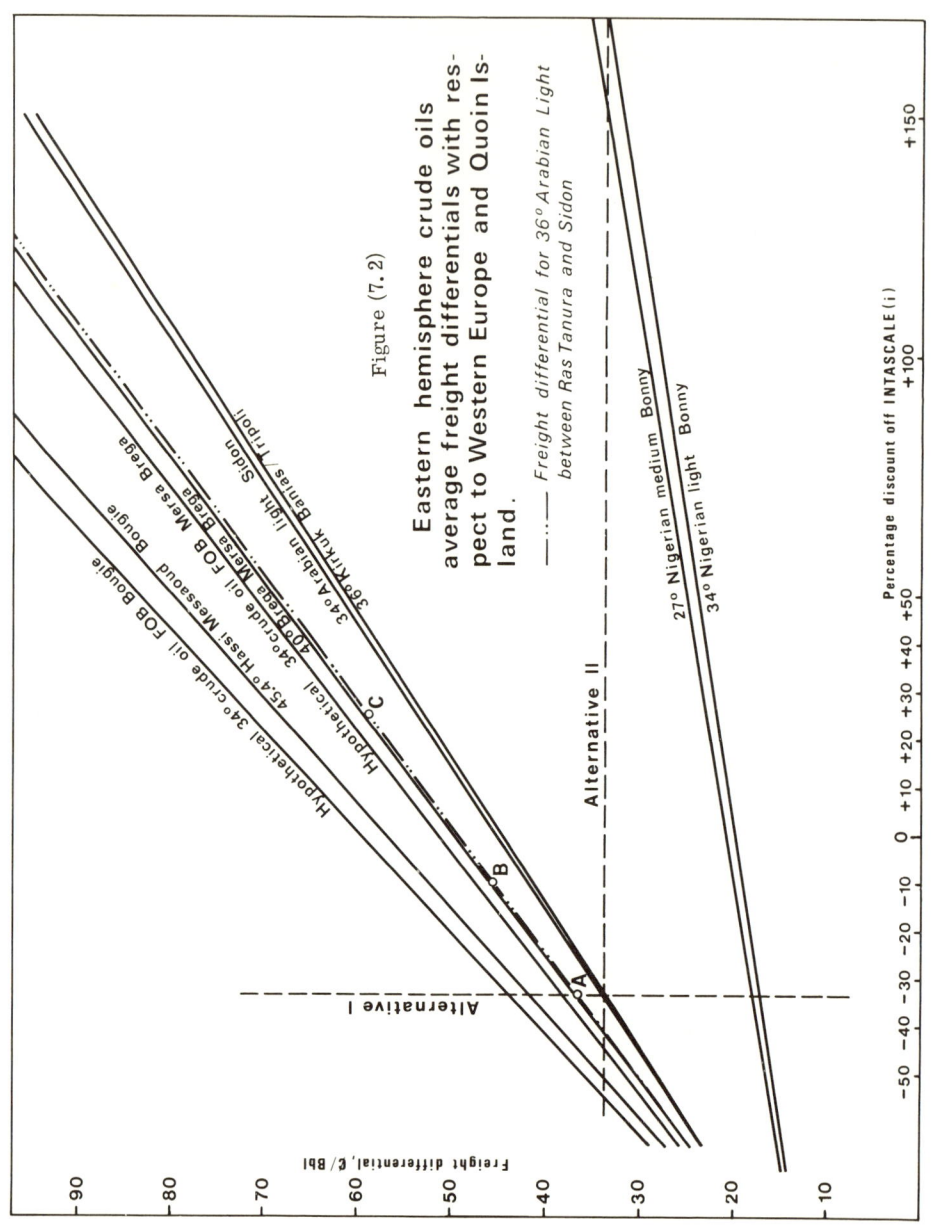

Figure (7.2)

Eastern hemisphere crude oils average freight differentials with respect to Western Europe and Quoin Island.

—— Freight differential for 36° Arabian Light between Ras Tanura and Sidon

In other words, the existence of an over-all pricing pattern in the Eastern Hemisphere is subject to the condition that the average freight rate relating the different reference prices should simultaneously have the same value, whatever its actual value is.

The impact of the variation of the average freight rate (i) on the value of the freight differential is illustrated in Figure 7.2. The equalization of CIF prices is achieved in Western Europe, for which the freight cost is taken as the average rates to Rotterdam and Marseille. The impact of gravity is also illustrated in the figure since freight differentials for the different export areas are shown for two sets of gravities: a common value of reference gravity (Go = 34°), which might correspond to hypothetical crude oils in some areas (Libya, Algeria), and the actual gravity value of crude oils exported.

Having thus determined the value of reference prices in such a way as to equalize CIF prices in a given market for the same reference gravity and the same average freight rate (i), let us check whether this equalization can be extended to crude oils of different gravities. This would be the case if the basic gravity differentials (a') of the different sets of CIF prices related to the different export areas would have the same value. Such an equalization would be possible only if the basic gravity differentials governing FOB prices would have different values from one export area to another.[6] In fact, this is untrue in the actual pricing structure in the Eastern Hemisphere, where FOB prices in all export areas are governed by the same (BGD) value of 2 cents a degree. It follows that the equalization of CIF prices under the conditions indicated above is valid only for crude oils having the same reference gravity (Go).

If we bear in mind that the selection of a common average rate (i) for the purpose of pricing is somewhat arbitrary and may vary with time, that the selection of a given marketplace where the equalization of CIF values is made is equally arbitrary, and that such equalization is limited to the single value of the common reference gravity, it follows that posted prices in the Eastern Hemisphere cannot fit into any over-all pricing pattern governed by the net-back formula concept. This theoretical conclusion will be confirmed in the next chapter, where it will be demonstrated that the actual posting levels of reference crude oils cannot be equalized in any given market for the same average freight rate (i). In fact, pricing patterns in the different export areas were brought about through an empirical process under different conditions and at different times. They were determined with respect to the Gulf postings, which display the "historical" value of 2¢/Bbl/degree with which they were aligned without any questioning of the real significance of the gravity differential or the validity of its extension. In practical terms, the gravity differential was regarded by producing countries as an established fact to relate prices and was accepted as such in order to determine whether prevailing prices were "fair" or not under their subjective "equal treatment" approach to the pricing issue.

NOTES

1. Petroleum Press Service, April 1951.
2. United Nations, The Prices of Oil in Western Europe (Geneva, March 1955).
3. Petroleum Press Service, April 1954.
4. The linear approximation of the conversion factor (φ) versus gravity can be expressed by the following relation:

$$\varphi = -0.00082\, G + 0.16177$$
$$20 \leq G \leq 45° \text{ API}$$

The pricing pattern of CIF prices can be expressed as follows:
$$Y = X + f = X + \varphi(1+i)R$$
$$Y = aG + b + (-0.00082\, G + 0.16177)(1+i)R$$
$$Y = [a - 0.00082\,(1+i)R]\, G + b + 0.16177\,(1+i)R$$

Say $\quad Y = \quad\quad a' \quad\quad G + \quad\quad b'$
where $\quad a' = a - 0.00082\,(1+i)\, R$
and $\quad b' = b + 0.16177\,(1+i)\, R$
so that $a - a' = 0.00082\,(1+i)\, R$

5. The freight differential relating Quoin Island and another export terminal outside the Gulf area, with respect to the same discharging port in which CIF values are equalized, can be expressed as follows:

$$f_o - f = \varphi_o . R_o\,(1+i_o) + 0.88\varphi_o - \varphi(1+i)R$$

The existence of an overall pricing pattern in the Eastern Hemisphere is subject to the condition that the average freight rate should simultaneously have the same value ($i = i_o$) so that the freight differential would be given by:

$$f_o - f = (\varphi_o . R_o - \varphi . R)(1+i) + 0.88\varphi_o$$

If we assume that reference crude oils in the different export areas have the same reference gravity, ($\varphi = \varphi_o$), then it follows that:

$$f_o - f = \varphi_o\,(1+i)(R_o - R) + 0.88\varphi_o$$

6. Reference posted prices would fit into an overall pricing pattern if they prove to be related in terms of freight differential as defined in the above mentioned formula, that is, CIF values of reference prices are effectively equalized in a given market. Historically speaking, the reference pricing pattern in the Gulf area is governed by a basic gravity differential of 2 cents, and the corresponding reference pattern of CIF prices is governed by a basic gravity differential (a^1) the value of which is listed in Table 7.2 as determined by the relation:

$$a^1 = 0.02 - 0.00082 \; (Ro \, [1 + i] + 0.88)$$

Now if we want the linear price scale of CIF prices governed by the above mentioned value of (a^1) to be equally representative of all pricing patterns in the different export areas, which corresponds to the equalization of CIF values of crude oils from all sources and for all gravities, then the value of the basic gravity differential of FOB prices in each export area (a) should be related to the predetermined value (a^1) by the relation:

$$a^1 = a - 0.00082 \; R \, (1 + i).$$

The combination of these two relations would give:

$$a = 0.02 - 0.00082 \, (1 + i) \, (Ro - R) - 0.00072$$

Since the overall pricing pattern is based on the same value of (i), and since each export area is characterized by a specific value of (R), it follows that the basic gravity differential in each and all export areas has specific values different from each other, and all different from the reference value of 0.02¢/Bbl/degree prevailing in the Gulf.

To illustrate these theoretical considerations, we have determined what should have been the pricing patterns in the different export areas in the eastern Hemisphere under the assumption that corresponding CIF values should be equalized at Rotterdam for all values of gravity. To achieve this, the pricing patterns would be governed by the following values of basic gravity differentials, which vary from one area to another and depend on the average freight rate considered (i).

The Gulf area	a = 2¢/Bbl/degree	
The Eastern Mediterranean (Sidon)	a = 1.72 when i =	0
	1.80	i = -0.35
	1.82	i = -0.50

Libya (Mersa Brega)	a = 1.68	i = 0
	1.76	i = -0.35
	1.80	i = -0.50
Algeria (Bejaia)	a = 1.61	i = 0
	1.72	i = -0.35
	1.78	i = -0.50
Nigeria (Bonny)	a = 1.82	i = 0
	1.85	i = -0.35
	1.88	i = -0.50

CHAPTER

8

ANALYSIS OF PRICING PRACTICES OUTSIDE THE GULF AREA

In the absence of any rational criteria for the elaboration, a priori, of an over-all pricing pattern in the Eastern Hemisphere, it is generally stated that crude oil prices have been set by the oil companies in accordance with the net-back formula concept, the equalization of CIF prices being made in Northwestern Europe and also occasionally in Southern Europe. In the present exercise, an overall figure will be considered for Western Europe and computed as the basis of the average freight to Rotterdam and Marseille.

Leaving aside the fundamental theoretical problems related to the impact of gravity on pricing patterns, we will examine in this chapter whether the actual pricing practices of the oil companies were consistent, in one way or another, with the above-mentioned conventional and simplified picture of the industry. The best way to conduct this investigation is to consider the chronological evolution of crude oil exports and pricing in the different producing areas. Posted prices in the Eastern Mediterranean are of special importance because they have been published since the early 1950s in association with crude oil postings in the Gulf area, and they also have contributed to the up-and-down movement of posted prices until their general freezing at their August 1960 levels. Moreover, the Eastern Mediterranean was the only exporting region outside the Gulf area until 1958, when crude oil exports began in Algeria and Nigeria, shortly followed by Libyan exports in 1961. The posted prices of crude oils available by the end of the 1960s are listed in Table 8.1.

However, it should be pointed out that the posted prices of Algerian and Nigerian crude oils are not significant for the purpose of our investigation. Algerian oil was produced, priced, and exported under special political conditions largely exceeding the limits of the oil industry; also, it was produced under French sovereignty until

TABLE 8.1

Eastern Hemisphere Crude Oils, Posted Prices Outside Gulf Area Prior to the Teheran and Tripoli Agreements

No.	Crude Oil	Terminal	Gravity (°API)	Posted Price ($/Bbl)	Sulfur (weight percent)	Pour Point (°C)	Conversion Factor (♦)
	Saudi Arabian						
1	Arabian light	Sidon	34-34.9	2.17a	1.63	−26	0.13310
	Iraq						
2	Kirkuk	Banias/Tripoli	36-36.9	2.21a	2.00	−39	0.13155
3	Bai Hassan-Jambur	Banias	34-34.9	2.07a	2.16	−14	0.13310
	Algeria						
4	Hassi-Messaoud	Bougie	Above 40 (45.4)	2.35b	0.12	−60	0.12480
5	Arzew	Arzew	Above 40 (43.1)	2.365b	0.10	−21	0.12660
6	Zarzaitine	La Skhira	Above 40 (41.4)	2.30b	0.13	−32	0.12770
	Libya						
7	Brega	Mersa Brega	40-40.9	2.23c	0.23	+2	0.12840
8	Libyan light	Es Sider	40-40.9	2.23c	0.30	+3	0.12800
9	Libyan light	Ras Lanuf	39-39.9	2.21a	0.37	+10	0.12925
10	Libyan light	Zueitina	Above 40 (42.1)	2.23c	0.25	+4	0.12725
11	Libyan high-pour	Ras Lanuf	36-36.9	2.10a	0.18	+18	0.13155
12	Serir	Mersa Hariga	37-37.9	2.10a	0.14	+27	0.13075
	Nigeria						
13	Nigerian light	Bonny	34-34.9	2.17a	0.26	+12	0.13310
14	Nigerian medium	Bonny	27-27.9	2.03a	0.25	+12	0.13900

a2¢/Bbl/degree gravity differential.
bFlat price posted from 1965 to 1969. Other sets of prices and story are detailed in the text.
c2¢/Bbl/degree differential below 40° API.

Source: Compiled by the author.

1962, and thereafter within the special context of cooperation between Algeria and France. Posted prices were introduced in Nigeria only in 1967 to align with the recently established pricing precedent in Libya.

In fact, the pricing of Libyan oil played a determinant role in raising the problem of an over-all pricing pattern in the Eastern Hemisphere since the posted price of 39° Zelten was first published by Esso and challenged by the Libyan Government. Posted prices published thereafter in Libya were aligned with the reference price of 39° Zelten, and later 40° Brega crude oil, after appropriate adjustment in terms of 2 cents per each degree (°API) gravity without questioning the significance or validity of this historically established gravity differential value. In other words, the pricing problem was raised in terms of comparative levels of reference crude oils: Zelten crude oil was posted at a time when posted prices were frozen in the Eastern Hemisphere, and the main question was whether the value announced by Esso was "fair and just" when compared to the existing precedent in the Eastern Mediterranean.

The evolution of crude oil postings in the Eastern Mediterranean during the 1950s will be analyzed in order to investigate the mechanism by which they were probably related to the Gulf postings. Posted prices in other exporting areas will then be compared with the reference price level thus established in the Mediterranean.

THE EASTERN MEDITERRANEAN POSTINGS

The highly productive oil field in Northern Iraq was discovered in October 1927 and rapidly proved to be one of the world's richest. The construction of a pipeline to evacuate the crude oil began in 1932 and was completed by the end of 1934. It was composed of two 12-inch lines each with a 2 million ton annual capacity: the northern line running across Syria and Lebanon to the Tripoli terminal and the southern line running across Jordan and Palestine to the Haifa terminal. A decision to more than double the pipeline was reached in 1938-39, but the construction was postponed because of World War II.[1] The construction of a 16-inch line began in 1946, parallel to the southern 12-inch line. It was almost completed when Israel occupied Palestine in 1948, prompting the Iraqi decision to stop oil pumping to Haifa so that the old and the new southern lines, with a combined capacity of 6 million tons, had to be abandoned. The construction of the 16-inch northern line began late in 1948 and was completed in 1950, thus bringing the total export capacity at Tripoli to 6 million tons a year. Another 30-32-inch line of about 8 million tons capacity at the Banias terminal was completed in 1952.[2] Expansion and improvement of the pumping facilities increased the aggregate transportation capacity to about 25 million tons by 1958 and to about 28 million by mid-1960. A fourth 30-32-inch line to Tripoli was completed in 1961, thus bringing the total capacity to slightly over 50 million tons a year.[3]

During the early development of export facilities, prices were of no real significance to the Iraqi Government since its income was based on a fixed tonnage royalty. On February 3, 1952, a supplementary agreement was reached with the IPC group allotting to the government 50 percent of net profits to be assessed by deducting productions costs from the border value of the crude oil.

The "Iraq border values" were defined as the "values of Iraqi crude oil at the points of exports from Iraq, having taken into account the geographical position of such points of export, applicable posted prices and average realizations from cargo and long term contract sales." The posted price of 36° Kirkuk crude oil was set at $2.29 a barrel in the Banias and Tripoli terminals, and the border value at the Iraq-Syria border was set at $1.65 a barrel.[4]

Saudi Arabian crude oil is evacuated by the Tapline crossing Jordan, Syria, and Lebanon to the Sidon terminal. The 30-31-inch pipeline was completed by the end of 1950, and the first shipment from Sidon took place on December 2, 1950.[5]

The posted price of the then 36°-36.9° Arabian light crude oil was the same as for Kirkuk crude oil, $2.29 a barrel FOB Sidon. This price was of true significance right from the beginning since the 50/50 profit-sharing agreement between Aramco and the Saudi Government was signed on December 30, 1950.

Posted prices at the Eastern Mediterranean do not display any freight disparity between the three terminals at Banias, Sidon, and Tripoli. They are related in terms of a gravity differential of 2 cents a barrel except for the 34° Bai Hassan-Jambur blend posted in June 1960 FOB Banias at $2.17 a barrel—that is to say 10 cents below its parity position with reference to 34° Arabian light posted at $2.27 and to 36° Kirkuk crude oil posted at $2.31. Two months later, all posted prices were uniformally reduced by 10 cents so that the price of the Bai Hassan-Jambur blend kept its 10 cents deviation, which seems to have been a policy-induced penalty illustrating the hostile climate between the IPC group and General Kassem, then engaged in a deadlocked negotiation. The same motivations were responsible for the underpricing of the 36° Basrah crude oil FOB Khor el Amaya, as already noted in Chapter 6.

We will use 36° Arabian light as the reference crude oil in the Gulf area and the Eastern Mediterranean. Since it is physically the same in Sidon and Ras Tanura, the price difference between these two export terminals is to be attributed solely to the advantage of the geographical location of the Sidon terminal, which is nearer to European markets. The freight advantage is a real and substantial one since it corresponds to a reduction of the transportation distance from the Gulf by some 3,500 miles and includes the savings of the Suez Canal charges. The difficulty is to assess this difference in economic terms—that is to say, to agree on the average freight rate to be applied to these 3,500 miles. The selection of such an average rate is a policy-motivated decision that does not necessarily have any direct relation to the actual conditions and performances of the tanker market. This statement is confirmed by the comparative evolution of 36° Arabian light postings FOB Sidon and Ras Tanura during the period 1948-60 as illustrated in Table 8.2. Price differences are converted to freight differentials in terms of presumed average freight rates (given as percentage figures over or below flat Intascale values) for which CIF prices might be equalized in Western Europe. These presumed average freight rates, effectively applied for the purpose of pricing, are compared to the actual performances of the tanker market, over the same periods, for both AFRA and spot rates. These figures are fairly representative because the tanker market was relatively simple in structure since a single AFRA value was published up to July 1959.

Table 8.2 clearly demonstrates that freight rates cannot be correlated either with AFRA or with spot rates, although they all vaguely follow similar trends of evolution. Therefore, it can be safely stated that neither AFRA nor spot rates have been retained by the oil companies to determine the relative levels of Eastern Mediterranean postings in comparison with the Gulf postings. Average pricing rates, effectively applied by the majors, should be considered as "policy-motivated historical values" the explanation of which is rather difficult to ascertain at this stage of technocratic investigation.

The comparative level of the Eastern Mediterranean postings seems to have stabilized at the average rate of Intascale minus 32.5 percent since January 1958, and this value was frozen as of August 1960 as a result of the general price freezing. It should be noted that this value is associated with a 36° gravity. If we consider that the 36° Arabian light began to be exported as a 34° crude oil in 1959, then the corresponding average pricing rate would become Intascale minus 33.5 percent instead of the above-mentioned value of minus 32.5 percent. However, the difference is not significant and is consistent with the over-all approximation of the pricing issue.

LOOKING FOR AN OVER-ALL PRICING PATTERN

Is there a single freight value for which the reference prices in each and all export areas could be related to the reference price

TABLE 8.2

Comparative Evolution of Posted Prices, 36-36.9° Light Arabian,
FOB Ras Tanura and FOB Sidon

Period	Posted Price, $/Bbl Ras Tanura	Sidon[a]	Price Difference $/Bbl	Presumed Freight Intascale Percent	Actual Freight (average) Percent Intascale AFRA[b]	Spot
4/48-7/48	2.22	2.95	0.73	+61	flat	+75
7/48-10/48	2.07	2.80	0.73	+61	flat	flat
10/48-4/49	2.03	2.76	0.73	+61	flat	-57
4/49-7/49	1.88	2.59	0.71	+56	-7	-45
7/49-10/49	1.88	2.57	0.69	+50	-31	-31
10/49-7/50	1.75	2.28	0.53	+9	-32	-26
7/50-10/50	1.75	2.25	0.50	+1	+3	+56
10/50-4/53	1.75	2.41	0.66	+43	+45	+80
4/53-7/53	1.75	2.29	0.54	+11	-19	-20
7/53-2/56	1.97	2.39	0.42	-20	+8	+1
2/56-3/57	1.97	2.46	0.49	-2	+34	+115
3/57-5/57	1.97	2.69	0.72	+58	+51	+6.5
5/57-9/57	2.12	2.69	0.57	+19	+51	+6.5
9/57-1/58	2.12	2.59	0.47	-7	+51	+6.5
1/58-2/59	2.12	2.49	0.37	-32.5	+24	-51
2/59-8/60	1.94	2.31	0.37	-32.5	+6/-10	-55
8/60	1.84	2.21	0.37	-32.5	—	—

[a]The same as for 36-36.9° Kirkuk FOB Tripoli before commencement of exports from Tapline.
[b]"Panel Award," before AFRA was introduced in 1954.

Source: Compiled by the author.

in the Gulf through the net-back formula mechanism, despite the fact that such an average value bears no direct relation with actual market conditions and performances? In other words, since such a value results from a policy-induced decision, did oil companies adopt similar pricing policies in the different export areas in the Eastern Hemisphere? A practical way to investigate this question would be to eliminate freight disparities by computing parity prices FOB Quoin Island by deducting appropriate freight differentials from actual FOB prices for a given freight rate value and to discuss how these parity prices would compare with the corresponding linear price scale of the Gulf postings FOB Quoin Island. An over-all pricing pattern would exist if price deviation were less than 2 cents a barrel for a given freight rate value.[6]

Parity prices FOB Quoin Island are plotted against gravity for three different values of freight rates in Figure 8.1, which also shows corresponding linear price scales. It is obvious that price deviations vary inconsistently from one case to another so that it may be safely stated that there is no over-all pricing pattern governing posted prices in the Eastern Hemisphere. This statement is better illustrated in Figure 8.2, which plots the price deviation for the different export areas against the freight variation. An over-all pricing pattern would exist if all these lines cut the zero line for the same value of freight rate, which is obviously not the case. Furthermore, Figure 8.2 clearly illustrates the impact of gravity since price deviation is nil at Intascale plus 85 percent for 27° Nigerian medium and at Intascale plus 95 percent for 34° Nigerian light crude oil, both posted FOB Bonny, as indicated in Table 8.1.

It is worth mentioning that Algerian crude oils were priced at flat values for 40° gravity and above, although actual gravities varied from 41.4° (Zarzaitine) to 45.4° (Hassi Messaoud), as indicated in Table 8.1. This raises a problem if we attempt to fit their parity prices FOB Quoin Island into the Gulf linear price scale; should we consider the 40° pricing value or the actual crude oil gravity? The two values are shown in Figure 8.1 for Hassi Messaoud crude oil.

THE PRICING OF LIBYAN CRUDE OILS

Libya emerged as a new source of supply in late 1961, and its production underwent fantastic expansion, exceeding 3 million barrels a day in less than 10 years. Libyan crude oils are generally of high gravity and of very low sulfur content; their pour point is relatively high, as shown in Table 8.1. The first crude oil to be exported was 39° Zelten, which later became 40° Brega crude oil, a blend of Zelten and Raguba fields. Its posted price was first published by Esso at

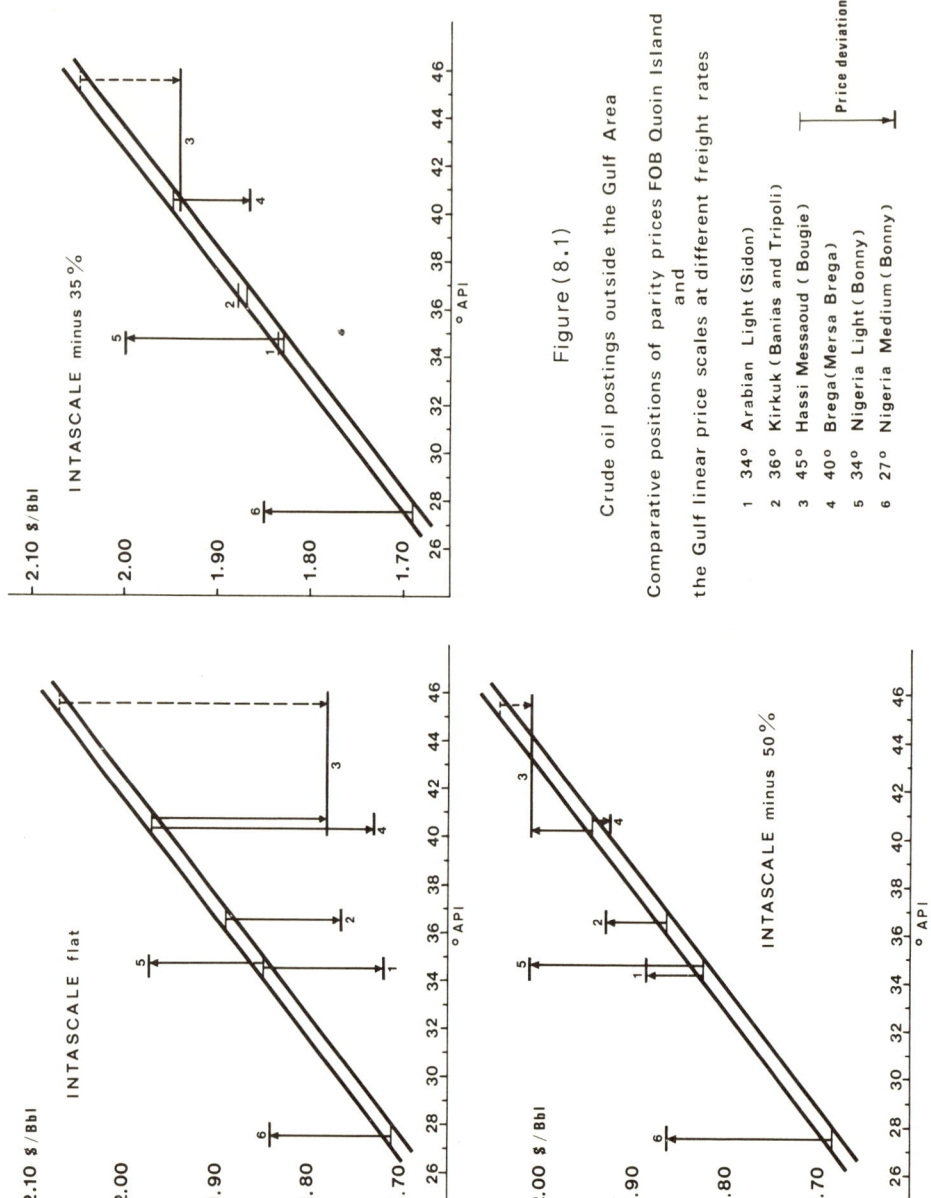

Figure (8.1)

Crude oil postings outside the Gulf Area

Comparative positions of parity prices FOB Quoin Island
and
the Gulf linear price scales at different freight rates

1 34° Arabian Light (Sidon)
2 36° Kirkuk (Banias and Tripoli)
3 45° Hassi Messaoud (Bougie)
4 40° Brega (Mersa Brega)
5 34° Nigeria Light (Bonny)
6 27° Nigeria Medium (Bonny)

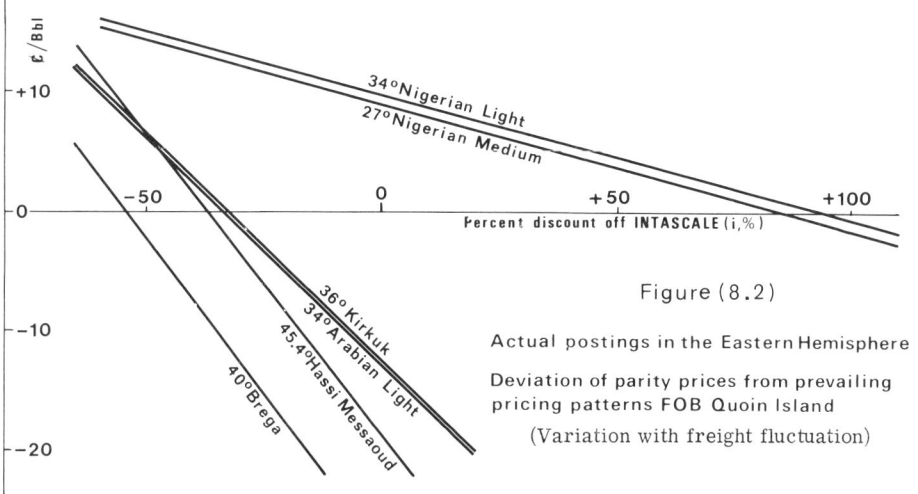

Figure (8.2)

Actual postings in the Eastern Hemisphere

Deviation of parity prices from prevailing pricing patterns FOB Quoin Island

(Variation with freight fluctuation)

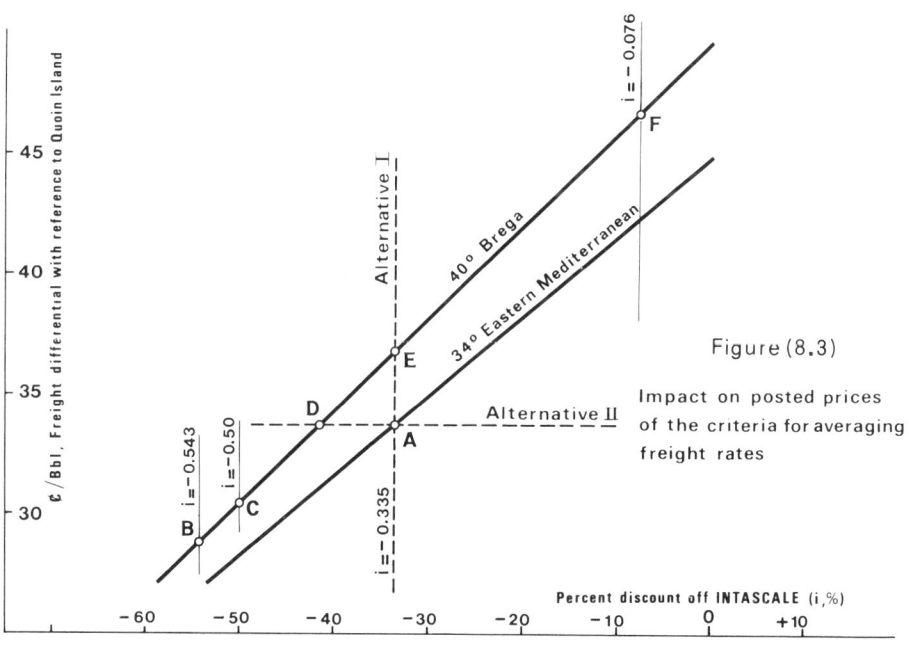

Figure (8.3)

Impact on posted prices of the criteria for averaging freight rates

$2.21 a barrel FOB Mersa Brega, thus creating the first pricing precedent in Libya and establishing the level of further postings by other companies.

Libyan postings derive from the reference price of 40° Brega crude oil according to the following pattern:

- There is no freight differential between the different export terminals located in the Gulf of Sirtica, except for Mersa Hariga near Tobruk, which is penalized by 2 cents a barrel to reflect higher freight rates.
- Posted prices are related in terms of a 2 cents/barrel basic gravity differential with a flat value at $2.23 for crude oils of 40° and above.
- High-pour crude oils are penalized by 5 cents a barrel.

Esso took the initiative of posting a price for its Zelten crude oil at a time when it was under no legal obligation to do so and despite the fact that such a move was of no real fiscal significance since payments to the government were to be made on the basis of realized prices. Nevertheless, the Libyan Government was not satisfied with the level of the Esso posting and protested the company's initiative, claiming that the price was unduly low and that it failed to give Libyan oil full credit for higher gravity and lower sulfur content. Esso International submitted a memorandum to the government explaining and justifying its pricing philosophy and action. This memorandum became a public document upon its publication in a specialized review,[7] and its full text is reproduced in Appendix A.

Government protests were of no practical effect though they were reiterated by successive governments on the occasion of each new posting. The companies' response was invariably to claim that they believed prices to be fair and consistent with the provisions of the law and that they enjoyed the exclusive right to set prices. Other serious problems were raised in 1966 when Mobil priced its 36° crude oil at $2.10 instead of $2.15, that is to say with a 5 cent penalty, alleged to be due to its high wax content. Government protests were equally ineffective.

A clarification of the legal background and framework of the pricing issue in general, and of the 1961 Esso move in particular, would be most helpful if we are to achieve a better understanding of the many problems and conflicts between the government and the oil companies that developed during the 1960s and culminated in 1970 in the harsh, direct confrontation that ignited the present major crisis in the international petroleum industry. The negative attitude of the oil companies generated deep feelings of frustration and suspicion, which led the Libyans to believe that they were the victims of premeditated injustice. The inefficiency of the protests of the monarchical regime appeared later to the young Libyan revolutionaries as suspicious collusion with "imperialism and monopolies" and prompted their early attacks against the oil companies to repair past injustice through an appropriate price increase. The following explanation will enable us to have a better understanding of the psychology of the Libyans and their emotional approach to the pricing conflict under the new revolutionary regime.

The activities of the oil companies operating in Libya were governed by the Petroleum Law of June 18, 1955. Under the provisions of Article 14, the government was entitled to receive 50 percent of the concessionaire's profits, which were arrived at after deducting the following items from the gross income: expenses, amortization, and a depletion allowance at the rate of 25 percent of the gross income provided it did not exceed 25 percent of the profits for the year. All fees, rents, royalties, and income taxes already paid were considered as a credit against total payment, and balance, if any, was payable as a surtax.

The law failed to give a clear definition of gross income, but it was generally understood to be equal to actual realization. Furthermore, the concession holders were under no legal obligation to post a price for their crude oils. The royalty at 12.5 percent was to be evaluated on the basis of a price to be agreed upon, but its value was considered as a credit against total payment and would not alter the company's fiscal obligations.

Shortly before the start of export operations, a basic modification was introduced into the Petroleum Law by a Royal Decree issued on July 3, 1961 and published in the Official Gazette of July 15, 1961. Among other important modifications, Article 14 was remodeled to clearly define the pricing procedures and the fiscal obligations of the concession holders. The gross income and the royalty were to be assessed on the basis of posted prices. The depletion allowance was entirely eliminated, the fees and rentals were mostly increased, the depreciation rate of physical assets was reduced to 10 percent instead of 20 percent, and the amortization of other capital expenditure was reduced to 5 percent.

The profits, to be shared at 50 percent, were to be obtained by deducting operating expenses, depreciation, and amortization from the gross income. Paragraph 5a of Article 14 defined the gross income in relation to crude oil export as "the posted price per ton of such crude oil less marketing expenses as defined by regulations, multiplied by the number of tons of crude oil so exported "—that is to say, realized prices in practical terms.

The procedure for setting posted prices was defined as follows in paragraph 5a of Article 14:

> Posted Price means the price FOB Seaboard Terminal for Libyan crude oil of the gravity and quality concerned arrived at by reference to free market prices for individual commercial sales of full cargoes and in accordance with the procedure to be agreed between the concession holder and the Ministry or if there is no free market for commercial sales of full cargoes of Libyan crude oil then posted price shall mean a fair price fixed by agreement between the concession holder and the Ministry or in default of agreement by arbitration having regard to the posted prices of crude oil of similar quality and gravity in other free markets with necessary adjustments for freight and insurance.

Although established concession holders were under no legal obligation to convert to the new terms heavily burdening their fiscal obligations, the government bargaining position was strong enough to induce "voluntary" alignment by all companies, which were keen to keep on the best terms with the administration in view of the tremendous potential reserves they began to visualize in the Libyan subsoil.

The amendment was made at the time Esso was preparing for its first crude oil exports. The initiative it took in August 1961 to post the price of 39° Zelten crude at $2.21 a barrel FOB Mersa Brega was achieved under, and within the framework of, the 1955 Petroleum Law, since the company had not yet aligned its operations to the just promulgated amendment. The company traded its acceptance of the new rules against further clarifications to be issued by the government to limit the financial charges occasioned by the Royal Decree of July 3. This was achieved in Regulation No. 6, issued on December 6 and published in the <u>Official Gazette</u> on December 23, 1961. It seems that Esso officially accepted the July modification only after the promulgation of Regulation No. 6, which defined the pricing procedure indicated in paragraph 5a.

Two of the provisions of Regulation No. 6 are relevant to the present issue. Article 15 defined marketing expenses, which may be deducted from posted prices to arrive at the company's income, as the "sum total of rebates, if any, from the posted price which the concession holder is obliged to grant for the purpose of meeting competition in order to sell Libyan crude oil." Therefore, there could be no confusion, and the company's income is to be assessed on the basis of realized prices.

Far more important was the opposition of the companies to the principles embodied in Article 14 with respect to the definition of posted prices. The unacceptable feature was that setting posted prices was to be achieved "in accordance with the procedure to be agreed between the concession holder and the Ministry." The companies considered the right to determine, at will, the price of their crude oils as a fundamental principle of their activities and as an inviolable privilege they could not afford to yield without alienating their strategies or weakening their ability to control world markets.

It seems that Libyan legislators recognized the soundness of this objection about principles and confirmed the companies' privilege to set posted prices unilaterally. Article 14 of Regulation No. 6 defined the "procedure" referred to in paragraph 5 of Article 14 of the amended law in such a way as to render that paragraph completely inoperative. In these conditions, Esso could align itself to the new amendments, as integrated and complemented by Regulation No. 6, which stipulated that crude oil pricing be done, legally, at its sole discretion.

Paragraphs 1 and 2 of Article 14 of Regulation No. 6 read as follows:

> Article 14—The procedure referred to in the definition of "posted price" in Paragraph (5) of Article 14 of the law and in Paragraph (5) of Clause 8 of the second Schedule to it shall be determined as follows:
> 1. The concession holder or its affiliates shall from time to time establish and publish its posted price for Libyan crude petroleum of the gravity and quality concerned, which shall be the price at which crude petroleum of that gravity and quality is offered for sale by the concession holder or its affiliates to buyers generally in cargo lots FOB Seaboard Terminal.
> 2. Posted prices, with reference to crude petroleum exported by the concession holder or its affiliates, shall be the prices FOB Seaboard Terminal posted by the concession holder or its affiliates, as quoted by Platt's Oilgram for the quality and gravity concerned. If Platt's Oilgram is discontinued or no longer quotes applicable posted prices, some other internationally accepted publication shall be used.

Thus, Esso pricing was legally covered in 1961. It was formally approved by the government on the occasion of the promulgation of the Royal Decree of November 20, 1965, which introduced a major amendment to the companies' fiscal obligations. Under the new provisions, payments to the government are to be made according to the so-called OPEC royalty expensing formula, which was outlined in Chapter 4. In particular, marketing expenses are to be limited to 0.5 cents per barrel; this element eliminated rebates and discounts off posted prices.

The independent oil companies fiercely opposed the new amendments, since their taxable income was based on realized prices as low as $1.60 instead of posted prices at $2.21 a barrel. They were forced into acceptance after one of the most dramatic episodes of the oil history in Libya. The major oil companies took the initiative in this respect and acted, on this occasion, as the allies of the government to force the independents into capitulation, in the hope of seeing them phased out of the market or at least of diminishing their aggressive competition by substantial cuts in their profit margins. In fact, during the period 1961-65, Esso paid taxes usually based on posted prices, although it could claim discounts in its transfer prices to affiliates. The remuneration it expected from this "voluntary overpayment" was to convince the Libyan officials of the harmful effect of letting independents in and the attractiveness of doing business with large integrated companies.

The companies, mainly Esso, were rewarded for their acceptance of the new fiscal arrangements by the addition of the following paragraph to Article 9 of their concession agreements:

> The basis which has been utilized by the Company for the assessment of the level of posted prices, discounts, rebates and allowances for the purpose of calculating total sums to be paid by the Company to the Government, including royalty, will be considered as fair and valid for the assessment of the Company's obligations towards the Government for all the periods prior to the date of enforcement of the amendment by which the present paragraph 5 has been added to the concession agreement.

In the light of the foregoing, it appears that the Esso pricing initiative was fully safe and cautiously covered by the laws and regulations in force. However, the apparent legal validity did not prevent the development on the part of the Libyans of feelings of frustration and could not dissuade them from challenging the fairness of the Esso price. In fact, the legality of the action does not obviate the necessity of our investigating the soundness of the arguments presented by Esso to support and justify the price it arrived at. This will be achieved by a thorough analysis of the Esso memorandum, which is fully reproduced in Appendix A.

This memorandum is of a real importance since it is almost the only document published by a major oil company that explains the philosophy of posted prices and outlines the mechanism of setting their values. The memo admits posted prices do not necessarily bear a direct relation to market realities,[8] and their level is shown to be derived from prevailing postings in other exporting areas by the net-back formula concept and from consideration of gravity, freight, and quality differentials.[9]

The $2.21 posted price of 39° Zelten crude oil is explained and justified as follows:

• The 2 cents per degree gravity differential was recognized in principle but not fully applied, on the alleged grounds that the gasoline yield is higher than the market requirement and that the high wax content means that "from a yield standpoint, Zelten performs more like a 35° to 36° crude than like a 39° to 40° API."[10]

• The low sulfur content was recognized as an advantage but of only limited significance. It was alleged to be outweighed by the high wax content, by the high pour point, and by a loss of about 10 percent in refining capacity because of distillation bottlenecking.

• The freight differential was recognized as the only significant advantage and was assessed on the basis of an average freight rate of Intascale minus 50 presented as the best appraisal of term charter rates for tankers in 1960-61.

In light of the foregoing, the posted price of 39° Zelten crude oil at $2.21 a barrel FOB Mersa Brega was arrived at as the arithmetical average of the posted prices of 34° Arabian light FOB Ras Tanura and Sidon, 34° Iranian light FOB Abadan, and 36° Kirkuk FOB Banias/Tripoli, adjusted for freight differentials.

Very few of the above arguments are actually valid, and one can wonder why they were stated in so confusing a manner to justify a result that, as we have seen, is fair and logical (if one accepts the policy-induced selection of Intascale minus 55 percent as a representative average freight rate). The price would have appeared fully in harmony with prevailing posted prices, had the memorandum not put it in a dubious position where arguments are in contradiction with the principles they are supposed to implement.

Without our going into great detail, the following remarks are essential for understanding of the situation:

• Zelten (Brega) crude oil is fully eligible for a basic gravity differential of 2 cents per degree, as will be demonstrated in Chapter 14. It might have been that some initial samples of Zelten crude oil proved to be waxy, but this is not confirmed in the quality of current commercial exports.

• The wax content and pour point of Brega crude oil and of the fuel oil derived therefrom are of the same magnitude as for 39-39.9° Murban crude oil (Abu Dhabi) was posted by Esso in 1962 at $1.88 a barrel FOB Jebel Dhanna without any penalty.

- The alleged distillation bottlenecking and the lower realization of high-gravity Zelten crude (an argument in contradiction with the nonrepresentativeness of API gravity) are features common to all light crude oils, and none of them (Qatar, Abu Dhabi, or others) had been penalized for that.
- The very low sulfur content is an effective advantage whose economic and market value cannot be neglected.

However, all these considerations dealing with quality differentials and market realizations would be controversial since there were no appropriate precedents to compare them with. This was not true for the freight differential valued by Esso on the basis of Intascale minus 50 percent, since there was the posting precedent at the Eastern Mediterranean, which was related to the Gulf postings by an average freight rate of about Intascale minus 33 percent. The validity of this precedent was recognized by Esso in these terms: "The prices posted at the Eastern Mediterranean reflect obsolete ocean transportation rates and should be adjusted downward to be in proper relationship with the Gulf postings. Nevertheless, since the Banias and Sidon postings have not as yet been so adjusted we used them in our study."[11]

We have already pointed out that the posted price of any crude oil might be deducted from the reference value for the same gravity FOB Quoin Island by adding the freight differential associated with the export terminal in question. The value of the freight differential might be somewhat arbitrary since it is determined by a "policy-induced" selection of an average freight rate. Figure 8.2 shows that such average pricing rates were Intascale minus 33 percent for the Eastern Mediterranean postings and Intascale minus 55 percent for the Brega posted price FOB Mersa Brega. The freight differentials of these two terminals are plotted against freight rates in Figure 8.3, and the actual Eastern Mediterranean posting is represented by point A.

Now, if we assume that the 39° Zelten crude oil was fully credited with the advantage of the 2 cents per degree gravity differentials, then its posted price in 1961 could have been set at any of the following values when one freight rate or another in the range of the actual rates prevailing in the early 1960s is retained:

$2.21   corresponding to Intascale minus 55 percent (point B), which corresponds to the actual value of the Esso posting;

$2.228   corresponding to Intascale minus 50 percent (point C) claimed by Esso to represent the average term charter rates but which actually represents the average rate of the spot market in 1960-61;

$2.265   corresponding to Intascale minus 41.5 percent (point D), for which the Libyan posting would not display any freight advantage with respect to the Eastern Mediterranean;

$2.30   corresponding to Intascale minus 33 percent (point E), which is the historical precedent set in the Eastern Mediterranean postings;

$2.41   corresponding to Intascale minus 7.6 percent (point F), which is the average AFRA value for the period 1960-61.

This range of normal price indetermination stretching over as much as 20 cents a barrel provides a striking illustration of the political nature of the pricing issue. Any value between $2.21 and $2.41 might be considered a fair price for 39° Zelten crude oil if the appropriate freight rate is admitted to be fairly representative for the purpose of pricing in conjunction with the Gulf area. We have already indicated that there are no objective or rational criteria for the selection of the "right and fair" average freight rate for the purpose of pricing. The selection is a policy-motivated action, and the value might be somewhat arbitrary. In the present Libyan case, the posted price of 39° Zelten crude oil set by Esso in 1961 at $2.21 a barrel would be "fair and just" if the average freight rate were admitted to be Intascale minus 55 percent. On the other hand, it would be underpriced by some 9 cents if it were aligned with the existing precedent at the Eastern Mediterranean and by some 20 cents if one considered the average AFRA rate of 1960-61 as representative of the tanker market.

THE PRICING OF ALGERIAN CRUDE OILS

Exploration, development, and production of crude oil in Algeria took place while that country was under full French sovereignty. First exports started from Bougie in 1958, and two other terminals were later brought into operation at Arzew and La Skhira (Tunisia). French companies held majority shares in Algerian concessions, with minor participation of some American and European companies.

Algerian crude oils are very light, with gravities exceeding 40°. Their sulfur contents, as well as their pour points, are very low, as indicated in Table 8.1. Their price structure is characterized by a flat value FOB each terminal. Prices are quoted for crude oils of 40 and above, with no provision for gravity differential. The price differences from one terminal to another reflect their relative freight advantages. In comparison with the most distant La Skhira terminal, Bougie enjoys an advantage of 4-5¢/Bbl, whereas Arzew enjoys 5.5-6.5¢/Bbl.

Crude oil pricing in Algeria is a very complex issue that can be understood only within its own peculiar context and specific conditions, which bear very limited similarities with, and are quite independent of, the conventional context and general framework of the international petroleum industry. A brief recapitulation of the historical background of Algerian oil will provide the key to a proper understanding of its recent developments.

Hassi Messaoud crude oil (45.4° API) was posted at $2.90 a barrel FOB Bougie in 1958, at a time when posted prices of Middle Eastern crude oils were at their highest. In February 1959, it was reduced to $2.77 a barrel along with the general price reduction of 13¢/Bbl in the Middle East. It was further reduced by 12¢/Bbl down to $2.65 in 1960, at the same time Middle Eastern prices were collectively reduced by 8-10¢/Bbl. It is obvious that the position of Algerian prices with respect to the Gulf and to the Eastern Mediterranean postings was not altered by the price modifications in the 1958-60 period.

During this period, and up to the Algerian independence in 1962, Algerian crude oils were actually of French nationality. Their pricing and the related fiscal arrangements were elaborated under French regulations intended to achieve the best safeguard of French interests. The price levels, which were determined under the Saharan Petroleum Code, were presumed to reflect the essentials of the market value of this "indigenous" crude oil as appraised from "within" by a major Western consuming country, as opposed to foreign crude oils, paid for in scarcely available dollars and competing for the supply of the domestic market. However, the income tax was paid by the oil companies on the basis of realized prices.

When Algeria became independent in 1962, Algerian oil became a "foreign" one, although most reserves remained French assets and crude oil continued to be produced mostly by French companies in the "franc" monetary zone. Therefore, it continued to enjoy a privileged position in the French market where the prevailing regulations enable the government to compel oil companies operating in France, including French subsidiaries of the majors, to cover a substantial share of their supply needs from sources located within the franc zone. Nevertheless, crude oil was no longer "national," and its pricing and related fiscal obligations were no longer governed by specific national considerations under the control of the French Government, which is, directly or indirectly, the major shareholder of French oil companies operating in Algeria.

This change in crude nationality induced a "political" cut in Algerian posted prices in 1963 by 30¢/Bbl, without any similar move anywhere else in the world, thus affecting the relative position of Algerian crude oils with respect to other producing areas in the Eastern Hemisphere.

The price reduction took place at a time when the Algerian Government was in a rather embryonic state and the general economic and oil policies were still in gestation, governed by complex and intimate ties with France. The oil interests were only a part of an overall package deal that progressively developed to set political and economic relations between the two countries in a special and privileged framework of cooperation. Algerian interests at stake included the marketing of Algerian wine, Algerian immigrant workers in France, financial assistance and further investments, and technical and cultural assistance; French interests at stake included the safeguarding of investments and assets in Algeria, strategic and political influence, and keeping Algeria in the franc zone. The oil matters symbolized this new spirit of cooperation and complementary interests. France needed Algerian oil produced by French companies and paid for in French francs, and the Algerians appreciated the privileged outlets secured for their oil in the French market.

This new and original approach to oil matters was materialized in the cooperation agreement signed by the two governments on July 29, 1965. It covered all aspects of the oil industry, and the exploitation of hydrocarbon resources was to be undertaken jointly by a cooperative association, ASCOP, grouping French companies and Sonatrach, the Algerian National oil company. The crude oil prices and the fiscal obligations of operating companies were an integral part of this overall agreement. The posted prices were confirmed at their 1963 levels but gained no fiscal significance with respect to the determination of income tax paid by oil companies. Taxation was to be applied to the gross profit obtained by deducting operating costs from tax reference prices instead of realized prices. No provision for royalty expensing was provided.

The main terms of the ASCOP agreement can be summarized as follows:

1. It increased the rate of income tax from 50 percent—as prescribed under the Saharan Petroleum Code—to 53 percent during the years 1965, 1966, and 1967, to 54 percent in 1968, and to 55 percent for subsequent years.

2. It discontinued the practice whereby the companies had the right to defer payment of taxes on the 27.5 percent depletion allowance.

3. Tax reference prices applicable to exports of Algerian oil from various terminals for the purposes of computing income tax and royalties were set at $2.095 a barrel FOB Arzew, whereas the posted price was $2.365 a barrel; $2.08 a barrel FOB Bougie, whereas the posted price was $2.35 a barrel; and $2.04 a barrel FOB La Skhira, whereas the posted price was, $2.30 a barrel.

4. Depreciation allowances were brought into line with the comparable provisions in the Middle East oil concessions, whereas previously they had been computed on the basis of negative balance results.

5. The Algerian Government agreed to grant nine new concessions to French companies, which, in general, were based on the Saharan Petroleum Code.

6. The French Government agreed to pay to the Algerian Government nonrepayable grants totaling 200 million francs ($40.8 million) during the five-year period following the effective date of the agreement.

7. The French Government undertook to grant the Algerian Government loans totaling 800 francs ($163.2 million) during the five-year period following the effective date of the agreement. The loans bear an interest at the rate of 3 percent a year and will mature in 20 years.

8. The French Government undertook to guarantee supply credits to an amount of 200 million francs a year for the five-year period following the effective date of the agreement, totaling about $204 million.

9. The French Government undertook to create, develop, and finance an Algerian Institute of Petroleum for research and training.

If we bear in mind that the real significance of posted prices in the Eastern Hemisphere, from August 1960, was that these provided a basis for assessing the tax and fiscal obligations of operating companies, it appears from the above that Algerian oil economics has specific characteristics of its own, quite different from prevailing rules in the Middle East and elsewhere. In particular, neither posted prices nor tax reference prices in Algeria are comparable to conventional posted prices. The further implications of oil economics with respect to other interests outside the framework of the profession (political and monetary relations, Algerian wine and labor in France, general cooperation), aside from the different aspects of the ASCOP agreement as noted above, play in fact the role of nonprice discounts. What would be comparable to posted prices in the Middle East is a hypothetical value exceeding tax reference prices by an amount, expressed in terms of dollars a barrel, corresponding to an over-all estimate of extra financial advantages granted by the French Government. A fair comparison would be based on total financial performances within the conventional framework in the Middle East and within the Franco-Algerian package deal (royalty expensing was not in effect by the time the ASCOP agreement was negotiated and finalized).

It is very difficult indeed to assess the equivalent value of nonprice discounts in terms of dollars a barrel, although the methods developed by Ching Chin Chen[12] may be applied to grants, loans, and credit facilities provided by the French Government. It is much more difficult to assess the value of new concessions granted to French oil companies or the substantial share in the French domestic market that provides a guaranteed outlet for Algerian production. It should be noted however that insisting on comparing Algerian oil economics with the conventional framework of Middle Eastern oil would constitute a profound misunderstanding and misjudgment of the real nature of the situation. The two contexts are not directly comparable, and Algerian oil economics, as embodied in the ASCOP agreement, should be assessed on its own specific merits.

The history of the oil scene in Algeria since the mid-1960s has been characterized by a continuous series of disputes and tension. The revolutionary regime in Algeria settled on a nationalist oil policy, aiming, in the long run, for direct exploitation of natural resources. In contrast with similar statements of political principle formulated by other producing countries, Algeria has deliberately chosen to travel the hard way of developing skilled labor and competent management and acquiring its own professional experience and technical know-how. The Algerians wanted to control effectively their natural resources according to the same standards as the Americans and Europeans. They meant business and knew that they would not be taken seriously unless they proved to be up to their responsibilities in all respects. They accepted the challenge, even provoked it, and were determined to win.

Sonatrach rapidly developed into a large, well-equipped company and won its first battle, politically and professionally, with the construction of Algeria's third pipeline to Arzew, which it owns and operates exclusively. This initiative was opposed by the oil companies on the grounds of principles and rights. A multitude of service companies were developed as joint-venture deals with specialized companies of international standing. These agreements were developed on sound professional grounds, and the Algerian good will and business-minded approach created enough confidence to attract American investments with a minority share of up to 49 percent in a "socialist" country. Partners include Southeastern Drilling Company (Alfor-Drilling), Independex (Algeo-Geophysics), Corelab (Alcore-Reservoir Engineering), Davis Mud and Chemicals (Al Fluid-Mud Logging), and Baker Oil Tools Company for field services.[13] Algeria proved to be a good partner, gaining the reputation of "respecting scrupulously all the agreements she has concluded independently."[14]

The major long-term objective of Sonatrach was to gain majority control of all hydrocarbon reserves in Algeria. French assets were protected by the special context of political ties and economic commitments between Algeria and France. No such bridles limited the

Algerian action aiming to recover most, if not all, oil assets held by non-French companies. After having arranged the purchase of the British Petroleum retail network in January 1967, the Algerian Government nationalized all other distribution networks in the heat of the Arab-Israeli war in June 1967. The conflict, and the alleged U.S. collusion with Israel, provided an easy political alibi for a long-matured decision of bringing American producing assets under governmental control. Mobil, Sinclair, El Paso, and Phillips were put in the hands of Sonatrach.

Confident in this privileged position of strength, the Algerians began to exert sustained pressure in order to induce "voluntary" modification of the concession agreements of these companies. The options proposed, both painful, appeared to be either to renounce the concession terms, which were to be replaced by minority participation with Sonatrach, or to face the prospect of gradual and total elimination in one way or another, with or without compensation.

Getty Oil had its own reasons for accepting the deal and entered into a joint-venture agreement with Sonatrach in October 1968. Getty agreed to grant Sonatrach 51 percent of its interests in Algeria, including producing fields and pipelines. It also agreed to invest a minimum of $16 million in the five years to come and to grant to nonrepayable "advance" of $2.25 million to Sonatrach from then on to be operator of the new association. Counteradvantages conceded to Getty included the lifting of governmental control and recovering 49 percent of assets, the granting of an 11,500 square kilometer exploration permit in a promising area, and relatively easy financial obligations. Posted prices were established at their pre-1963 level, but income tax was to be calculated at the rate of 55 percent on the basis of tax reference price set at $2.195 a barrel FOB Bougie, substantially lower than general Algerian claims. No provisions for royalty expensing were made. This agreement was intended to set a model for further Algerian action. It was the first time that a producing country acquired majority participation in existing production, and the agreement was claimed to set profit-sharing at 88/12, in the favor of Algeria.

El Paso reached another arrangement with Sonatrach in mid-1969 that provided for the supply of 10 billion cubic meters per year of gas, over a 25-year period, the largest deal of its kind ever signed. The total cost of the project would be around $1 billion, and Sonatrach would provide about two-thirds of the financing.

But other non-French oil companies did not have reasons similiar to those that motivated Getty to accept the governmental deal and vigorously resisted Algerian pressure and intimidation. On April 29, 1969, a decree was issued canceling Sinclair's oil concession rights in Algeria on the grounds that Sinclair's merger with Atlantic Richfield in March, without the prior approval of the Algerian authorities, constitued a violation of the Saharan Petroleum Code and the provisions of the concession agreement. Sinclair assets were put up for public auction under such conditions that only Sonatrach could win. It actually did get Sinclair's properties in spite of the company's appeal to the International Court of Justice at the Hague.

The last round took place in June 1970, when four non-French companies were unexpectedly nationalized: British-Dutch Shell; American Phillips along with its subsidiary Drilling Specialists; Italian AMIF; and German Elwerath-Sofrapel. Apart from Mobil assets under sequestration (the company is presumably inclined to reach an agreement), the only crude oil not yet under Algerian control is held by those French companies with which Sonatrach's dealings were confused and made inoperative by the political context of the complex relations between the Algerian and French governments commanding the oil activities of French companies.

In the late 1960s Algeria was not prepared for a full confrontation with France so that French assets seemed to have been immunized against the Algerian assaults by the 1965 agreement. Therefore, Algerian objectives were concentrated on the improvement of the financial terms of their deal with the French, especially since the ASCOP agreement was to be negotiated in 1969. Algeria took several initiatives to strengthen its bargaining position in the coming negotiations: Posted prices were restored to their pre-1963 level, that is, $2.65 a barrel FOB Bougie (renamed Bejaia); the French companies were informed that tax reference prices were "temporary and provisional" as of January 1, 1969; several measures were taken to reinforce the control over money transfer; and so on. Algeria also secured full support from OPEC in its negotiations with France.[15]

Negotiations with the French started in November 1969 and were concerned with technical and economic aspects; the meetings were a confrontation of experts. The Algerians "demonstrated" that their crudes were worth $2.65 a barrel and the French "counterdemonstrated" that they were rather overpriced even at actual tax reference prices because of a world oversupply capacity. The technical confrontation could lead nowhere because of the immense gap between the philosophy and real objectives of both parties. French negotiators could find Algerian arguments "academic although intellectually interesting" but far from realistic. The doctrinal background of the Algerian attitude and arguments was clearly the negotiation of what the Algerians termed the "colonial pact" dictated in 1965, but political pressure and intimidation could not scare the French because French retaliation possibilities were at least as potentially effective. The real importance of Algerian oil to France is that it is located in the franc zone and is paid for in French francs. There is a maximum price France is prepared to pay for the consolidation of its balance of payments (alternative crude should be paid for in dollars), and obviously this price is far below Algerian ambitions.

Expectedly the negotiations failed to achieve any positive result and were discontinued in June 1970. The Algerian demands, rejected by the French, were as follows:

- Posted prices to be confirmed at actual values (pre-1963 levels)—$2.665 per barrel FOB Arzew; $2.65 per barrel FOB Bougie; $2.61 per barrel FOB La Skhira—for crude oils of gravity in the range 40-44.5° API. Prices are to be reduced by 1.5 barrels per each degree above 44.5° and below 40°;
- Taxes to be applied, at the rate of 55 percent, based on posted prices;
- Royalty expensing according to the OPEC formula; and
- These conditions to be effective as of January 1, 1969.

However, the over-all situation and atmosphere had changed by that time, and the "political" protection of French assets seemed less and less effective. The Libyan revolutionaries were engaged in a direct confrontation with the oil companies over the pricing issue and coordinated their plans and action with Algeria. It was felt that the international petroleum industry would undergo deep modifications and that the producing countries would enjoy preponderant bargaining positions. It seems as if the Algerians finally settled on an over-all confrontation with the French.

On July 27, 1970, the Algerian Ministry of Industry and Energy unilaterally dictated new tax reference prices to replace the old ones declared as provisional since January 1, 1969. The new reference prices to which taxes would apply at the rate of 55 percent were as follows: $2.870 a barrel FOB Arzew; $2.855 a barrel FOB Bougie; $2.815 a barrel FOB La Skhira.

This was the starting point of a dramatic episode that led to a complete break between Algeria and France, and to the nationalization of the French oil companies as will be indicated in Chapter 16.

THE PRICING OF NIGERIAN CRUDE OILS

The first commercial discovery in Nigeria was made onshore in 1955 by Shell-BP, which remained the sole producer until 1963, when Gulf made the first offshore discovery. Other commercial discoveries both onshore and offshore by SAFRAP (French), AMOSEAS, Tennessee, Mobil, AGIP/Phillips, and Phillips helped set Nigerian production in early 1967 over 550,000 barrels a day (b/d) and confirmed the emergence of this Western African country as a major new supply area.

In June 1967, Biafra declared its secession from the Nigerian Federation and called for its recognition as an independent country. Shortly thereafter, civil war flared up throughout the country and brought petroleum activities and exports to a standstill, since the major oil fields and export terminals, under Biafran control, were blockaded by the federal government. Exports were resumed in late 1968 and recovered their prewar level in April 1969.

Peace returned in December 1969, and Nigeria's remarkable recovery from the effects of its civil war exceeded all expectations and forecasts. In May 1970, production passed the 1 million Bbl/day level, nudging out Algeria as the 10th-biggest producer in the world. The 2 million Bbl/day mark was expected to be attained no later than the end of 1972.

Until 1967, there were no posted prices in Nigeria since the fiscal obligations were based on realized prices, and the royalties were credited against income tax liabilities. Nevertheless, the concession agreements included a "most favored African nation clause" by which the fiscal obligations of the companies would have to align with any improved terms conceded to any other African country. Upon the amendment of the Libyan Petroleum Law in November 1965 eliminating discounts off posted prices and implementing the OPEC's royalty expensing formula, the Nigerian Government claimed that it was entitled to similar treatment and requested that the oil companies align with the Libyan precedent. A series of negotiations followed, mainly with Shell-BP, and on January 5, 1967 a decree was issued, with a retroactive effect as of January 1, 1966, amending the Tax Act of 1959. The new Petroleum Profits Tax Act provided that posted prices, to be established in prior consultation with the Nigerian Government, would be the basis for tax and royalties payments and that royalty would be expensed. Oil companies had no chance to escape the new provisions of the law, and their protestations could not alter its enforcement.

Shortly afterwards, the following posted prices for Nigerian crude oils were published by BP-Shell FOB Bonny terminal: $2.17 a barrel for 34-34.9°. Nigerian light, and $2.03 a barrel for 27-27.9°. Nigerian medium, with a gravity fluctuation differential of 2 Bbl/degree above and below indicated gravities.

A new Petroleum Law was issued on November 27, 1969, which became effective immediately. It did not specifically alter the existing financial provisions but confirmed the 1967 amendments. According to the new law, posted prices are to be set by the concession holder "after agreement with the commissioner as to the procedure to be followed for the purpose." The price is to bear a "fair and reasonable relationship" to established posted prices for Nigerian crude of comparable quality and gravity, or if there are none, to those at "main international trading export centers," with due regard, in either case, to freight differentials "and all other relevant factors."

Although posted prices were published only for two crude oils in Nigeria, the pricing pattern governing them has its own specific characteristics. Its main feature is a basic gravity differential of 2 Bbl/degree, but in contrast to the Gulf postings, it does not include a penalty for "investment discontinuity" when shifting from 34° Nigerian light to 27° Nigerian medium.

Let us examine how the $2.17 posting of 34°. Nigerian light, taken as a reference crude oil, would compare with other postings in the Eastern Hemisphere. The impact of gravity on the freight differential relating Nigerian prices to the Gulf FOB Quoin Island is outlined in Figure 8.2. If we consider the freight differential for a 34° gravity between Bonny and Quoin Island with respect to Western Europe (average freight rates to Rotterdam and Marseille) and consider actual freight rates prevailing in 1966-67 when Nigerian postings were determined, the posted price of 34° Nigerian light FOB Bonny could have any of the following alternatives values: $1.962 a barrel, corresponding to Intascale minus 65 percent, which was the average rate of the spot market; $1.976 a barrel, corresponding to Intascale minus 55 percent, which was the value upon which the Libyan postings were based, as indicated above; $2 a barrel, corresponding to Intascale minus 36 percent, which was the average value of AFRA rates; or $2.004 a barrel, corresponding to Intascale minus 33 percent, which was the value upon which the Eastern Mediterranean postings were based.

It is obvious that the "normal" range of freight fluctuation was very limited, and the posted price of 34° Nigerian light crude oil FOB Bonny would have been "technically" justified anywhere in the $1.962-2.004 range. A "fair" price would have been most likely set at $2 a barrel corresponding to Intascale minus 36 percent, which meets three important criteria simultaneously: It represents the prevailing and prospective average value of AFRA (the six-day war of 1967 could not be foreseen); it almost represents the basic rate (Intascale minus 35 percent) governing posted prices in the Gulf; and it almost coincides with the historical precedent of the Eastern Mediterranean at Intascale minus 33 percent.

Consequently, setting the posted price at $2 a barrel FOB Bonny would have been in full harmony and accordance with an over-all pricing pattern in the Eastern Hemisphere. Now, BP-Shell actually set the price at $2.17, thus granting Nigerian light a premium of 17¢/Bbl, which can hardly be accounted for in terms of freight differential. Figure 8.2 shows that one should consider an average rate of Intascale plus 94.6 percent in order to bring the $2.17 price into harmony with the Gulf postings. Obviously, retaining such a value in the conditions prevailing and foreseeable in late 1966 is absurd and cannot be justified on any grounds whatsoever.

The only possible explanation is that the $2.17 price has a built-in quality premium over and above the prevailing price rules in the Gulf, mainly gravity differential. Checking the crude assay of Nigerian crude oils against those available in the Gulf would suggest that the 17¢/Bbl premium would account for the balance of the penalty for high pour point, premium for better valorization in terms of products yields, and premium for low sulfur content.

The economic evaluation of these three items will be undertaken in the following chapters.

It is worth noting that this substantial shift from the prevailing pricing patterns in the Eastern Hemisphere, inside and outside the Gulf, combines with another important shift, which fails to penalize crude oils of 32° API and below by 9-11¢/Bbl in terms of investment discontinuity. The initiative of these pricing innovations were taken by BP (and Shell). BP also introduced the first pricing innovation in 1953, when it brought the investment discontinuity penalty into the Gulf pricing pattern.

NOTES

1. Stephen H. Longrigg, Oil in the Middle East (London: Royal Institute of International Affairs, Oxford University Press, 1968), p. 77.
2. Ibid., p. 179.
3. Ibid., p. 361.
4. Zuhayr Mikdashi, A Financial Analysis of Middle Eastern Oil Concessions, 1901-65 (New York: Praeger Publishers, 1966), p. 307.
5. Longrigg, op. cit, p. 206.
6. Price deviation is computed by the following formulas:
CIF price: $Y = Z + f_o = X + f$
parity price: $Z = X - (f_o - f) = X - \varphi (1+i)(F_o - F) - 0.88 \varphi$
price deviation: $\Delta X = X_o - Z$
$$\Delta X = X_o - X + (\varphi(1+i)(F_o - F) + 0.88\varphi)$$
where X is X FOB price at the export terminal for a given gravity value (G); X_o is the Gulf reference price FOB Quoin Island for the same gravity (G); and ($f_o - f$) is the freight differential with respect to Quoin Island.

It is obvious that the price deviation ΔX varies with the crude oil gravity (G) through the conversion factor (φ); the freight rate is (1 + i); and the geographical location of the export terminal through is (F).

7. The Review of Arab Petroleum and Economics, August 1968, p. 28.

8. The Esso memorandum reproduced in the Appendix A.

9. Ibid.

10. Ibid.

11. Idem, pp. iii-iv.

12. Ching Chin Chen, "Crude Oil Prices and the Postwar Japanese Refining Industry" (Ph.D. Dissertation, Massachusetts Institute of Technology, February 1967).

13. Petroleum Press Service, "Sonatrach Increases Its Stake," December 1968, p. 453, and "Sonatrach's Oil Empire," February 1972, p. 53.

14. Oil and Gas Journal, April 20, 1970, p. 88.

15. The OPEC Resolution XIX, 105 of December 16, 1969 states that

"The Conference,

having heard the statement of the Head of the Algerian Delegation concerning the opening of the negotiations for the revision of the fiscal terms applicable to the French oil companies,

considering that Algeria is notably aiming, through this action, to secure proper consideration of the advantages Algerian crudes enjoy due to their quality and their geographical proximity to the markets,

recognizing that Algeria is acting in full accordance with OPEC principles and objectives and economic conditions prevailing in the international oil industry,

expresses its full support for any appropriate measures taken by the Algerian Government to safeguard its legitimate interests."

CHAPTER

9

SULFUR DIFFERENTIAL

Sulfur compounds are present in different forms and in various proportions in almost all crude oils. There is no advantage whatever to their presence. On the contrary, they are responsible for serious hazards that cannot be allowed to continue and whose elimination is fairly expensive. The noxious effects of sulfur can be located in three domains: corrosion and hazards in refining operations, the sulfur contents of white products, and the sulfur contents of residual fuels with respect to air pollution.

The technical and economic aspects of almost all the problems involved have been thoroughly investigated and assessed in the literature. Higher-sulfur crude oils are more costly to process in response to given market demands and specifications than comparable crudes of lower sulfur content. The incremental costs involved can be fairly assessed on the basis of the actual facts, performances, and economics of the industry. The findings are normally given in terms of incremental costs as a function of increasing sulfur contents. In other words, sulfur-related economics concern penalties for higher sulfur contents and not premiums for relatively lower sulfur contents.

Serious problems arise when one questions the relationship between sulfur differentials and crude oil pricing patterns. The evaluation of such differentials on technical and economic grounds is most intricate, but still it can be achieved within reasonable ranges of approximation. The real problems are related to the basic concepts of pricing philosophy and to the political dimensions of pricing policies and strategies. Higher sulfur contents mean inevitably higher expenses with respect to maintenance, desulfurization, sulfur pollution abatement, and other factors. The main question is to determine who should share in meeting such extra expenses and to what extent. The magnitude of these expenses can be very high indeed both in terms of per barrel and total costs, and many approaches could be visualized by which total or partial costs might be allocated to one or another of the parties involved. The interests at stake are mostly conflicting, and any outcome of such indirect confrontation would reflect relative bargaining positions rather than any actual fact of the industry.

SULFUR COMPOUNDS IN CRUDE OILS

Elemental sulfur may exist as such in some crude oils, but generally it is present as more or less complex molecules in the whole range of petroleum fractions. On the other hand, the different refining processes may modify the nature and distribution of sulfur compounds.

General correlations between sulfur content and the main characteristics of crude oils are most difficult to establish. Nevertheless, qualitative indications are generally accepted as fairly representative. Sulfur contents vary with geographical location from one sedimentary basin to another. Generally speaking, Middle Eastern crude oils have a relatively high sulfur content, African crude oils a very low sulfur content, and Venezuelan crudes lie in between. Middle Eastern crude oils may be subdivided into two ranges: Iranian crude oils at the lower range and Iraqi and Kuwait crude oils at the higher range, and other crude oils in between. American crude oils display a very wide range of sulfur content, according to W. L. Nelson.

It is generally stated that higher °API gravity crude oils contain less sulfur. However, the relationship between gravity and sulfur content, although qualitative, has significance only in relative terms. As indicated by W. L. Nelson,[1] there is no accepted definition of what constitutes a low-sulfur crude oil. Obviously, a 40° API crude oil that contains 0.3 percent sulfur is certainly not a low-sulfur oil, whereas a 20° API crude oil that also contains 0.3 percent sulfur would be certainly a very low-sulfur oil. Accordingly, American crude oils have been classified as being "low-sulfur" oils if they contain less than the maximum amounts given in Table 9.1 for the gravities indicated.

These maximum percentages have been retained by Nelson at about 60 percent of the average sulfur content of major crude oils produced in the United States of the same °API gravity. It is questionable whether these standards can also apply to the classification of crude oils produced in other areas.

Total sulfur content of crude oils is not important as such; it is rather the distribution of sulfur in the various fractions obtained when processing the crude that is responsible for refinery corrosion and for the resulting incremental costs of maintenance, whereas the sulfur content of finished products is responsible for incremental costs of desulfurization.

TABLE 9.1

Maximum Sulfur Content of "Low-Sulfur" Oils

Gravity (°API)	Sulfur (percent of weight)	
12°	under	0.93
15°	"	0.88
20°	"	0.75
25°	"	0.60
30°	"	0.46
35°	"	0.35
40°	"	0.25
45°	"	0.15

Source: W. L. Nelson, Oil and Gas Journal, Vol. 68 No. 33, (August 17, 1970), p. 78.

Hydrogen sulfide is a readily recognizable constituent of many crude oils and is generally present in high percentages in the gases associated with oil production. Thermal decomposition of some sulfur compounds such as mercaptans, sulfides, and disulfides generally yields hydrogen sulfide. This occurs in distillation or in cracking operations. Hydrogen sulfide is a smelly and toxic gas responsible for serious corrosion of processing units.

Corrosion cost American refiners about $560 million in 1956— that is, about 17.9¢/Bbl of processed crude oil.[2] Total sulfur corrosion (high sulfur with sourness) amounts to about 16.5¢/Bbl, or 92.4 percent of the total, whereas high sulfur corrosion alone amounts to about 12.7¢/Bbl, or about 70 percent. These 1956 figures were confirmed later when Nelson stated that "most of the maintenance expenses of a refinery can be traced to the corrosion caused by sulfur in the crude oil."[3]

Gasoline and naphtha specifications stipulate that such products should be sweet in order to be marketable. Therefore, sweetening processes, or the elimination of mercaptans, are a must for practically all refineries. They apply to straight-run streams from distillation and result in increasing the refining cost of higher-sulfur crude oils. Such an increase is generally expressed as a penalty affecting the value of the crude oil incriminated. The value of the penalty, as compiled by Nelson, depends on two factors: the sourness of the crude oil, and its total sulfur content.[4] Average figures applicable to Middle Eastern crude oils correspond to a sulfur differential penalty of about 4-5 cents for each 1 percent sulfur over and above an average sulfur content of about 1.5 percent. The magnitude of corresponding penalties ranges between 12 and 20 cents per barrel of crude oil.

The sulfur content of straight-run middle distillates has generally the same value as for the crude oil itself. In diesel oils, sulfur compounds increase wear and may contribute to the formation of engine deposits. The sulfur content of heavy distillates used as heating fuels contributes, to some extent, to air pollution. Current sulfur specifications for diesel oil are around 0.5 percent maximum, but it is expected that they may go as low as 0.25-0.30 percent of total weight. To meet these specifications, adequate desulfurization facilities are needed, and in fact, such facilities have become a common feature in almost all refineries. The incremental cost thus induced varies from one case to another, and there seems to be no general rule to estimate its value. Direct operating costs of desulfurization are estimated at about 10 cents per barrel for straight-run distillates and above 20 cents per barrel for cracked distillates. Total costs, including investment-related capital charges might be double the costs indicated above.

The largest part of sulfur compounds is concentrated in fuel oils utilized mainly as a source of primary energy. The sulfur content of fuels is responsible for corrosion during combustion by exhaust gases containing sulfur oxides at low temperature, and this corrosion is included in maintenance expenses; it is also largely responsible for air pollution in heavily industrialized areas.

The economics of sulfur pollution and its impact on the international petroleum industry were investigated by this author in a paper presented to the OPEC seminar held in Vienna in June-July 1969.[5] Unlike the preceding two items—sulfur corrosion, and the desulfurization of distillates—which correspond to actual extra expenses incurred by refiners in their current operations, sulfur pollution deals with potential problems that are most likely to materialize to different extents in the future, and the related expenses may be incurred partly or totally by the petroleum industry.

The main factor that will govern future evolution is a limitation on sulfur in residual fuels, which may be imposed by competent authorities in industrialized countries. Such drastic limitations have already been implemented on the East Coast of the United States, where low-sulfur fuel oils enjoy an established market premium, and it is expected that increasing volumes of low-sulfur fuels will be drained to the U.S. market from different Eastern Hemisphere sources. The situation is completely different in Western Europe, which is the natural market for Eastern Hemisphere crude oils. To the average observer, it might appear that, because of the absence of adequate regulations and the infrequency of alerts or hazards, there is no serious sulfur pollution there. This assumption is superficial and misleading, for the problem is actually as serious there as elsewhere. However, the high availability of naturally low-sulfur crude oils from North and West Africa has helped to ease the situation and to postpone the inevitable time when the seriousness of air pollution reaches a crisis level.

In order to reduce the sulfur content of fuel oils refiners prefer to process low-sulfur crude oils in as large quantities as possible. However, due to the limited availability of such crude oils, the second alternative would be to get the most advantage out of the operational flexibility of existing facilities to desulfurize distillates (which is a relatively simple and not too expensive process) and to blend them back to the fuel pool. However, this possibility has limited production when it is applied to Middle Eastern crude oils.

The third alternative would be the total desulfurization of the atmospheric residuum, bringing its sulfur content to less than 0.5 percent for most Middle Eastern crude oils. Some processes are available, but their reliability seems to be somewhat doubtful. This alternative is very costly: The investment is estimated to range between $2 and $3 per barrel of fuel, and the total cost of producing one barrel of 1.5 percent fuel oil might range between 45 cents (Abu Dhabi crude oil) and 90 cents (Kuwait crude oil). The total cost of producing one barrel of 0.5 percent fuel oil would exceed $1 in the most favorable case of processing the least sulfurous crude oils (Qatar or Abu Dhabi).[6]

SULFUR DIFFERENTIAL: REFINING

It is an objective fact of the industry that processing higher-sulfur crude oils is more costly than processing lower-sulfur crudes, and thus the latter would represent an equivalent savings for the refiner. Considering the impact on crude oil economics in terms of sulfur penalty or sulfur premium is essentially a matter of philosophy and policy. Sulfur-related expenses are actual costs incurred by refiners and represent a full part of the integrated over-all economics of crude valorization. The main problem is to know how to allocate such expenses—or savings—between the parties concerned, that is, oil companies and producing countries. On the contrary, sulfur pollution represents more or less potential expenses—or savings—that could be shared in different proportions not yet determined, by producing countries, oil companies, and other parties such as consumers, groups, consuming countries, and others. These two aspects of sulfur differential derive from different philosophies and will be discussed successively below.

During the 1950s when pricing patterns were elaborated in the Eastern Hemisphere, crude production was mainly located in the Gulf area where crude oils were characterized by fairly high sulfur content except for limited production of 1.05 percent Qatar crude oil. In other words, the sulfur impact on refining economics materialized only in terms of a sulfur penalty, that is, in terms of effective and actual incremental expenses. They were reflected in crude pricing patterns in terms of a sulfur differential penalizing high-sulfur crude oils in terms of about 3¢/Bbl for each 1 percent of sulfur above 1.6-2 percent weight. This penalty differential may be explained in connection with the sourness penalty, total sulfur corrosion, and distillates desulfurization, as discussed earlier in this chapter.

The situation gradually changed with the advent of low-sulfur African crude oils. Processing such crudes resulted in a substantial reduction, if not total elimination, of these expenses or penalties. The development of their relative share in over-all European supplies was presumably accompanied by a relative reduction in the desulfurization capacity of new refineries. Corrosion problems—not related to sourness—were reduced drastically. Therefore, refiners who could dispose of such crude oils as an alternative to conventional Middle Eastern supplies achieved effective savings equivalent to the incremental expenses that would have been incurred otherwise. Lower sulfur content thus represents a real and genuine economic advantage over high-sulfur crude oils, and such an advantage is a permanent one and is not subject to any significant fluctuation.

Oil companies and European refiners enjoyed and exclusively monopolized this economic advantage throughout the 1960s, which witnessed the extraordinary expansion of African production, since the advantage was not reflected, in any way whatsoever, in their prices. Low sulfur content is an effective quality worth a substantial premium, but it was secured free, to the detriment of producing countries. Without prejudging the real motivations of oil companies in denying this advantage to low-sulfur crude oils, one can understand their attitude in the historical context of the early 1960s. First, the production of low-sulfur crude oils took place in Algeria, then under full French sovereignty. These crudes were outside the framework of international oil economics and should be considered as captively produced and utilized by an industrialized European nation according to its internal economic standards. However, these crude oils were posted at $2.65/Bbl, whereas parity value with Middle Eastern pricing patterns would have given only $2.35/Bbl (See Chapter 8). Thus the French granted the high-gravity low-sulfur Algerian crude oils a 30¢/-Bbl premium without apparent justification, but one might suspect that their very low sulfur content actually played a significant role in this respect. This premium was eliminated upon Algerian independence, when posted prices were reduced to $2.35/Bbl, which aligned with the Libyan precedent established by Esso at Mersa Brega, where 40° Brega crude oil was posted at $2.35/Bbl without its enjoying any sulfur premium. However, this reduction was not significant to the Algerian Government since its income was based on a tax reference price of $2.08/Bbl and royalty was not expensed.

The production of low-sulfur crude oil soon followed in Nigeria, where the prevailing regulations did not provide for any postings, since taxation was based on realized prices. Posted prices of Nigerian oil were announced in 1967. They displayed a premium of about 17¢/Bbl over parity prices, a fact not explained or justified by the companies concerned. However, higher yields of middle distillates could account for a large part of this premium, although the very low sulfur content might also be suspected to share in the responsibility for this premium.

The question was really raised on the occasion of the first Libyan posting by Esso in 1961, which has been discussed in detail in Chapter 8. Although the company was not under any legal obligation at that time to post a price for its Libyan production, it elected to do so for reasons indicated elsewhere in this study. These reasons caused the setting of the price below its parity position with respect to the Gulf and Eastern Mediterranean crude oils, and, indeed, Brega crude oil was effectively underpriced by about 8-10¢/Bbl. In these conditions, one can easily imagine that no company was actually prepared to introduce, voluntarily and spontaneously, an original innovation in the established ethics of the industry and to increase prices by a sulfur premium. In fact, the introduction of such a premium, even if it were very modest, was in contradiction with the majors' philosophy and policies; they were eager not to create a precedent that could lead to uncontrollable developments throughout the world.

Furthermore, two other important factors militated against the recognition of the economic advantage of low sulfur and its introduction in crude pricing patterns. On political grounds, crude prices were frozen as of August 1960 after a period of continuous erosion and successive reductions. The factors commanding further price deterioration were still active, and their inhibition was mainly due to political considerations. In other words, should prices have been allowed to move freely again, such a move could be only downwards, especially in Libya, where realized prices were substantially lower then posted prices.

Even if political circumstances had favored an increase in prices, oil companies would still have been tempted to oppose the implementation of such an increase being presented as a sulfur premium. On financial grounds, processing low-sulfur crude oils would result in substantial savings in processing costs, which add to the companies' profits; in the meantime, the savings would represent potential losses should the companies have to pay back to producing countries the value of the sulfur premium they enjoyed. Such a potential loss cannot be recovered from consumers by reflecting it, in one way or another, in product prices; it would be a net potential loss to the companies, or better, a net cut in the incremental profits they enjoyed exclusively.

Let us here examine whether the recognition of a sulfur premium for low-sulfur crude oils would be justified or not. First of all, sulfur content is as genuine and permanent a feature of crude oils as gravity, and its actual economic value is an intrinsic one, not subject to market fluctuation. In other words, it is of the same essence and nature as posted prices themselves, and the introduction of a sulfur premium or its denial would be a long-term policy-induced move and not a tactical one. The argument that use of low-sulfur crudes would make it possible to meet market specifications by blending these lower with higher-sulfur crudes is not relevant. Blending requirements would influence, to some extent, the makeup of crude supply patterns from different possible sources. However, one cannot deny crude oils the economic advantage of this specific feature; otherwise the same question would arise for gravity and the whole pricing system would collapse. Market demand patterns for petroleum products are currently met by processing blends of crude oils of different gravities yielding different proportions of products that balance each other—excess gasoline of light crudes being balanced by excess fuel oil of heavy crudes, for example—and yet gravity does enjoy a differential in crude pricing patterns. In fact, a sulfur differential has been effectively recognized in prevailing pricing patterns in accordance with the integrated approach to crude oil economics, but only in terms of a penalty for higher-sulfur crudes. The question raised here is to extend the sulfur penalty differential applicable to sulfur content over and above 2 percent into the range of low sulfur below 1.5 percent, where it would be converted into a sulfur premium differential.

On philosophical and moral grounds, denying this intrinsic economic advantage to low-sulfur crude oils would mean its monopolization by the oil companies (since it does not reflect onto consumers) to the detriment of producing countries, whereas the basic concept and philosophy of posted crude prices is to share the advantages and disadvantages between the two parties in a fair and equitable way.

SULFUR DIFFERENTIAL: POLLUTION

It is obvious that the sulfur pollution phenomena did not play a significant role in the formulation of crude pricing patterns throughout the 1950s, since the existence of such phenomena at that time was merely potential. In fact, it was only during the second half of the 1960s that such problems emerged in the forefront of public and official concerns and contributed to increased pressure for the promulgation of stringent regulations. Therefore, arguments in this respect cannot be related to, or justified by, market evidence or realities. They can be presented only on psychological, moral, and political grounds and would result in inducing substantial changes and segregation in the marketplace.

The only objective facts that can be taken as a common basis for discussion are that sulfur-pollution abatement has become a priority necessity for heavily populated areas of industrialized countries and that such abatement is very costly. Technical solutions have been developed, but their economics seems to be prohibitive, or at least, cannot be supported wholly by only one party, and the real problem actually concerns the ways and means of having the total costs shared by all the parties concerned. Sulfur pollution, and more generally air pollution, is a specific problem of industrialized countries and can be considered a by-product of high standards of living and high per capita consumption of energy, services, and commodities. It is normal and moral to expect these countries to pay for these privileged advantages, but who should actually pay? Citizens are the ultimate beneficiary, and, as usual, they could be burdened with taxes and levies; but this would be effective only for that fraction of pollution related to individual domestic utilization (higher prices for lead-free gasoline or minimum-sulfur heating oil, for example). Unfortunately, this would have only limited effect on sulfur pollution, which results mainly from industrial activities and power generation. The costs would be passed indirectly to citizens by a general increase for manufactured goods, but this would proceed by a general increase in the cost of primary energy, which modern industrialized nations, for political and economic reasons, cannot afford to undertake. Furthermore, the economic performances of some industries might be marginal so that they could not survive the shock of such a brutal increase. Similarly, fossil fuels, which are directly responsible for sulfur pollution, cannot bear the full burden of pollution abatement; such a move could be a fatal blow to European coal industries, which can survive only with the help of government subsidies.

To quote the tentative conclusions of our paper on sulfur pollution, it is obvious that

> the real battle lies on legislative grounds where an overall framework and specific regulations have to be designed in order to ensure efficient sulfur pollution abatement, to conciliate the conflicting interests involved and to allocate

the high cost between them. . . . Should flue gas desulfurization be imposed on a large scale . . . then the largest part of necessary investment will have to be supported by fuel-consuming industries and the oil companies would be affected marginally. On the contrary, should severe restrictions be imposed on fuel sulfur content, then the major [question] to be answered would be . . . who will share in the high cost which obviously exceeds the possibilities of the industry.[7]

The situation has worsened rapidly in Japan and the United States before technical questions have found satisfactory answers. In the face of the doubtful and still unreliable performances of alternative solutions, severe limitations on fuel sulfur content have been enacted and implemented progressively. Because of the limited availability of naturally low-sulfur crude oils, desulfurization facilities have started to develop both in Japan and the Caribbean, the costs of which have been shared, directly and indirectly, by all parties concerned. The governments contributed by fiscal and taxation facilities, and the companies put down the investment effort and recovered it partly from the consumers by adding a sulfur premium to the price of low-sulfur fuel oils.

The situation has been completely different for Western Europe, the economics of which controls, to a large extent, the economics of Eastern Hemisphere crude oils. The issue seems to have been practically ignored, and most European countries are proud to state publicly that they do not suffer from serious sulfur pollution and that therefore there is no necessity to enter the debate over a sulfur premium. Fuel oil supplies are abundant in the market where low-sulfur fuels do not enjoy any significant premium; they are even available in surplus quantities, which are occasionally exported to the East Coast of the United States. It is understandable that European governments do not dramatize the situation, because of its serious repercussions for the fragile coal industry. It is true that sulfur pollution has not developed as acutely and seriously as in Japan and the United States, but does this mean that European countries are less industrialized and civilized than these two pollution leaders? Certainly not, and the explanation may be found in the fact that the process of extensive development of sulfur pollution, which invaded Japan and the United States throughout the 1960s, was, for Europe, naturally, continuously, and progressively curtailed and eroded by the availability of increasing supplies of naturally low-sulfur crude oils from North and Western African sources. This advantage was obtained at no cost since a sulfur premium has never been recognized for these crude oils or reflected in their prices. And by the very logic of things, moderately polluted air inhibited any tendency of the market to recognize any specific advantage for low-sulfur fuel oils. Here we find an inescapable vicious circle that would perpetuate the indirect subsidizing of sulfur pollution abatement at the expense of low-sulfur crude oils, that is to say, to the detriment of their producing countries.

In other words, the claim for a sulfur premium cannot be put within the framework of actual facts, since low-sulfur crude oils would be the victims of their own specific quality. It can only be forwarded on political grounds, and the arguments can be formulated as follows: Should European countries have been denied the advantage of getting naturally low-sulfur crude oils, then sulfur pollution would have developed similarly as in other industrialized areas and desulfurization would have become a necessity. Low-sulfur crude oils actually made it possible to avoid such a problem and, therefore, should be credited with the advantage of the marginal cost of alternative solutions that they have indirectly made possible to avoid up to now.

The debate is a highly complex one, especially because it has to take place between two parties, consuming and producing countries, that have no direct contact with each other. Approximate compromises will have to be elaborated by the oil companies, which will strive not to be the victim of such an indirect confrontation.

Consuming countries would be most reluctant to start paying for something they have been enjoying free. They would aim to perpetuate the present situation by which low-sulfur African crude oils pay for high-sulfur Middle Eastern crude oils so that European air remains sulfur-free with acceptable pollution standards.

Countries producing sulfurous Middle Eastern crude oils are not concerned as long as their prices remain unaffected by any penalty intended to enforce sharing in the over-all desulfurization costs. They have been protected by the stability of posted prices that provide the basis for the calculation of their income at a time when realized prices of high-sulfur crude oils have been progressively deteriorating (especially for exports to Japan).

Countries producing low-sulfur crude oils are directly concerned and are calling for the introduction of a sulfur differential, which would break the vicious circle. They will not continue subsidizing the progressive abatement of sulfur pollution and thus paying for European well-being and high standards of living. Their argument is rather simple but cannot be neglected or ignored: If the Europeans are not prepared to recognize the economic value of the advantage inherent to low-sulfur oil—that is, if they do not accept the existence of a specific quality that makes African crude oils more valuable than those from the Middle East—then they should be prepared and willing to dispense with it. Therefore, they should not mind this advantage being denied to them (unless they pay for it) in order to valorize it in other markets, such as Japan and the United States, where it will be paid for. Producing countries would even be prepared to face the risk of withholding their production momentarily so as to make consumers realize the advantage they have been enjoying free of charge and accept the principle of paying for it. Their criterion of judgment would be the cost of alternative solutions (desulfurization), which are far too prohibitive.

Oil companies have been responsible for unconscious and involuntary complicity with European countries throughout the 1960s, since they chose not to recognize the advantage of the low sulfur content of African crude oils and not to reflect its economic value in their prices. In particular, Esso bears a direct responsibility in this respect for establishing the first precedent by posting Zelten crude oil in 1961. The argument that market realities at that time could not justify such an innovation in prevailing pricing patterns is not relevant, for the reasons elaborated earlier in this chapter. The determination of crude prices is a long-term strategic move, intended to induce a predetermined over-all evolution of the market, rather than a move demanded by actual realities that are the result of recent past history and performances. This strategy is governed by political considerations and by a genuine evaluation of interests and the balance of forces. All conditions in the early 1960s militated to bind together the interests of the majors and those of and Europe, and every move towards lower crude prices was welcome. Should oil companies have elected to adopt a different strategy at that time, for any reason of their own, and consequently should they have introduced a sulfur premium in posted prices of African crude oils—and this would have been largely justified on maintenance and distillates-desulfurization grounds—then such a premium would have become a natural fact of life and the over-all economics of the industry would have developed accordingly. It is reasonable to expect that such an innovation would not have created insurmountable difficulties for consumers since the majors had the industry under tight control, and several historical precedents, to be outlined in Part III, have induced important changes in crude pricing patterns although in apparent contradiction to the prevailing professional and economic conditions.

In the confrontation over prices during the 1970-71 crisis, the arguments supporting the views of the countries producing low-sulfur crude oils were preponderant. A sulfur premium was formally recognized and introduced in the Tripoli Agreements, despite the technical arguments of the oil companies based on the fact that the actual market conditions did not show any premium for low-sulfur fuel oils. However, such a premium was readily displayed by the industry thereafter and has become a permanent feature of the market since then.

NOTES

1. W. L. Nelson, "Sulfur Contents of U.S. Crude Oils," Oil and Gas Journal (OGJ) 68, 33 (August 17, 1970): 78.
2. W. L. Nelson, Petroleum Refinery Engineering (New York: McGraw-Hill, 1958).
3. W. L. Nelson, "Sulfur and Sourness," OGJ 62, 13 (March 30, 1964): 89.
4. W. L. Nelson, "Revised Sulfur and Sourness Penalties," OGJ 67, 22 (June 2, 1969): 96.
5. T. Rifai, "Sulfur Pollution and Its Impact on the International Petroleum Industry," OPEC Seminar, Vienna, June-July 1969.
6. Ibid.
7. Ibid.

CHAPTER
10
FREIGHT DIFFERENTIAL

    The basic concept of crude oil pricing in the Eastern Hemisphere originated with the 1928 Achnacarry Agreement, which, among making other decisions that shaped the international petroleum cartel (see Chapter 12), decided that crude oil prices in a given market should be the same whatever the sources of the oil. In other words, the majors' pricing strategies and their concerted plans for coexistence would jointly set the level of CIF prices, and the corresponding FOB values would be then derived by a net-back formula in which the freight element would govern directly the ultimate results.

    This mechanism was universally applied through the Gulf-plus pricing system prior to World War II, during a period when Middle Eastern production was very limited indeed. The development of a specific pricing pattern in the Eastern Hemisphere was equally based on the same concept: The price level in the Gulf area was initially set by equalizing its delivered value at the U.S. East Coast with the CIF price of crude oils from the Gulf of Mexico. Later on, crude oil prices in other export centers (Africa and the Mediterranean) were linked to those prevailing in the Gulf area by equalizing CIF values somewhere in Western Europe.

    The freight differential appears as a major component of the mechanism of price formation. However, this statement embodies a fundamental contradiction since crude oil prices are intended to be universal in scope and stable in space and time, whereas the freight component is essentially fluctuating in nature and magnitude. Ultimately, each case has its own characteristics, and any attempt to elaborate an average cross-section of the transportation economics would not correspond to any past, present, or future reality.

    This fundamental conflict is irreconcilable on professional or economic grounds, and it would be misleading to attempt obstinately to relate consistently crude oil prices with actual transportation economics. Still, such interrelation is indispensable to the universal scope of any pricing pattern. The way out of this dilemma is to ignore the actual detailed facts of tanker market economics and to transform the tactical microeconomic freight component into a strategic macroeconomic freight differential based on a policy-induced and stable average freight rate that cannot be consistently correlated with actual freight rates at any moment. This process has been detailed in Chapter 7, where it was presented as the political dimension of pricing patterns.

The definition of average freight differentials for the purpose of price formulation was relatively simple in the 1950s and the early 1960s because the structure of the tanker market was not too complex. There was a single category of tankers with respect to size and a single route from the Gulf area to Western Europe via the Suez Canal. The situation changed dramatically in the late 1960s as a consequence of the rapid increase in tanker size, the prolonged closure of the Suez Canal, and the development of different export centers outside the Gulf area. The increasing complexity of transportation economics proved to be troublesome during the Tripoli price negotiations in 1970-71, since it has become very difficult to agree on a single definition of the freight differential. Because of the market segregation with respect to tanker size and the alternative routes via the Suez Canal or around Africa, different models could be considered to define the freight differential. In particular, according to the size range of the tanker considered, one might consider three main tanker categories based on whether they were able to pass through the canal loaded or not (trip from the Gulf to Western Europe) and whether they could do the same in ballast (return voyage).

Considering the same couple of export and discharging terminals under identical conditions, different values of the freight differential would be obtained should one consider the three main alternative routes, Suez/Suez, Cape/Suez, and Cape/Cape. The freight arrangements included in the Tripoli Agreement of March 20, 1971 represents a compromise between the contradictory evaluations of the Libyan Government and the oil companies, which cannot be fully appraised without a prior investigation of the recent evolution of tanker transportation economics.

CONSEQUENCES OF A MAJOR CRISIS

The crisis built up surprisingly in a few weeks' time in the spring of 1967 and burst rather unexpectedly before anyone could really foresee what was going to happen and take the necessary precautions. The international oil industry was taken by surprise, and all of a sudden Gulf exports had to be routed around the Cape. It was estimated that an extra tanker capacity of about 13 million tons was needed in continuous employment to fill the gap.[1]

In the early days of the crisis, there were doubts about the continuation of Arab oil supplies because of the stoppage of exports from the main Arab producing countries and because of the fear of irremediable action or sabotage in the intense emotional atmosphere then prevailing, where the oil was regarded as the ultimate weapon against "Western imperialism supporting Israeli agression." However, apprehensions rapidly cooled off when it became clear that the ban on exports was limited to a few countries (Britain and the United States) alleged to be involved indirectly in the conflict and that the ban was far from efficient. The restrictions were lifted after the Khartoum Arab Summit, and production again started to soar upwards. In the meantime, Iranian production was pushed to an all-time high and the Bandar Mahshar terminal was reactivated. Western Hemisphere sources contributed largely to filling the gap; Venezuelan production increased by 500,000 Bbl/day and U.S. allowables in Texas and Louisiana were raised substantially so that by early July shipments of crude oil from the U.S. Gulf ports were 650,000 Bbl/day above normal.

Therefore, the continuity of oil flow to Europe was not threatened by a lack of adequate emergency supplies. The real bottleneck was the possibility of providing the necessary extra tonnage to move oil continuously over the new longer routes. By flexibility and promptness in redeploying tanker transport and stretching its capacity to the utmost, Europe was supplied from worldwide sources thanks to the efficient and prompt efforts of the international oil companies, which in a few weeks' time, managed successfully to meet the new demands. While rearrangements and rescheduling were taking place and while awaiting the oil en route, which took a longer time to arrive than usual, Europe was drawing on its oil stocks, built up and prudently maintained at a high level since the 1956 Suez crisis. All this occurred at a time, the summer, when refinery runs were at their seasonal lowest levels.

The essential objective in crisis periods is to maintain the continuity of supplies at all cost. Once it became obvious that no dramatic situations would occur, attention could focus again on the economics of such emergency supplies and on the prospects of its future evolution. Expectedly tanker rates shot up, as shown in Figure 10.1, and the total bill was further increased by the conjugated effect of longer routes around the Cape. The impact was greatest and rather instantaneous on the spot market, which covered the badly needed available extra tonnage. Up to May 1967, tanker market performance was very poor indeed. AFRA was as low as Intascale minus 46.6 for large

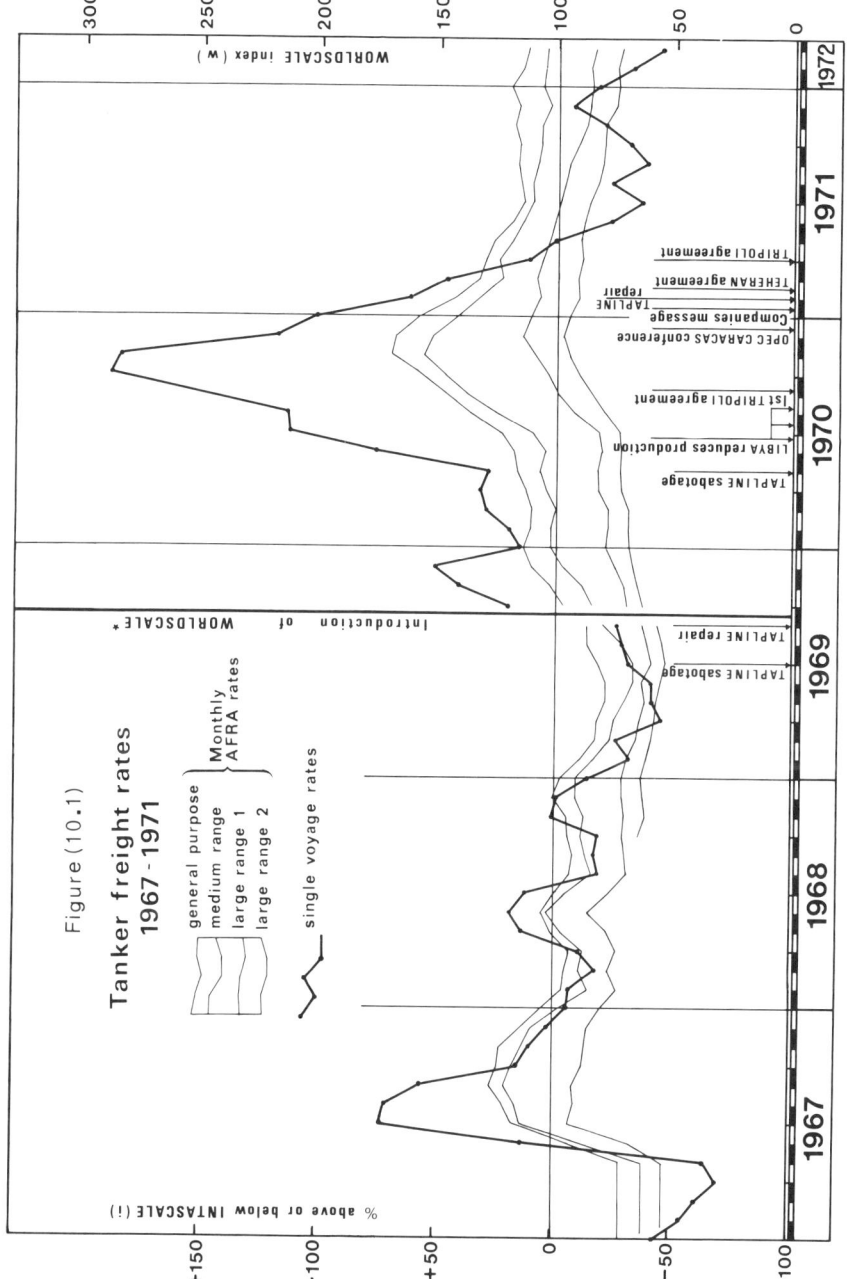

tankers in the first half of 1967, whereas spot rates reached Intascale minus 70 in April. A good deal of excess tonnage has been available in recent years, causing orders for new tankers to slow down and enhancing the trend for multipurpose ore-oil and cereal-oil combined vessels. The Canal shutdown completely changed the situation, and the tremendous demand for tonnage caused over 3.5 million tons ex-layup and grain trade to be added to the carrying capacity of the world tanker fleet within a month. Obviously, any other tonnage that could be freed to join the oil trade was added to the world tanker fleet after the end of June. Thus the oil industry was facing the winter season in the Northern Hemisphere with very little tanker tonnage in reserve, and for a time it could be feared that severe winter weather, during the next six months, would cause further trouble.

These fears proved to be unjustified, and the tanker market continued its smooth downward movement for both spot and AFRA rates, and by mid-summer 1969, the market was already approaching its pre-Suez level. This was largely due to rapid growth in new tanker tonnage, particularly in the mammoth range. The largest ship afloat in 1967 was still the 210,000-dwt Idemitsu Maru, but by mid-1969, there were some 35 tankers in the 200,000-to-326,000-dwt range in service and almost 200 more on the way.[2] In fact, this phenomenon began well before the six-day war, but the shift to super-tankers was accelerated by the crisis.

In face of market uncertainty and the necessity of following its evolution as closely as possible, the London Tanker Brokers' Panel replaced the half-yearly AFRA publication by monthly average assessments. AFRA figures were further sophisticated by the introduction of a new category of tanker size (Large II) in October 1968 to cope with the new trend in the market. The Worldscale schedule was introduced in September 1969 to replace both Intascale and ATRS and was in use until the end of 1970, when its rates were revised upwards to take into account the changes in market conditions, and particularly the increase in the price of bunker fuels. Another revision was made on January 1, 1972, and it is expected that further revisions will be made periodically every year.

The freight market started to deteriorate again in late 1969 with the sabotage of the Tapline in Syrian-occupied territories and Israel's long-standing refusal to allow Aramco to make the necessary repairs. This mini-crisis was a prelude to the explosive situation that built up throughout 1970, when freight rates soared to unprecedented levels under the conjugated impact of the Tapline sabotage in May and production cuts in Libya. The political background to the over-all confrontation between oil companies and producing countries, mainly Libya, will be discussed and analyzed in detail in Chapter 16. The freight differential and the Suez allowance played a major role in the

Tripoli negotiations in the readjustment of prices in September 1970 and in the setting of new prices for Mediterranean oil in March 1971.

The prolongation of the Suez Canal closure and the absence of any serious prospect for a peaceful and durable settlement of the Arab-Israeli conflict pushed the oil companies into a feverish rush to diversify their actual and potential sources of supply. Offshore exploration was intensified in the North Sea as well as in the most promising continental shelf areas around the world. Indonesia recovered its long dormant potential and Alaskan oil was contemplated as a possible alternative to risky Middle Eastern oil. The diversification of sources, although still embryonic, coupled with the deep process of modification affecting the tanker market irresistibly engaged in its race towards gigantism, induced a subtle but substantial change in the concept of freight differential so that the debate about the Suez allowance that took place in late 1967 was somewhat different from the rough discussions of 1970-71 in regard to the new prices for Libyan crude oils. Freight differentials will be calculated and discussed in both cases, and the impact of super-tankers will be thoroughly investigated.

FREIGHT DIFFERENTIAL: THE SUEZ ALLOWANCE, 1967

The freight differential concept was utilized to correlate crude pricing patterns in the different export areas in the Eastern Hemisphere with reference to Gulf postings. Its assessment was based on the assumption that oil flow from the Gulf area to Western Europe would actually take place through the Suez Canal, so that the freight advantage enjoyed by a different export area nearer to European markets (Eastern Mediterranean, Libya, Nigeria, for example) was taken as the full difference between the transportation costs of the two routes considered. Freight differentials at flat Intascale rates could be unquestionably calculated, and the main difficulty was the "political" selection of a constant average freight rate, in terms of the percentage discount of the established flat differential, for the purpose of pricing as opposed to the ever fluctuating value of actual rates.

The pricing of Eastern Mediterranean oil during the 1950s raised no special problem in this respect. Total oil movement from the Gulf was made through the Canal to Western Europe and the average freight rate finally settled at the historically established level of about Intascale minus 35. When Libyan oil had to be posted for the first time in 1961, the shift to larger tankers was just beginning. The first 85,000-dwt vessel was launched in 1956, followed towards the end of

1958 by the first 100,000-tonner. Then there was a pause until the first half of 1962, when a ship of about 130,000 dwt was launched by a Japanese oil company.3 In other words, vessels that could not go through the Canal in the early 1960s represented a rather insignificant proportion of total tonnage. Furthermore, these marginal tankers were utilized for the transportation of oil east of Suez to Japan rather than around the Cape to Europe even though they could go through the Canal partially loaded and there was no problem for them to come back through in ballast. In other words, the freight differential concept was the same when applied to the Libyan postings by Esso in 1961. In fact the disparity between the Libyan posting and the historical precedent established at the Eastern Mediterranean was mainly due to the "political" selection of a different average freight rate, as outlined in Chapter 8.

The shift to super-tankers, in orders made to shipyards, developed slowly throughout the early 1960s in a period when the logistics and economics of mammoth tankers were maturing in the board rooms. In the first half of 1967, the average deadweight tonnage of new tankers delivered was about 53,771 tons and the average deadweight of the existing fleet was no more than 33,021 tons. A single 170,000-dwt tanker was in operation, and the percentage of the tonnage routed around the Cape to Europe was still marginal indeed. Furthermore, because of the available surplus tonnage, some oil companies were routing some of their smaller tankers anyway around the Cape in order to keep crews and vessels occupied. Under these conditions, the freight differential for the purpose of pricing could still be derived from the Suez-Suez concept, especially since its "political dimension" immunized it against actual market conditions.

The closure of the Suez Canal caused a formidable rush to order super-tankers in numbers as great as shipyards could tolerate all around the world. In the second half of 1967 an average of 10 orders for for new tankers were made every month, and by the end of the year 200,000-dwt tankers represented about 64 percent of the total tonnage under construction or on order. However, deliveries could be only progressive, and it would take several years before super-tankers would represent a substantial part of the total tonnage in use, and in fact, tankers of more than 80,000 dwt represented only 10.6 percent of the total tonnage in use as of July 1, 1969, two years after the Canal closure; the percentage climbed to 15 percent by January 1, 1971. The percentage was no more than 1 percent in the second half of 1967, so that the freight premium enjoyed by Mediterranean oil over Gulf supplies—that is, the Suez allowance—derived from the same concept as the freight differential and could be calculated by comparing the economics of Suez/Suez routes (Canal open) and Cape/Cape routes (Canal closed). This concept was valid at least as of late 1967, when negotiations took place between the oil companies and the producing countries concerned. This concept will be utilized below to calculate the Suez allowance in 1967/68. Other concepts involving large tankers going through the Canal in ballast and so forth would have to be taken into consideration for later periods along with the substantial increase in the percentage of super-tankers in the worldwide oil movement.

The closure of the Suez Canal caused the cost of transportation of Gulf oil to increase as a result of the conjugated effect of two distinct factors:

1. The longer route around the Cape, which affected flat Intascale rates as follows (dollars/long ton):

	Before June 1967 (Suez/Suez)	After June 1967 (Cape/Cape)
Quoin Island—Rotterdam	6.38 + 0.89	9.40
Quoin Island—Marseille	4.77 + 0.89	9.00
Quoin Island—Europe	5.575 + 0.89	9.20

2. The higher rates, given in terms of discounts of flat Intascale (i), resulting from the dramatic shortage of necessary tonnage to compensate for the closure of the Canal and the occasional stoppage of pipelines (Tapline). The impact on spot and AFRA rates is illustrated in Figure 10.1.

In the meantime, the routes of African and Mediterranean oil supplies were not altered, so that Intascale flat rates were the same before and after June 1967. However, the actual transportation costs rose with the general increase of freight rates (i). The freight differential, or the Suez allowance, enjoyed by such supplies would then correspond to the net balance of the increase in transportation costs for the two routes considered with reference to Gulf oil.

The impact since June 1967 of the closure of the Suez Canal on freight differentials as evaluated above represents the technocratic approach to real and effective savings in transportation costs when disposing of Mediterranean or African oil in comparison with alternative supplies from the Gulf routed around the Cape. Reflecting these savings on crude prices in some way or other is essentially a political decision that, in terms of financial benefits, would designate who would or should get or share in the relative advantage associated with crude supplies outside the Gulf area.

The producing countries concerned (Eastern Mediterranean exports of Iraqi and Saudi oil, Libya, Algeria, Nigeria, and so on) could legitimately claim that the freight advantage—or the Suez allowance, as it is called above—is the consequence of, and directly related to, the geographical location of their export terminals so that they should enjoy this advantage alone without sharing it with any other party. In other words, the Suez allowance should be added totally to posted prices, which, in fact, means that producing countries accept implicitly that oil companies are entitled to an equal share of this advantage because the fiscal arrangements are based on 50/50 profit sharing.

<u>Consuming countries</u> in Western Europe were the direct victims of the crisis since the cost of their oil supplies from the Gulf area (and to a lesser extent from other sources, because of the general increase in freight rates) increased substantially. The increase was a real and inevitable one, but sustained efforts were made to keep it to a minimum. The best way to achieve this, at any given moment, would be to obtain as much supply as possible from the nearest sources, while keeping prices at precrisis levels—in other words, to make the producing countries concerned share in the total bill to the largest extent possible. When considering the conditions that prevailed after June 1967, it clearly appears that Libyan oil represented the main target for European aspirations since Eastern Mediterranean oil was limited by the capacity of IPC and Tapline pipelines and by the uncertainty of their operation; Algerian oil was largely hampered by the developing conflicts with France; and Nigerian oil was stopped by civil war in Biafra. In fact, Libyan oil production was pushed at unprecedented high rates and largely contributed to ease the pressure on European supplies. But, at the same time, it increased the relative dependence of European countries on Libyan oil, a fact that played an important role in the 1970-71 confrontation.

<u>Oil companies</u>, mainly the majors, were the real beneficiaries of the crisis since they were the arbitrator of the situation. The producing countries were concerned with FOB prices and the consuming countries were concerned with CIF prices, and the oil companies were in between, with the quite exclusive privilege of setting the levels of both prices at will. The fact that the companies successfully managed to avoid a real crisis in European supplies gave them a great "moral" advantage vis-à-vis the European countries and enabled them to present and justify higher CIF prices. The general increase in freight rates, while benefiting independent shipowners because of the sharp jump of spot market rates, actually produced substantial extra profits for the oil companies, even for their Gulf supplies, because a large part of crude transportation was made by tankers either company-owned or under long-term charters, the operating cost of which was relatively unaffected and much lower than the new higher level of rates invoiced to their affiliates in the consuming countries. Extra profits were made on supplies from export terminals outside the Gulf area by reflecting only a part of the Suez allowance in posted prices, whereas the balance was added to their integrated profits or partially shared with the consuming countries by delivering Mediterranean

and African oil below CIF parity prices. By manipulating the total or partial allocation of the Suez allowance in one way or another, they could keep the largest part to themselves while subdividing the balance between producing and consuming countries to the limits dictated by the safeguarding of their political positions and strategic interests in these countries.

<u>Independent shipowners</u> certainly benefited from the crisis to a very large extent, but it is very important to note that they are not concerned with the present debate about the allocation of an extra freight advantage—or Suez allowance—to posted prices of Mediterranean and African oil in comparison with Gulf supplies. In fact, they do benefit from the increase of the transportation costs for each and all routes considered, but they have no connection with the relative differential between the routes of the two alternative sources of supplies. To be more precise, if we assume that a concordance of interests with producing countries would have induced a "political" decision by the major oil companies to reflect the Suez allowance totally on posted prices in order to restore and maintain the parity of CIF prices, the consequences would have been a larger bill for consuming countries and a relatively lesser increase in Libyan production, but independent shipowners would have enjoyed the same extra profits from their tankers whether on one route or another.

In fact, the problem was readily raised by the producing countries concerned—namely Iraq, Libya, and Saudi Arabia—which called for full benefit from the extra freight advantage enjoyed by their Mediterranean exports. However, this specific question happened to occur at a special phase in the relations between the major oil companies and the OPEC member countries, which were engaged in fiscal negotiations in connection with the elimination of the allowances provided for in the royalty-expensing agreement reached in late 1964, as already noted in Chapter 4. Yet the producing countries were not in a very strong bargaining position and were successful in achieving a partial and progressive elimination of the allowances over 10 years. The bargaining power of producing countries with Mediterranean exports became relatively weaker over the additional issue of the Suez allowance that rose up unexpectedly. In reality, the situation was very complex and varied from one country to another. Nigeria was out of the picture because of its civil war, and Algeria had no initiative by that time since its French partners denied Algeria any advantage resulting from the closure of the Canal on the grounds of the cooperative agreement. Iraq was in an inextricably troubled situation with the IPC group resulting from the confiscation of 99.5 percent of its concession under Law 80 and from the fact that it was the only OPEC member country that did not ratify the royalty-expensing agreement.

Negotiations took place in October-November 1967 with only Saudi Arabia and Libya, which reached a compromise with the oil companies. The basic claim for the Suez allowance was dropped, and as a compensation the companies accepted complete temporary elimination of the allowances and discounts on Mediterranean exports as long as the same companies agreed to accept the continuation of "the extraordinary circumstances" prevailing in the Middle East consequent to the closure of the Canal. The agreement was to be implemented retroactive to the date of resumption of exports from Mediterranean terminals after stoppage of oil flow in June 1967 (July 6 for Libya and September 20 for Saudi Arabia). Shortly afterwards, an over-all agreement was reached with OPEC about the gradual elimination of allowances and discounts over a period extending until 1975, the year to which the temporary full elimination for Mediterranean exports would apply. Under these conditions, the compromise increase granted to Mediterranean Saudi and Libyan crude oil prices as a compensation for the Suez allowance would amount to the figures given in Table 10.1. These agreements came to a premature end as a consequence of the 1971 price agreements. Their performance during the short period of their validity is compared in Table 10.1 with a tentative evaluation of the Suez allowance computed under the two different concepts outlined above. It is obvious that the elimination of the allowances represented only a part of the freight advantage enjoyed by Mediterranean oil.

THE ECONOMICS OF SUPER-TANKERS

This subject became a very popular one by the mid-1960s, and enthusiastic visions of the future were displayed in the literature with the publication of numerous articles and studies pointing out the "irresistible" attraction of the economic advantages of larger and larger vessels. Among other publications, the paper presented by Michael Hubbard to the Institute of Petroleum in London on December 1966[4] had a large audience. Psychological considerations were partially responsible for this feverish enthusiasm, especially at a time when the tanker market was very depressed indeed. The June 1967 crisis, and the closure of the Suez Canal, added further political dimensions to the debate and provided more arguments to emphasize the ineluctable race towards gigantism so as to minimize dependence on the Canal.

The early enthusiasm of technocrats contributed to creating a confused situation that did not help in the development of realistic ways out of the crisis. The confused situation that prevailed in the late 1960s was best described by W. H. Ashford as follows: "I am

TABLE 10.1

1967 Compromise Agreement,
Suez Allowance Against Elimination of Royalty-Expensing Allowances

		1967	1968	1969	1970	1971	1972	1973	1974	1975
Rates of Allowances										
Percent off posted price		6.5	5.5	4.5	3.5	2.0	—	—	—	—
Gravity, ¢/Bbl per degree over 27°		0.2647	0.2647	0.2647	0.2647	0.2647	0.2647	0.176467	0.088234	—
Gravity, id. (G >37 as if 37°)		—	0.059559	0.119118	0.178677	0.238236	0.297795	0.198530	0.099265	—
34° Lt. Arabian, $2.17/Bbl FOB Sidon										
Percentage allowance	¢/Bbl	14.10	11.93	9.76	7.60	4.34	—	—	—	—
Gravity allowance	¢/Bbl	1.85	2.27	2.69	3.10	3.52	3.94	2.62	1.31	—
Total	¢/Bbl	15.95	14.20	12.45	10.70	7.86	3.94	2.62	1.31	—
Suez allowance: economic	¢/Bbl	37.50*	28.10	22.00	—	—	—	—	—	—
political	¢/Bbl	49.60*	34.00	23.50	—	—	—	—	—	—
40° Brega, $2.23/Bbl FOB Brega										
Percentage allowance	¢/Bbl	14.50	12.26	10.03	7.80	4.40	—	—	—	—
Gravity allowance	¢/Bbl	3.45	4.04	4.64	5.23	5.83	6.42	4.28	2.13	—
Total	¢/Bbl	17.95	16.30	14.67	13.03	10.23	6.42	4.28	2.13	—
Suez allowance: economic	¢/Bbl	37.00*	27.20	20.00	—	—	—	—	—	—
political	¢/Bbl	50.70*	34.00	21.90	—	—	—	—	—	—

*2d half of 1967.

Source: Compiled by the author.

inclined to think that size has tended to become something of a fetish, and imagine that the Torrey Canyon disaster will give rise to some furious rethinking, even if it does not act as a deterrent."[5]

Tanker economics and the mechanism of its market are very complex when considered on professional grounds as a major branch of the international oil industry. The matter becomes unintelligible when viewed as a major segment in the mechanism of price formation. It played a "hot" role in the post-Suez crisis, which culminated in the 1970-71 price confrontation. It has been demonstrated throughout this study and was recalled earlier in this chapter that there are no objective or rational criteria for expressing tanker economics in simple formulas and reflecting them in pricing patterns, even when the tanker market is relatively simple, as during the 1950s. The impact of super-tankers, as visualized in early 1967, on company investment decisions, on international supply patterns, on the over-all supply costs, and on crude oil prices was far from maturing in board rooms. The prospects of super-tankers and the evaluation of their impact could be worked out and manipulated at will to serve specific political or professional interests: to show that the Canal is worth reopening or that it should remain closed, to justify or to deny an extra freight premium for producing countries, to reveal or to camouflage the huge profits gained by shipowners, to justify or to reject projects for new transit pipelines, and so forth. Conflicting political interests largely transcended the framework of the petroleum industry; the reopening of the Canal became a "hot" issue in the Arab-Israeli war, whereas the continuation of its closure was highly appreciated by the U.S. Department of Defense because it cut Russian supplies to the Far East and diverted them to the costly route around the Cape.

Under these conditions, the investigation of the economics and impact of super-tankers is most difficult to undertake on professional grounds and can be only controversial and speculative in the context of the then prevailing political environment. Although simple in appearance, the technocratic approach as displayed by some oil company economists is misleading because it is generally used to demonstrate a "political" case, not to search for an objective answer to the real problems.

The optimistic enthusiasm about super-tankers resulted from technocratic studies and extrapolations that were published in the period 1966-68 while such vessels existed only as blueprints. Their attractiveness resulted from the economies of scale, which showed substantial savings in capital and operating costs. The charts published by M. Hubbard[6] in late 1966 showed that capital costs would drop from about $160 a ton for a 25,000-tonner to $100 a ton for a 85,000-tonner and further to only $80 a ton for a 150,000-dwt vessel. This would help in the securing of funds necessary for financing ambitious developments foreseen for tanker construction. Savings in operating costs supposedly would result from several factors that were optimistically enumerated by several authors.[7]

The dramatic shortage in tanker capacity and the need to fill the gap that resulted from closure of the Canal in June 1967, combined with the enthusiastic psychological atmosphere outlined above, gave an extraordinary impetus to the construction of super-tankers. This "revolution" was reported and commented on in the literature.[8] The evolution of the size structure of the world tanker fleet is illustrated in Table 10.2.

The oil companies actively participated in this move in an average proportion of one tanker for every three tankers ordered. By the end of 1967, Esso had on order 40 tankers totaling more than 5 million tons capacity, Shell on order had 25 tankers totaling about 5 million tons capacity, and total tonnage ordered by the oil companies was about 18 million tons as against 22 million for independent owners. The oil companies looked upon the economics of super-tankers differently from the approaches indicated above. They were interested in the actual costs of transporting their crude oil supplies rather than in absolute tanker performances. In this respect, the Suez Canal played a central role in differentiating the various categories of tankers and in comparing their relative performances. Just before its closure, the Canal could accommodate tankers with 38-foot draft, to be increased to a 40-foot draft by the end of the year. Under these conditions, the upper size limit of fully loaded tankers able to move through the Canal would be about 80,000 tons, although specially designed shallow-draft 90,000-tonners could be ultimately received. With a depth of 40 feet, 170,000-200,000-tonners would be the limit for vessels able to pass through the Canal when in ballast (unladen) while routing around the Cape loaded.

However, an over-all economic evaluation should take into consideration the impact of gigantism on other factors, independent of the vessel itself, such as the ratio between the size of the tanker and the size of the refinery it is intended to supply and the accessibility of ports and navigation routes.[9] Summing up most of the technical and professional arguments, we find that tanker size seems to be related to the normal storage capacity of the refinery it supplies. Therefore, the economics of larger tankers in the over-all consolidated accounting of a major integrated oil company should add the cost of extra storage. On the other hand, super-ports are necessary to receive super-tankers, and huge sums of money would have to be spent in order to provide adequate facilities for the large tankers both at export and discharge ports. Heavy investments have been made in the main Gulf export terminals, but the situation in the congested European ports is more delicate. In this connection, the specific geographic situation of

TABLE 10.2

Evolution of Size Structure of World Tanker Fleet
(excluding government-owned vessels)

Size Group (dwt)	December 31, 1965				June 30, 1967				December 31, 1967			
	Existing Fleet		On Order		Existing Fleet		On Order		Existing Fleet		On Order	
	No.	Thousand dwt	No.	Thousand dwt	No.	Thousand dwt	No.	Thousand dwt	No.	Thousand dwt	No.	Thousand dwt
10,000- 29,999	1,685	31,797	46	822	1,606	30,411	62	1,236	1,608	30,500	55	1,130
30,000- 59,999	879	36,329	24	1,182	898	37,472	6	240	895	37,826	12	438
60,000- 99,999	203	15,068	144	11,045	316	23,847	63	5,192	349	26,707	48	3,996
100,000-149,999	15	1,669	28	3,213	41	4,719	48	5,582	59	6,753	37	4,250
150,000 and above	—	—	19	3,221	3	520	75	15,577	7	1,203	142	30,285
Total	2,782	84,863	261	19,483	2,864	96,969	254	27,827	2,918	102,989	294	40,100
Average Tonnage	30,504		74,647		33,021		109,553		35,298		136,394	

Size Group (dwt)	December 31, 1968				December 31, 1969				December 31, 1970			
	Existing Fleet		On Order		Existing Fleet		On Order		Existing Fleet		On Order	
	No.	Thousand dwt	No.	Thousand dwt	No.	Thousand dwt	No.	Thousand dwt	No.	Thousand dwt	No.	Thousand dwt
10,000- 29,999	1,572	29,930	63	1,362	1,507	29,204	110	2,533	1,482	29,018	—	2,428
30,000- 59,999	909	38,404	13	457	911	38,452	21	722	916	38,784	—	1,702
60,000- 99,999	385	29,613	42	3,501	411	31,851	27	2,198	429	33,372	—	1,541
100,000-149,999	82	9,417	30	3,601	96	11,338	33	4,114	110	12,773	—	3,154
150,000-199,999	17	2,996	23	4,037	30	5,338	9	1,571	34	6,010	—	1,194
200,000-249,999	15	3,103	125	26,827	53	11,213	125	27,347	117	25,257	106	23,821
250,000-299,999	—	—	41	10,439	2	510	72	18,713	8	2,051	123	32,537
300,000 and above	2	654	7	2,322	6	1,960	6	2,000	6	1,960	10	3,597
TOTAL	2,982	114,117	344	52,548	3,016	129,581	403	59,198	3,102	149,225	432	70,254
Average Tonnage	37,254		152,758		41,816		146,892		46,897		162,625	

Source: World Tanker Fleet Review (London: John I. Jacobs and Co., December 31, 1972).

refineries would dictate the philosophy of major oil companies. Until now, only Gulf has taken the initiative of building a deepwater discharging port at Bantry Bay (Republic of Ireland).

The typical tanker built and ordered in 1967-68 was actually around 200,000 dwt. Some builders call such ships the "Shell-type," because in the early rush to large tankers, Royal Dutch/Shell decided on 200,000-tonners as the new backbone of its fleet. They are large enough to haul crude oil economically around Africa to European refineries, and small enough to return in ballast through the Suez Canal, when it is open. They are also able, assuming some harbor improvements, to enter a number of major European ports. The decision reached by Esso to adapt 250,000-tonners was based on the fact that extra investment in the case of trans-shipment terminals for 300,000-plus-tonners is prohibitive, whereas the advantage of transiting back through the Canal in ballast is rather insignificant.

Although some enthusiastic views continued to be stated occasionally about a further increase in tanker size, the industry in practice settled at the size levels indicated above. The industry needed to check the optimistic forecasts of technocrats and to see how effectively large tankers would actually perform in comparison with theoretical forecasts. This pause was needed to digest the unprecedented revolution in tanker transportation and was virtually imposed by the fact that shipyards all around the world were congested with orders running up to 1974-75, with the largest part of the orders concentrated in the 250-280,000-dwt class.

In fact, the over-all economics of super-tankers in the early 1970s appeared more gloomy than the bright prospects described by the early commentators. From 1967 to 1971, the economics of tanker transportation was deeply and adversely affected by several factors. The savings resulting from the economies of scale were largely offset by a fantastic increase in the cost of construction, as indicated in Table 10.3. This was due to an initial underestimation of the construction cost of the early 200,000-tankers, to the increase in steel prices, and to general inflation, which affected the cost of materials and labor. Furthermore, the sharp demand for tankers helped make the tanker construction market a seller's market, whereas it had been a buyer's one for many years.[10]

Operating costs were equally and heavily affected. The insurance premium for mammoth tankers has continued to increase in recent years.[11] Furthermore, the cost of repairs are unusually high due to the scarcity of dry docks, towing facilities, and related installations of adequate size. Ocean pollution problems will add further expenses. Generally speaking, all items making up the total operating costs have increased in one way or another, but the increase has been greater for larger tankers. It is currently estimated that the transportation

TABLE 10.3

Approximate Construction Cost of 250,000-Tanker

Date of Order	Cost (millions of dollars)
end 1967	15
end 1968	19
end 1969	22
end 1970	32-34
end 1971	38-40

Source: Revue Française de l'Énergie, no. 237 (December 1971), p. 138.

cost from the Gulf terminals to Rotterdam increased by 31 percent for the 30,000-tonners, by 80 percent for the 70,000-tonners, and by 120 percent for the 250,000-tonners over the period 1967 to end-1970.[12]

The uncertainty about the economic performances of supertankers discussed above progressively became of secondary importance, and rising costs appeared as an established fact of life in a world ravaged by inflation. The acute tanker shortage that took place in spring 1970 subsequent to the Tapline sabotage and the reduction of Libyan production was largely responsible for the over-all confrontation between producing countries and oil companies that resulted in the Teheran and Tripoli Agreements. Since then, an oppressive psychosis about energy shortage has swept over the world and the continuity of supply emerged as a determining factor in strategy planning for both oil companies and governments of consuming countries. In the meantime, hopes for a prompt reopening of the Suez Canal vanished along with the prospects of a political settlement in the Middle East, and the everlasting instability in the area contributed to make pipeline transportation most unreliable. Under these conditions, disposing of enough tonnage to transport oil from the Gulf area via the Cape under all circumstances was viewed as an insurance premium for continuity and security of supply that should be paid for at any cost.

In fact, the rush to tanker orders kept on at a high pace and shipyards all over the world are currently fully booked four to five years ahead. The feverish rush to order is motivated by the normal increase in demand, by the strategic necessity of having enough tonnage available to cope with emergency conditions by which Mediterranean supplies would be largely reduced or even cut off, and by the growing import needs of the United States. According to a USMA estimate, a fleet of 500 "very large crude carriers" (VLCC's, of 250,000 dwt each) will be needed by 1980, supported by some 5,000 coastal vessels to cope with final delivery and local distribution under the assumption that about half of U.S. needs would then have to be imported from the Gulf area.[13] However, inflation should be considered as a contributing factor to the relentless rush to orders despite rising costs. John I. Jacobs reports that with available VLCC construction berths for 1974/75 deliveries virtually all committed by the last quarter of 1972, price escalation for new tanker orders was then running at about 1 percent a month.[14]

Total newbuilding tonnage rose to the record level of 120 million dwt by the end of 1972, more than the entire fleet in service just four years earlier. The capacity of the existing commercial tanker fleet increased by almost 11 percent in 1972 to a year-end total of over 187 million dwt. VLCC's of 200,000 dwt and above represent about one-third of this total but account for more than 75 percent of the tonnage in order. A new size category (70,000-90,000 dwt) seems to emerge in new orders. These vessels are viewed as indispensable secondary carriers to transport oil from VLCC transhipment terminals to consumption centers without deepwater ports. This size ship will be ideal to deliver growing volumes of crude oil from Atlantic transshipment terminals into the United States, which is still without a single deepwater port.

The other newly popular category is that of mammoth carriers larger than 350,000 dwt. Total orders jumped from 7 in mid-1972 to 22 by the year's end. Some of the newbuildings will have an almost revolutionary design with a relatively shallow draught. The size record set by Shell's 533,000 dwt newbuildings have already been surpassed by Globtik's letter of intent for a 706,000 tonner, and Shell is also reported to be considering more ships of up to 650,000 dwt. It seems that the possibility of the 1-million-ton tanker is becoming more and more feasible, and several yards should be able to build such ships within a few years.

NOTES

1. *Petroleum Press Service*, August 1967, p. 282.
2. *Petroleum Press Service*, November 1969, p. 419.
3. W. H. Ashford, "The Economics of Large Tankers," *The Journal of the Institute of Petroleum* (London) September 1967, p. 290.
4. M. Hubbard, "The Comparative Cost Oil Transport," *The Journal of the Institute of Petroleum*, January 1967, pp. 1-23.
5. Ashford, op. cit., p. 291.

6. Hubbard, op. cit.

7. E. G. C. Colley, "Economics of Large Tankers," Petroleum Times, April 15, 1966, p. 470.

8. See in particular, Petroleum Press Service, June 1967, April 1968, August 1968, October 1968, October 1970, April 1972; Oil and Gas Journal, August 19, 1968; Oil and Gas International, February 1971; Petroleum Intelligence Weekly, November 20, 1967, September 30, 1968.

9. See in particular, an article published by the French subsidiary of British Petroleum, "Tendances et perspectives à long terme de la construction des pétroliers," l'Industrie du Pétrole, Gaz et Chimie, November 1967, p. 63.

10. P. Dietrich and Y. Poulizac, "Le Prix du transport maritime du pétrole," Revue Française de l'Énergie, no. 237 (December 1971), pp. 127-146.

11. "Mammoth Problems," Petroleum Press Service, May 1970, p. 167.

12. Dietrich and Poulizac, op. cit., p. 138.

13. Petroleum Press Service, April 1973, p. 133.

14. World Tanker Fleet Review (London: John I. Jacobs and Co., December 31, 1972).

CHAPTER

11

**RELATIVE VALUES
OF CRUDE OILS**

The evaluation of the relative values of crude oils has been a fashionable concept and a popular approach to crude oil prices in recent years. W. L. Nelson has been compiling, for many years, extensive information and data on the subject in his "Process Costimating" series, currently published in the Oil and Gas Journal, and has presented a method for "Standard Crude Oil Realization" in 10 articles published between November 18, 1963 and December 14, 1964. The author supervised a comprehensive unpublished study of the subject in 1966, which this chapter is largely inspired by. Another approach was suggested by F. C. Turner in his paper "Methods of Crude Oil Evaluation" presented at the OPEC Seminar held in Vienna in July 1969.[1] The most recent contribution came from P. H. Frankel and W. L. Newton in their paper "Comparative Evaluation of Crude Oils."[2] On the other hand, it is well known that OPEC has been investigating the subject in order to relate, if not explain, the posted prices of crude oils in terms of their relative values. The results have always been disappointing whatever the method used for crude evaluation actually was.

The idea of evaluating relative and/or comparative values of crude oils is very attractive in itself. Among the multiple aspects of its popularity, psychological considerations are certainly the most important, especially when one keeps in mind that none of the methods proposed actually succeeded in reaching the professional objectives they aspired to. Crude oil pricing has been always considered a taboo issue, where oil companies, especially the majors, were suspected of concentrating their most vital forces, and the regulatory mechanism enabling them to keep the industry under tight control.

Oil companies were generally very cautious when dealing with the pricing issue; they refrained from entering open debates and preferred to see public attention diverted to such intellectually stimulating exercises as the evaluation of the relative values of crude oils, which was undertaken by some outside observers or critics of the industry.

The general public was enthusiastic about such a challenge to the muteness of the majors, and the challengers were highly stimulated by their apparent success in exploring the unknown and explaining its mysteries. Their satisfaction with their own findings concealed the weakness of their approach and the failure of their achievements. The methods developed appeared to be easy to understand and to apply, especially because they relied on general concepts accessible to almost everybody, and most of the necessary information and data are generally available. In the face of this apparent simplicity, it could be questioned why the issue was held to be so complex and incomprehensible. Yet the mechanism relating the available data was lacking and the new intellectual approach seemed to be able to provide it. Naturally, the oil companies were most satisfied with this fashion and encouraged pioneers with moral rewards and professional encouragement so that the evaluation of the relative values of crude oils would confuse more and more the vital issue of crude oil prices.

Encouraged and supported by the psychological environment outlined above, intellectual contributions were further confused by their very scientific approach and economic nature. Simple and attractive formulas and models were successfully developed to represent the industry, and they proved to be flexible and full of promising possibilities for further investigation. Because of such intellectual stimulation, the temptation could be very strong for further and more sophisticated juggling with models and formulas at the expense of actual significance or representativity. Such an academic approach is actually the victim of its own apparent facility since it appears to succeed in imprisoning the industry in simple and accessible formulas, whereas the issue under consideration is a most complex one. It is very dangerous to supply simple explanations for complex phenomena, but it is still more dangerous to accept and adapt such explanations without discerning their real scope and significance and without recognizing clearly their specific limitations.

We do not intend to deny the merits and advantages of the different methods for the evaluation of the relative values of crude oils. On the contrary, we aim to clarify their nature and to delimit their scope so that they may be usefully and effectively utilized to achieve their intended objectives. In particular, we would like to dissipate any misunderstanding between the concepts of relative values of crude oils and their prices so that investigators may pursue further their efforts within a clear framework that lays out those objectives that they cannot, and should not, hope to attain.

PRICES VERSUS RELATIVE VALUES

All methods for evaluating the relative values of crude oils derive from the same process of thinking that will be tentatively summarized in the following. Crude oil, which is a physical mixture of a multitude of hydrocarbons, has no economic value as such (except when directly consumed as fuel), although it is traded at commercial prices. Its value derives from the sales of finished products obtained from it and sold in a given market where the products respond to effective needs, with respect to quantity and quality. Therefore, the economic value is attached to products as such, irrespective of their crude oil of origin. From technical and economic viewpoints, two crude oils could perform differently through the processing stage but meet the same market demand pattern; and meeting such patterns is the commanding factor and the raison d'etre of the petroleum industry. Since the market considers petroleum products an anonymous commodities and does not recognize their specific crude of origin, the ultimate economic performances of all crude oils is measured by the same entity, that is, market valorization corresponding to a given demand pattern at prevailing products' prices. Therefore, checked against this common and predetermined criterion, the relative values of crude oils would automatically be set in terms of their relative economic performances in achieveing, jointly or separately, common objectives.

In economic terms, the differentiation between crude oils takes place at the refining stage, where the crudes shift from specific entities into anonymous products. In all the other stages prior to their arrival at the refinery gate, their technical and economic performances are, more or less, governed by the same laws and mechanism related to their physical nature as fluids. The economics of exploration, development, production, and transportation are rather independent of crude oils themselves, and the small differences resulting from specific physical characteristics (complexity of field stabilization facilities, impact of viscosity on pipeline transportation and of gravity on tanker transportation economics) are fixed facts independent of market requirements and can be assessed, once and for all, on objective professional and economic grounds. The comparative evaluation of crude oils is to be performed in terms of, and checked against, market realization, which actually generates the cash flow of the industry, and not in terms of production and transportation costs, which play a role, in internal company accountings, in setting tax-paid costs. The relative values of crude oils are intended to provide a general framework for crude evaluation where a large number of "equivalent" patterns could be indifferently selected according to given criteria of comparison and judgment. The minimization of total tax-paid costs would provide necessary criteria for optimizing plans of action within the large framework indicated above.

Therefore, although the oil industry is essentially an integrated one, the relative values of crude oils are evaluated in terms of refining economics considered as an isolated and autonomous sector. This concept is not in contradiction with the very integrated nature of the industry but points out the real scope and limitations of the values thus elaborated. They have no absolute significance in themselves but are important in the context and framework within which they are computed. In particular, the same set of relative values of crude oils would have different significances to an independent refiner, to an independent producer, and to an integrated company.

Although different methods have been presented, or can be suggested, for crude oil evaluation, they all derive from the same basic concept, which may be characterized by the following major items:

1. Market demand to be satisfied defined by
    a. products' specifications;
    b. products' demand pattern, that is, their relative share of the market;
    c. products' prices;
2. Characteristics of crude oils as defined by their crude assays.
3. Refining models that would enable the different crude oils to meet, technically, market demands under given basic assumptions;
4. Elaboration of over-all economics of refining models for different crude oils.
5. Definition of the specific criteria of comparison and judgment in order to compare the economics of individual performances and therefore to elaborate their relative values according to a given concept of hierarchy.

The different items enumerated above have to be investigated when evaluating the relatives values under the different methods for crude evaluation. However, it is obvious that the results and their significance are very sensitive to the specific value retained for each of the multitude of factors and parameters indicated above. For example, the relative values of two crude oils, elaborated under given sets of conditions, would naturally evolve because

• The market demand pattern would change from one area to another, from one country to another in the same area (as in Western Europe), and even from one refinery to another, both belonging to the same company and located in the same country.

• Products' specifications would change in place and time.

- The concept of refining models could be elaborated under different assumptions (leading to different methods), and the same concept could change differently as a consequence of the change of the two preceding factors.
- The economic conditions of refining could vary largely in time and place and even from one company to another within an identical external environment.
- The economic criteria retained for comparison and judgment would vary according to the parties and conflicting interests involved.

In other words, there could be different sets of relative values for given crude oils, just as there could be different sets of conditions and parameters governing the industry in time and place, at a macroeconomic level. They may, as well, affect the internal structure of the industry for a given market, at a microeconomic level. In the face of this fundamentally aleatory and fugacious concept, crude oil prices are intended to be universal and stable indicators of commercial values, insensitive to the variation of the factors and parameters commanding the industry at a given moment. Should prices ever change because of professional, economic, or political reasons, they would move as a group without their internal structure being altered; that is to say, their relative positions are established once and for all. It is obvious that there is a fundamental difference and an intimate divorce in philosophy and concept between the relative values of crude oils and their prices. The distinction between the two concepts should be made and explained on different grounds as follows:

1. Crude prices are the expression of political forces and strategic considerations governing the international petroleum industry, whereas the relative values reflect tactical differences inherent to detailed specific conditions of the moment.

2. Crude prices are the expression of the integrated nature of the industry by which different crude oils <u>contribute collectively</u> to the meeting of global market demand whereas relative values ignore integration since it is assumed that crude oils would <u>compete individually</u> to achieve the same. It follows that crude pricing patterns are independent of market demand patterns (See Chapter 14), whereas their relative values depend directly and intimately on relative market demand for petroleum products.

3. Crude prices are intended to be homogeneous entities of universal scope, whereas relative values are, by nature and essence, specific measurements of individual cases and conditions. The stability of prices and the permanence of their structure provide an indirect evaluation of the degree and efficiency of control of the international petroleum industry by the major oil companies, whereas

the relative values reflect the extremely large range of flexibility of operation of the same majors within the universal framework they have set for political and strategic reasons.

4. Relative values of crude oils are determined on the assumption that they are commanded by competitive forces of supply and demand in a free open market, whereas both the supply of crude oil and the demand for products are highly inelastic to prices.

In terms of philosophy, crude prices derive from a "subjective" concept of an over-all framework of the industry, set a priori with no direct connection to the actual facts and conditions of the moment, whereas relative values would be an "objective" measurement of comparative performances corresponding to specific sets of conidtions within the above-mentioned over-all framework. The two concepts are irreconcilable, and it would be a glaring blunder to confuse them and to attempt to explain or evaluate the former by the superficial and intellectually attractive methods of the latter.

Different methods for evaluating the relative values of crude oils will be elaborated below. Their domains of validity and application should be always kept in mind for each specific case. They can be valuable tools of economic analysis and/or commanding factors in policy or decision-making in such cases as:

- Comparing supply offers to individual, independent refineries;
- Explaining or correlating realized prices, over a given period of time, for arm's-length transactions in the so-called open, nonintegrated market;
- Elaboration and optimization of flexible operational plans for a given company to meet a specific market demand in order to take into consideration its internal structure and the extent and complexity of its processing facilities and to adjust smoothly to market variations and fluctuations;
- Measuring possible discrepancies between over-all integrated plans of the group (majors), materialized by transfer prices and lifting programs and optimized operational plans of individual companies that make up the group and are generally of different nationalities and/or specializations.

Confusion between posted prices, realized prices, and the relative values of crude oils has overshadowed the industry throughout the 1960s, and especially in the late 1960s. Companies' attitude provided indirect encouragement to this confusion, which was pushed into the heart of the debate by OPEC's attempts to "understand" the pricing structures and to find adequate means to influence them. All those who have been in touch with the organization could have realized to what extent technocratic approaches to crude prices through the investigation of their relative values can be misleading in the absence of any doctrine or philosophy about the nature and significance of prices. The confusion was enhanced by astute propaganda by oil companies and consuming countries that focused attention on realized prices and pointed out their continuous erosion far below posted prices, thus suggesting that the latter have become anachronistic relics of the past that should be replaced by a more realistic pattern, naturally at a much lower level. A representative sample of such campaigns is provided by an article entitled "What Price Crude Oil?" published in Petroleum Press Service, December 1969, displaying a large array of agruments that ignore the essence of the industry's integration and the "political" nature of prices as defined in this study and suggesting a reshaping of the whole pricing structure in the image of arm's-length transactions. One cannot refrain from having a legitimate feeling of suspicion when reading such "candid" statements as

> These postings have long been divorced from commercial realities ... in the sense that they do not correspond to the prices at which crude oils change hands in arm's length transactions. ... Ideally, it may be suggested, the whole corpus of posted prices should reflect as closely as possible the realities of the market. ... Greater realism would be introduced if these fictitious prices were swept away and market forces allowed to do their work unimpaired.[3]

The article suggests that the method proposed by Frankel and Newton for the "Comparative Evaluation of Crude Oils" be taken as a basis for elaborating such "realistic" new prices.

Such an article is a challenge to the intelligence of any ordinary observer of the industry who may have had the opportunity to read that "most of the crude produced in the world is never sold at all,"[4] that best estimates put arm's-length transactions at no more than 15 percent of total world trade, and that open market prices have no direct commercial significance because of nonprice discounts. The article did not even attempt to camouflage its underlying political intent—to reduce payments to producing countries and, consequently, to lighten the burden of foreign exchange disbursment of consuming countries. Even Frankel and Newton could not accept such an extrapolation of their thinking and addressed Petroleum Press Service a letter, published in the January 1970 issue, in which they stated that the article

> goes further and asks whether it would not be appropriate to substitute the current framework of "posted or of

Tax-Reference" prices by a system based on realization of sorts. We do not believe that the long overdue self-contained rethinking of comparative valuations should be made on a more fundamental review, because such wider scope would make it wellnigh impossible to reach such a target in the foreseeable future. . . . The "free market" prices, as far as they are known, cover a sometimes small and thus erratic part of total oil movements whose visible monetary element is often only a part of a package deal and thus not relevant on its own. In fact, the armslength market for crude oil has much in common with the spot market for tankers which has long since been acknowledged as being an altogether unsuitable indicator of the tanker market at large.[5]

Comparative evaluation of crude oils is no more than evaluation of their relative values in given and well-defined conditions, and the significance and validity of the results are applicable only within the framework and limitations set by these conditions. Over-all generalizations and extrapolations would be meaningful only within the limits of validity of the generalization and extrapolation of the basic assumptions and conditions assumed to represent a cross-section of the industry instead of a specific actual case.

RELATIVE VALUES

The evaluation of relative values of crude oils is a very complex process that relies on a large range of information and data fed into one model or another intended to represent how the industry arranges to meet market demand through the refining stage.

The main factors making up any method for crude oil evaluation may be grouped as follows:

1. Crude oil supplies represented by representative crude assays.
2. Market structure, covering
    a. Products' demand patterns;
    b. Products' specifications;
    c. Products' prices or valorization.
3. Refining models enabling crude supplies to meet market demands for products.
4. Refining economics.
5. Criteria for comparing relative values of crude oils.

These factors will be thoroughly investigated. Emphasis will be put on Western European conditions that are believed to govern the economics of Eastern Hemisphere oil or at least to provide the professional background for political factors within which American interests play a leading role.

Crude Oil Supplies

One of the underlying assumptions governing the different approaches to the relative values of crude oils is that supplies are available in adequate quantities under competitive conditions in a free open market. Some authors recognize the shortcomings and inadequacy of such a basic assumption, but even though serious doubts could be cast on the representativeness and validity of their results, such authors usually make their objections in a rather academic way, and their suggested method for crude evaluation is elaborated on the same old, distorted basis. Competition between crude oils is, in fact, admitted implicitly in the elaboration of refining models, whose concept and economics generally ignore the integrated oligopolistic nature of the industry.

The fundamental contradiction outlined above has never been emphasized because of the general confusion between the concepts of crude prices and their relative values. In fact, crude oil assays were considered as the only objective data known with high precision since they do not deal with supply conditions but with routine analysis of the different crude oils considered. Crude analysis or processing handbooks are necessary in order to elaborate processing schemes that make it possible to meet market demand. Crude assays are generally available in the literature in a very simplified form (Oil and Gas Journal, W. L. Nelson's Petroleum Refinery Engineering, P. Wuithier's Raffinage et genie chimique), but they are far from being sufficient for the necessary detailed calculations of processing schemes and their economics.

Highly elaborated crude assays are currently prepared by oil companies for the purpose of their own refining operations as well as for their sales to third parties. Such documents have always been claimed to be highly confidential, although they are available to most people directly or indirectly involved in refining activities in industrialized countries.

Market Structure

The ultimate goal of the petroleum industry is to meet market demands to the satisfaction of consumers who will pay for all the expenses, taxes, and profits incurred in all the successive stages of the industry. However, the actual prices charged to consumers are misleading for the purpose of an over-all evaluation of oil economics because they embody a large share of levies and taxes imposed by local administrations in consuming countries. The prices of real significance are those that generate the cash flow of the industry considered as a whole. The most representative figures are ex-refinery prices, which are generally kept confidential. In fact, the total integrated chains of the oil industry still include marketing and distribution, where a further portion of the cash flow may be generated. Nevertheless, a large part of these activities is in the hands of independent operators (wholesale and retail) who are tied to one company or another to sell their products under their trademark. It has been often claimed that the profits (to companies) generated from this sector are very marginal, if they exist at all, especially if we take into consideration advertising costs. In fact, even if substantial profits were made at the marketing and distribution stage, they would not be significant to the elaboration of the relative values of crude oils since they derive from anonymous products, irrespective of their crude oils of origin. On the contrary, ex-refinery products' valorization has a direct impact on the relative values of crude oils in accordance with the refining and economic model considered.

This analysis provides a further confirmation of the concept of relating the relative values of crude oils to their comparative performances (considered as momentary autonomous sections) through the refining stage. Another consequence is that market demands should be met exactly, with respect to quantity and quality; otherwise, not only would there be a shortage in the cash flow generation but also, and most important, any inadequacy of supplies as provided by the model under consideration would be a highly unrealistic distortion of the real picture of the industry, which strives to adapt to and to meet all market needs and which realizes these elementary facts in a most satisfactory way. The key factor of these successful achievements is the integration of the industry, which is actually ignored in models designed to simulate the industry. Therefore, these models should aim to meet market demand as closely as possible, that is to say, with a minimum deficit and/or surplus production. The gap between refinery production as simulated by the model and market needs provides an indirect measurement of the nonrepresentativeness of the model considered.

Having clarified the general framework within which the market structure has to be defined, let us consider the three market components: demand patterns and products' specifications and prices.

<u>Products' demand patterns</u>, which normally should correspond to the refinery production pattern in the model intended to simulate

industry performance, are largely published in the specialized literature, especially the OECD publications. We will concentrate here on a discussion of the difficulties inherent in the elaboration of an overall pattern that can be considered as fairly representative for the purpose of evaluating the relative values of crude oils.

The first difficulty relates to the definition of the market being considered, and this immediately raises the question of the validity of retaining relative values to represent crude prices. The latter are, by essence, universal in scope and are valid on a countrywide, or at least regional, basis, irrespective of internal market variations, whereas each set of relative values is specific to a given and well-defined market; this concept is most difficult to materialize in practice.

If we start from the smallest scale possible of a single refinery, the demand patterns of the market it is destined to supply may be assessed differently. The most suitable case would be a single isolated refinery owned by independent operators and having to feed a specific, not very sophisticated market. This is generally the case in developing countries and even in developing regions in some industrialized countries. The market structure and conditions can be well defined, and the related relative values can provide valuable and efficient tools for assessing and comparing the commercial prices of alternative available supplies. Such a procedure takes place in evaluating bids for crude supplies to some countries in Latin America (Argentina, Brazil) and in Asia (India, Pakistan), and such groups as independent refiners in Italy. In these cases, market demand can be taken at the yearly demand averages after the necessary adjustments for seasonal fluctuations are made. It should be noted that the limited operational flexibility of refineries in such cases can introduce a permanent distortion between actual market needs and possible production patterns.

The situation becomes much less clear in the highly industrialized European countries. Regional segregation is a common feature that can be further complicated by the availability or lack of local transportation and distribution facilities. Seaside refineries generally have large bunkering activities unknown to inland refineries. The proximity of special types of industries or economic activities (petrochemistry, large power stations, urban centers, and so on) would encourage the creation of specific micro-markets for each refinery. The production of a given refinery could be deliberately different from the market demand pattern representing the nationwide optimized plan of the group it belongs to and depending on the specific share of the company in the regional market considered (commercial policy of some companies might exclude some products although such products are in great demand). The situation is often complicated further by inter-region and cross-region products' exchanges between different companies. Under these conditions, it is very difficult indeed to talk about the market demand pattern of a country, which cannot be representative of any real situation. The general procedure is to consider over-all statistical figures, in percentages, as representative of the average demand pattern. It is obvious that the relative values of crude oils elaborated on the basis of such an average pattern would be significant and valid only within the limits of significance and validity of the averaging considered. In particular, they cannot be of practical significance to any single refinery, and they might even be of contradictory significance to the same company when applied to coastal or to inland refineries.

This difficulty does not exist in reality for crude oil prices, which are universal in scope. This point will be expanded in Chapter 14, where it will be shown that crude pricing patterns are independent of market demand patterns, which command only crude supply patterns and not their prices.

The situation becomes inextricably complex when one considers the market demand for a large area such as Western Europe. All the above-mentioned difficulties are overshadowed by the diversity and contradictions of legal frameworks and administrative regulations. The difference in industrialization levels and the availability of alternative sources of energy contribute to making the different average demand patterns so dissimilar that one can really wonder how such patterns can be averaged further together into a unique over-all pattern. Coal-rich industrialized countries such as France, Germany, and Britain are different from coal-lacking Italy, which, in turn, traditionally disposes of large excess refining capacity destined for export. All are quite different from other less industrialized European countries. Another major difficulty in elaborating such an over-all average derives from the diversity and contradictions of data and information (quantity and quality) and of products' specifications from one country to another. Actual market demands are generally unknown since such information is considered as highly confidential by the operating companies and is accordingly kept secret. Official statistical figures are generally available from different administrative sources, although with some delay; their interpretation and correlation are inevitably marked by a nonnegligible margin of error. We may mention conversions from a weight to a volume basis, and vice-versa, since products' gravities are only infrequently stated precisely and their specifications generally allow for a wide range of gravity variation. Disparities of specifications from one country to another, and even within a given country, and their loose ranges often make it quite difficult to compare products from different countries in order to itemize them collectively. Finally, we should emphasize the very large diversification of end products, whereas the refining models are generally elaborated on the basis of a limited number of large pools of products. Such a shift from complex reality to oversimplified average schemes embodies a sizable margin of arbitrary approximations, a fact that should be kept in mind when one is manipulating the relative values of crude oils.

Products' specifications are large in number and often contradictory in the limitations they induce in refinery operations. Meeting them is the sine qua non of refining since, otherwise, products would not be marketable. They are very seldom stated in rigid values but are given in terms of ultimate limits. There are several specifications for each products, but only a few of them are limiting factors in processing and blending schemes. Bottleneck limiting factors for the same product in the same market may vary with operational flexibility, with the crude oils considered, and with the availability of alternative and complementary sources.

A detailed investigation of products' specifications would be far beyond the scope of this study. It should be noted, however, that specifications for automotive fuels and light products are generally comparable in different European countries. The situation becomes confused for heating and industrial fuels, which vary in name and characteristics from one country to another. This is most troublesome because actually a large number of end products are grouped into the pool of heavy fuel (occasionally of distillates) and there are no general common rules for such grouping.

Products' prices should be taken as the ex-refinery prices for the multiple reasons discussed earlier in this chapter. The main reason is that the largest part of the total cash flow of the industry is generated at the refinery stage, which stage materializes the transformations of crude oils into individual products of established commercial values. Once market needs have been physically satisfied, the industry is interested in products' valorization or total realization rather than in individual prices. Similarly, the elaboration of relative values of crude oils depends directly on products' valorization calculated by combining the refinery production pattern with individual prices.

Products' prices and valorization are very difficult to determine, and the reliability of published data is often doubtful. Products' valorization is composed of two components—namely, refinery production related to market demand pattern, and products' prices. We have pointed out earlier in this chapter the large degree of imprecision prevalent in determining demand patterns and the large margin of indetermination it casts on relative crude values. Products' prices are still responsible for a larger part of the indetermination and arbitrariness. Briefly, some of the main reasons are

• Retail prices are generally well known, but they include taxes and dealers' margins, whose extent cannot be determined with satisfactory precision.

- Actual ex-refinery prices are generally kept highly confidential, and the reliability of available data can be questionable because of the disparity of sources of information.
- Ambiguity of specifications and lack of precision of actual characteristics of marketed products, especially gravity, would cause serious trouble in checking and correlating available figures, mainly for conversion from dollars per ton to dollars per barrel and vice-versa.
- There is often a more or less large segregation in prices in the same market according to the specific conditions of the parties directly involved in commercial and/or transfer transactions. Products' prices are far from being homogeneous.
- Furthermore, prices are regularly subject to variations and fluctuation in time and space in unpredictable and uncontrollable ways.
- All the above-mentioned factors vary differently and independently from one country to another.

It should be emphasized that most European markets are governed by the over-all integrated pattern of the international petroleum industry. Only a minor part of products' movements take place within independent circuits in a so-called open market. Quotations of such marginal spot transactions are currently published in Platt's Oilgram (European bulk and European barge quotations). This part of the market could be compared, in nature and mechanism, to the spot market of tanker transportation. The corresponding figures cannot be taken as representative of the industry either in magnitude or in relative structure.

Refining Models

After one has selected the crude oils to be compared and has determined the market pattern to be met, an adequate model has to be elaborated to simulate how the industry may effectively make the link between crude oil supplies and consumer needs for specific products. This exercise would be rather simple for well-determined markets generally fed by a single isolated refinery (developing countries) but becomes inextricably complex and difficult for highly industrialized countries. The different approaches to this complex issue will be reviewed below before we discuss their limitations and validity.

Generally speaking, an "average" refinery is considered as representative of the whole refining industry in a given area. It is defined by its "average" size and its "average" processing scheme, and therefore it cannot be considered as really representative of any situation for the same reasons encountered in "averaging" market demand and prices. The very concept of an average refinery meeting average market demands is, in reality, a further deviation from the integrated nature of the industry, and it would be tedious to elaborate any more on this fundamental shortcoming of the comparative values of crude oils.

The different approaches that are found in the literature are presented briefly below.

I. Methods not restricted by specific market demands:

(a) The simple composite value of crude oil is figured on the basis of major straight-run products as obtained by distillation (gasoline, kerosene, distillates, and fuel oil) to which standard prices are applied. It is obvious that such a scheme does not correspond in any way whatsoever to reality and that its oversimplification would largely reduce its credibility.

(b) Source refineries in producing areas are generally concerned with exports and, therefore, are not committed to any specific market. Different products, however, could be advantageously marketed in different areas where they are specifically in high demand. A highly elaborated processing scheme could be optimized for each crude oil corresponding to specific investment and operating costs per barrel. All products would be valorized at ex-refinery export prices.

II. Methods restricted by specific market demands:

It is generally assumed that each crude oil is processed individually in order to meet market demands as closely as possible. Eventual excess products are disposed of at lower distress prices (ultimately at fuel value) and deficit products would be compensated for by imports at occasionally higher prices.

(a) The Nelson standard and modified methods have been published in different series in Oil and Gas Journal. To quote the original statement of the method: "Relative values of crude oils delivered to the refinery would be determined by:

1. Determining yields of products (derived from standard laboratory analysis) as compared with other crude oils, so that the total yields from all crude oils match the demand for each of the products throughout the world;
2. Determining the value of the products using prices posted for the products at the point or points of consumption;
3. Subtracting the cost of processing the crude oil for the products of item 1 above, at a refinery situated at (or in direct route to) the points of consumption of item 2 above.

The figure obtained in item 3 is the value of crude oil delivered to the refinery or region contemplated in item 2. To translate the value of delivered crude oil into the value of the home port, such costs as handling, transportation duties charges should be subtracted."[6]

P. H. Frankel and W. L. Newton[7] described the Nelson method as going "into considerable detail and yet tending to generalize in a somewhat arbitrary fashion. . . . It is largely based on American conditions which are different from those of the Eastern Hemisphere." In order to overcome differences between the United States and other countries, Nelson introduces complexity formulas and other adjustment factors (modified method).

(b) <u>Average refining methods</u> tend to reflect the actual refining situation in the market or area considered in the average refinery model, which model can be determined by statistical investigation. These methods correspond to what we may call the "refiner's approach" since refineries are actually designed with a given degree of built-in flexibility in order to be able to process a wide range of crude oils or blends. Refineries are actually what they are, and the different crude oils have to go through the predetermined average processing scheme. Products' valorization from one crude oil or another can be affected differently and adversely since the limitation of excess or deficit production cannot be kept under tight control. These methods include Turner's approach in his "net-back economics" scheme[8] and the more "liberal" approach suggested by Frankel and Newton.[9]

(c) Optimized refinery methods assume that each crude oil would be processed in a specific refinery especially designed to meet market demands. Under these conditions, the investment and operating costs would differ from one crude oil to another. This approach is an extrapolation of the concept of the source refinery mentioned above. It differs because the processing scheme is alienated to meet a specific demand pattern, which would result in the necessity of upgrading some streams for some crude oils (for example, cracking to produce more gasoline from heavy crude oils) or in the undervalorization of some streams produced in great excess (naphtha blended into refinery fuel pool). Turner's replacement economics" approach is an example of such methods.[10]

Other models might be thought about, but none of them would be satisfactory and representative. Each would give different sets of relative values and none can be taken as a basis for the elaboration or comparison of crude prices. This is because the basic philosophy of all such models is fundamentally erroneous. Each crude oil is supposed to meet, individually, the whole spectrum of market demand, whereas in reality, the integrated nature of the industry means that crude oils collaborate to meet market requirements collectively and do not compete individually to achieve the same. Under these conditions, a given crude oil may have to yield a specific pattern of products that might be completely different from the total average market demand pattern. Relative values of crude oils have no significance beyond the limits of validity of the models they are derived from.

Refining Economics

Once the refining model has been determined, its economics has to be worked out in terms of investment and processing costs. Here again we find further disparities in methods and data, which introduce further restrictions on the representativeness of relative crude values. Refining economics involves a large number of concepts that are far from having unique definitions. Furthermore, they may have different values according to sources of information and methods of calculation. General economic conditions in the area considered are of primary importance, but their impact cannot be easily ascertained. Figures are different for independent and major integrated companies, as outlined by Nelson, who continuously keeps updating data in the Oil and Gas Journal.

It is beyond the scope of this study to go into further detailed discussion of the different methods of industrial economics. However, it should be kept clearly in mind that any findings or results are valid only within the limits of the economic conditions retained for the model considered. Unfortunately, these conditions are very rarely stated, so that the figures that can be found in the literature have no real significance because one can never know which cost items they include and which they exclude. A more troublesome result is that these figures can be neither compared nor correlated significantly, and yet many authors compare and correlate them and thus enter into inextricable and meaningless juggling of figures. To illustrate the above in an illuminating way, one can state that investment figures may include some or all of such items as erected cost battery limits, offsites, land cost and site preparation, engineering costs related to different types of contracts, royalties, and financing charges. A major ambiguity, which often shadows operating costs, is whether they represent direct costs or whether they include capital-related charges (amortization, profits, interests, and so on). Furthermore, each individual cost item could have been assessed on a different (and unknown) basis.

Nelson's method is internally consistent since it is based on standard itemized costs that he tabulates and continuously updates so that all relative values elaborated by this method would be directly comparable. However, this type of data has very limited practical value since it is determined by a statistical compilation of adjusted information stretching very far in time and space and mainly inspired by U.S. conditions. Methods suggested by other authors, however, are confined to an academic presentation and do not explain how they would apply in practice.

Criteria for Comparative Evaluation

Once all the above calculations have been made, the economic balances of all cases under study can be elaborated in detail where all components and values are known except crude value (x, $/Bbl) considered as a parameter. Under a given set of conditions and assumptions and within the framework of the refining model considered, the economic balance for each crude oil (B) can be determined as a function of its value (x):

$$B = F(x)$$

Now, in order to relate the values of different crude oils under consideration, it is necessary to adopt specific criteria for comparing their relative economic performances. In practical terms, the criteria would define the nature and the formula of the economic balance so that the values of two crude oils would be equivalent, in relative terms, if they have identical performances measured by their economic balance (B):

$$B = Bo$$

or
$$F(x) = F(xo)$$

Bo and xo correspond to a reference crude oil retained as a basis for comparison.

It has been noted throughout this chapter that the scope and validity of relative values of crude oils are limited since all the technical and economic components contributing to the model are known with a certain degree of approximation or are calculated within a given range of assumption and simplification. The validity of the results cannot exceed the combined interaction of all these assumptions, approximations, and simplifications. Now, the selection of one criterion or another introduces a new disturbing factor, everything else being equal. In other words, having agreed upon a given model and having selected the most appropriate information and data, one can elaborate two or more different sets of relative values of crude oils considered by selecting two or more different criteria for comparison, each set being significant only within the framework of the political and economic concepts of the corresponding criteria.

To clarify the above in more specific terms, we will present different types of criteria that might be retained for the evaluation of comparative values of crude oils.

In those models where each crude oil is to be processed in a specific refinery especially designed and optimized for it, each refinery would correspond to a different investment cost, and the main implicit assumption of this approach would be that such a refinery would have to be constructed at that cost. Therefore, the economic context of the model considered would have mainly to deal with investment decisions both in terms of capital outlay and operating costs. The first criterion (investment cost as such) is not directly concerned with crude values, and it would provide some useful guidelines for the technical optimization of the processing schemes being considered. For example, one may consider selecting the processing scheme that would minimize total investment (availability of necessary funds) or minimize per barrel investment (profitability and economic efficiency of an investment) or minimize foreign exchange outlays (developing countries). The second criterion is concerned with operating costs, which include a substantial part of investment-related components (such as amortization, interest, overheads, taxes, maintenance), direct costs (labor chemicals, and so on), and crude values. Therefore, any formula that relates comparative values of crude oils through corresponding operating costs would include specific investment figures. The formula can be defined in different ways and thus lead to different sets of results. For example, one may consider equalizing profits, or equalizing profitability of investments, or equalizing the foreign exchange components of operating costs, and so forth.

In those models where all crude oils would have to be processed in the same average refinery, investment is the same for all crudes in all cases, and the implicit assumption of this approach is that the refinery does exist already. Under these conditions, investment-related components in operating costs would eliminate each other in any formula comparing the relative performance of any two crude oils, and the profitability would be measured directly by the profits derived therefrom and including only the direct operating costs, to the exclusion of investment-related costs.

NOTES

1. F. C. Turner, "Methods of Crude Oil Valuation," OPEC Seminar, Vienna, June-July 1969.
2. P. H. Frankel and W. L. Newton, "Comparative Evaluation of Crude Oils," The Journal of the Institute of Petroleum 56, 547 (January 1970).
3. "What Price Crude Oil," Petroleum Press Service, December 1969, p. 443.
4. J. E. Hartshorn, Oil Companies and Governments, 2d ed. (London: Faber and Faber, 1967), p. 133.
5. "Letter to the Editor," Petroleum Press Service, January 1970, p. 21.

6. W. L. Nelson, "Standard Crude Oil Realization," <u>Oil and Gas Journal</u>, November 18, 1963, p. 149.
7. Frankel and Newton, op. cit.
8. Turner, op. cit.
9. Frankel and Newton, op. cit.
10. Turner, op. cit.

PART

III

A STRATEGIC APPROACH
TO CRUDE OIL PRICES

CHAPTER 12

THE GENESIS OF PRICING PATTERNS

PRICING ECONOMICS IN WESTERN EUROPE IN THE EARLY 1950s

In previous chapters we have briefly outlined the nature of the dynamic forces and the extent of the conflicting interests that confronted each other in the postwar period as part of the shaping of a new framework for petroleum economics and crude oil pricing in the Eastern Hemisphere. The major oil companies, which controlled the largest part of free world reserves outside the United States, have been under converging and sustained pressure for lower crude prices. During World War II, the British Government initiated a move to secure lower-cost fuel supplies for its fleet east of Suez, and efforts continued thereafter to lower prices in general and to substitute "sterling crude" for "dollar crude" to as great an extent as possible. The most effective pressure was exerted by the U.S. Government through ECA and the Marshall Plan. Similar efforts were made by European governments, although their potential for pressure or retaliation at that moment was very limited. Nevertheless, it was quite obvious that, sooner or later, the influence of the U.S. Government on European affairs would progressively diminish during the recovery period and that the impact of the nationalistic feelings of European governments would be more and more heavily felt by the majors.

The new pricing pattern in the Eastern Hemisphere progressively developed in the early 1950s, at the same time Western Europe was emerging as a specific economic entity with genuine political maturity and unitary objectives. The European Economic Community (EEC) became a reality, paving the way to the spectacular success of the Common Market. Early attempts were concerned with the coordination and optimization of energy sources and costs as well as with heavy industries. The organizational setup for the realization of these objectives was provided by the European Coal and Steel Community.

The over-all strategy on which the majors settled in developing actual pricing patterns has certainly taken all these factors into consideration at the consumers' end, as well as the political and economic factors at the producing end in the Middle East. It was only natural to expect the maturing European community to scrutinize the majors' plans and actions in order to assess the extent to which they could be reconciled with its own objectives for security of supply at lowest cost in foreign exchange and in total value. Such an extensive investigation was undertaken by a special study group of the United Nations Economic Commission for Western Europe. Its findings were published in March 1955 in a report entitled The Price of Oil in Western Europe under the reference E/ECE/205. The present chapter relies heavily on this exceptional document, which will be referred to as the Report.

The primary intent of the Report was to decide whether the crude and products' prices charged to Western European countries were "fair and just." Although it concluded that there was no clear answer since "fairness is a versatile multi-facet concept," the Report endeavored to provide all the necessary information and data for an understanding of the mechanisms of price formation and for an appraisal of the dynamic forces and conflicting interests governing its evolution. Emphasis was placed on consumer interests, which called for market forces of supply and demand to play a larger role in pricing patterns, with the expectation that Middle Eastern prices would drop because of a large potential oversupply capacity at very low cost. The background of energy and petroleum economics in Western Europe was described and situated within the framework of the international oil business; then the price formation of Middle Eastern crude oil was analyzed and the pricing structure and price levels of petroleum products in Western Europe were investigated. The Report ended with a critical analysis of the prevailing system without committing itself to any specific judgment.

BACKGROUND TO PRICING PATTERNS IN THE EASTERN HEMISPHERE

In the early 1950s, petroleum products accounted for about one-eighth of Western European energy supplies, but their relative importance was about half the average annual increase since 1937. Total world consumption of petroleum products has been steadily increasing at an average rate of 7 percent since 1900. Future needs are expected to develop at still higher rates because of greater energy consumption and the higher-share percentages of hydrocarbons.

During the period between the two world wars, high rates of expansion were centered on white products because of the extensive development of motorized transportation for which there is no substitute. Emphasis shifted thereafter to residual fuels along with extensive industrial development, the increasingly declining economics of coal, and transportation restrictions due to war. Further expansion was equally expected in all sectors with an ever increasing percentage of petroleum products in over-all energy supply. These forecasts and expectations proved valid, as evidenced by Table 12.1.* European prosperity inevitably was to depend on petroleum imports. Continuity and security of supplies were the dominant underlying guidelines of European energy policy, with an immediate objective of bringing the total bill down to a minimum.

In the dollar-lacking Europe of the early 1950s, the foreign exchange component was relatively more important than the total bill. The Report indicates that the total ex-refinery value of petroleum products in Western Europe was about $2.2 billion, of which about one-half was FOB purchases of imported crude oil, one-quarter tanker transportation, and one-quarter refinery margins. The dollar component accounted for more than half of the total bill and constituted a heavy burden on European trade balances. Refining operations were primarily located in Europe, and a large part of the transport was made under European flags, so that the largest part of the dollar component was due to crude oil importing. Lower FOB prices were the overwhelming factor in shaping relations between consuming countries and U.S. oil companies. Furthermore, substituting "sterling crude" for "dollar crude" was a specific objective for British-controlled interests.

The postwar period witnessed two basic changes in the international oil business, changes that were emphatically concretized in Western Europe. First, refining operations shifted from producing centers to consuming countries, thus reducing the relative share of imported refined products from about 65 percent 1937 to 25 percent in 1953. Storage and transportation economics favored such evolution; but the determining factors were strategic considerations dealing with security, determination to reduce the dollar fraction of the bill, and the drive for intensive industrialization. Consequently, European

---

*The relative importance of petroleum products is actually larger than shown in Table 12.1, which deals only with primary energy. Other uses, for which there are generally no substitutes to petroleum products (engines, petrochemistry, domestic uses, and so on), are not included.

TABLE 12.1

EEC Consumption of Primary Energy

Fuel	1950	Weight Percentage				
		1960	1967	1968	1970	1980
Coal	74	53	31.3	29.8	27-32	9-16.5
Lignite	9	7	4.9	4.6	5	3.5
Petroleum	10	28	51.5	52.6	49-54	64-50
Natural gas	—	3	5.4	6.4	7	12-15
Primary electricity	7	9	7.3	7.1	7	12-15
Total	100	100	100	100	100	100
Total*	193	307	434	457	495	755

*In million tons equivalent of oil.

Source: J. Masseron, L'Économie des hydrocarbures (Hydrocarbon Economics) (Paris, Ed. Technip, 1969), p. 9.

countries became much more crude-price-conscious than before World War II.

The second change was a shift in interregional trade in crude oil and petroleum products from the Western to the Eastern Hemisphere. In 1952, Western Europe derived 93.2 percent of its crude oil imports from the Middle East as against 39.1 percent in 1937. This development was not a temporary phenomenon; it reflected a dramatic deterioration of the balance between reserves and production consumption in the United States and the development of fabulous reserves of low-cost crude oil in the Middle East. In the early 1950s, the United States became a net importer of both crude and products, as shown in Figure 12.1. Apart from the United States and Eastern Europe, which are producers but not exporters of crude oil, the largest part of crude oil produced in the world (mainly in the Middle East and the Caribbean) is not consumed in the producing areas but moved to the main consumption centers located in highly industrialized countries. The quantity of crude oil traded between the two hemispheres is relatively limited, with a net balance of exports from the Middle East to the United States. Crude oil production in the Eastern Hemisphere (that is, the Middle East) is disposed of in the hemisphere itself, where imports to Western Europe account for more than half of total production.

Figure (12.1)

Inter-regional trade in crude oil and petroleum products in 1952

Millions of tons

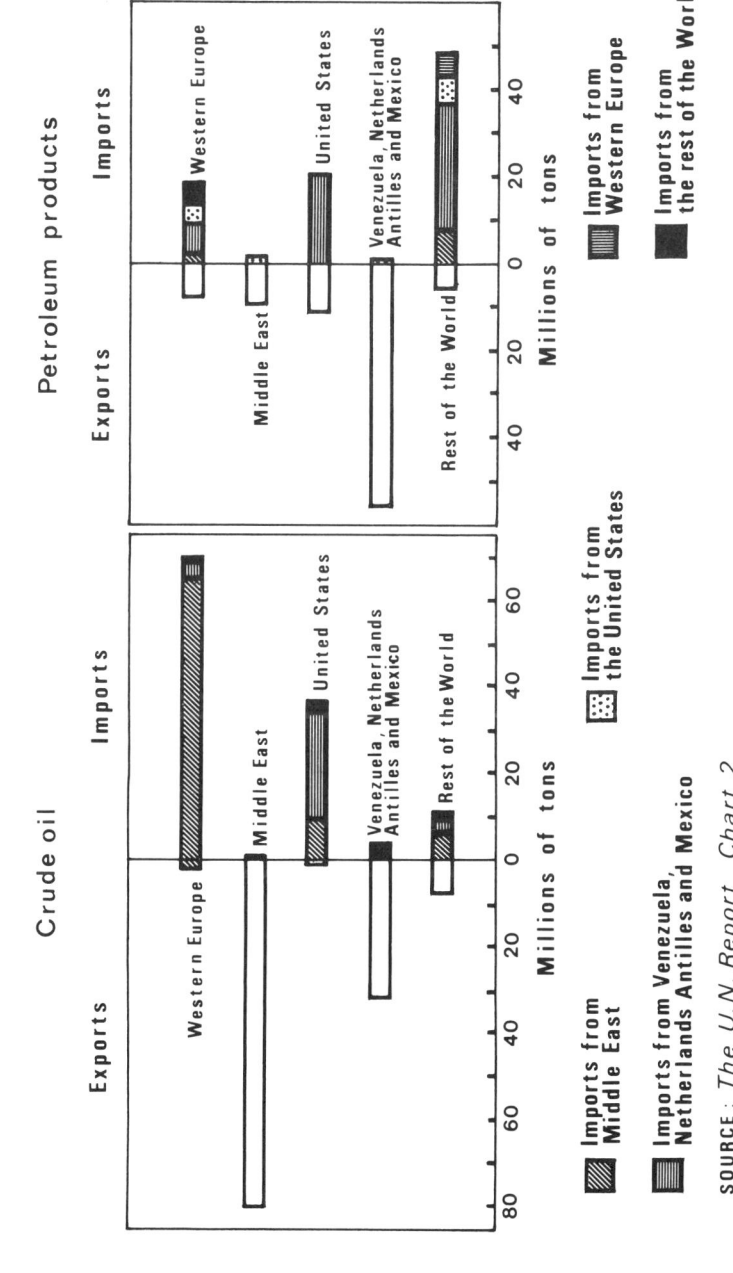

SOURCE: *The U.N. Report, Chart 2*

Because of the extensive development of refining capacity in consuming countries, the trade of petroleum products comprises relatively small volumes. The major exporting center is the Caribbean area, where exports are essentially made up of residual fuel sent to the East Coast of the United States. Western Europe imports include mainly special products from both the Caribbean and the United States. It should be pointed out that products' trade between the two hemispheres is expected to diminish with further expansion and sophistication of refining in Europe as well as with the resumption of operations in the Abadan refinery.

Before World War II, Western Europe was chiefly dependent on Western Hemisphere sources for its supplies of crude and products, and thus it came under U.S. pricing patterns, which applied throughout the world. During the postwar period, the two hemispheres were rather insulated from each other, and specific and separate pricing patterns developed. The "legitimate" aspiration of European countries is to get the best advantage out of low-cost Middle Eastern crude oil in the economic sense of supply and demand under competitive conditions. The realities of the international oil business were at odds with this simplified approach, and the Report provides a thorough and far-seeing approach to petroleum economics in Western Europe in the early 1950s.

A EUROPEAN LOOK AT CRUDE OIL PRICES

In the early 1950s, crude oil imports to Western Europe were invoiced at posted prices as determined by major oil companies controlling the largest part of crude reserves in the Caribbean and the Middle East. The Report investigated the evolution of posted prices in the postwar period and recalled how CIF prices of reference crude oils from the Eastern and Western Hemispheres were equalized, for 36° API gravity, at a line that progressively shifted westward before settling at the East Coast of the United States. FOB prices in the Gulf area were determined by a net-back formula in which the freight element plays a major role, a formula extensively investigated earlier in this study. The Report envisaged different criteria for the selection of the proper rate to be applied for the purpose of pricing but was unable to suggest a definite figure.

Now, if we consider that the Eastern Hemisphere is an autonomous entity, from the viewpoint of supply and demand as demonstrated above, the very low cost of production of crude oil in the Middle East should bring prices down from their high level initially determined as a function of high-cost U.S. crude oils. The Report estimated production cost in the Gulf Area at about 10-25¢/Bbl against an average of 76¢/Bbl in the United States and about $1.50/Bbl on the East Coast.

Profits derived from Middle Eastern operations are higher than in the Western Hemisphere. More significantly, profitability, in terms of return on invested capital, is incommensurately higher because of the extraordinary productivity of wells in the Middle East.

Should normal economic forces and laws come into play, FOB prices in the Gulf would drop, eventually to the ultimate level of the marginal cost of production plus taxes and royalties paid to host governments. In fact, the Report indicated that economic forces and laws were not related to the mechanisms of crude price formation. The fundamental factor of pricing patterns in the Eastern Hemisphere was the equalization of CIF prices at the U.S. East Coast so that only marginal quantities could penetrate the U.S. domestic market in so far as the U.S. market is not specifically protected by import control or restrictions. Lower prices could not be tolerated because of the inevitable invasion of domestic markets, which would result in deteriorating home prices and profits. Pricing strategies in the Eastern Hemisphere were designed to protect the conservation objectives of the United States and the stability of domestic prices. This could be effectively achieved because of the tight and over-all oligopolistic control of the international oil business by the majors and the strong support provided them by the U.S. Government. Indirectly, the commanding necessity of setting crude prices in compliance with U.S. interests was favored by producing countries since it resulted in the maintaining of crude oil prices, and consequently payments to host governments, at high levels.

The key element to a genuine understanding of pricing patterns in the Eastern Hemisphere would be, therefore, a thorough investigation of the U.S. domestic oil industry and its impact on international oil business.

PRICES OF PETROLEUM PRODUCTS IN WESTERN EUROPE IN THE EARLY 1950s

As indicated earlier, petroleum products traded in the prewar period were far more important than crude oil, and high-capacity refining operations were installed in producing areas—namely, the Gulf of Mexico, the Caribbean, and the Gulf area. Although their relative importance largely diminished after the war, the posted prices of petroleum products prevailing in these areas remained of the utmost significance to Western Europe. Posted prices FOB both the Gulf and the Caribbean were aligned with those prevailing in the Gulf of Mexico. Ex-refinery prices of petroleum products in Western Europe were determined neither by market realities and economics nor in relation to the structure of crude oil prices. They were aligned with import parity prices from the Caribbean and the Gulf area, that is, posted prices plus freight. This system is still operant in the official French pricing system. Although ex-refinery prices are generally unknown, the Report indicates that "in spite of some recent price adjustments, the largest part of the transactions takes place at import parity prices. The determination of a specific price structure, if ever it has actually occurred, has been only superficial and on a limited scale."

The postwar evolution of crude oil pricing in the Eastern Hemisphere, with the equalization line finally settling at the East Coast of the United States, benefited Western European countries by lowering the delivered prices of Middle Eastern crude oils. Another specific change in pricing patterns was to base them on a basic gravity differential of 2¢/Bbl/degree. In contrast to this marked shift from the prewar situation, the pricing of petroleum products in Western Europe remained unchanged in spite of the deep and permanent modification of the supply pattern, which became largely dependent on Eastern Hemisphere sources.

The impact and consequences of this pricing philosophy in Western Europe should be appraised from two viewpoints: the general level of prices based on import parity from the Caribbean and the Gulf, and the structure of products prices versus market demand patterns.

In light of the foregoing, the prices of petroleum products in any given region in Western Europe are arbitrarily related to its geographical location and not to actual refining economics, as might be legitimately expected. Everything being equal, the relative profitability of refining is governed by accidents of geography rather than by the verdict of costs. Starting from the same level, products' prices increase when going westward from the Gulf and eastward from the Caribbean along with the increase of the freight cost. At prevailing freight rates, prices reach their peak at an equalization line, which is the eastern coast of Italy. Since delivered crude prices increase continuously when going westward to the U.S. East Coast, it follows that refining profitability is highest at the equalization line (Italy) and diminishes progressively in both directions, although at higher rates when going westward, as shown in Figure 12.2. Here lies the explanation of the extensive development of refining installations in Italy (and to some extent in Southern France), with their substantial overcapacity for exports.

Another important consequence is illustrated in Figure 12.2. Should the freight cost change, which it actually does continuously, refining profitability (margin) to the east of Italy would not be significantly affected. On the contrary, the effects of cost changes are combined to the west of Italy so as to make refining economics very sensitive to freight fluctuations. Maximum amplification of this phenomenon is found in Northwestern Europe, where the largest part of refining capacity is located.

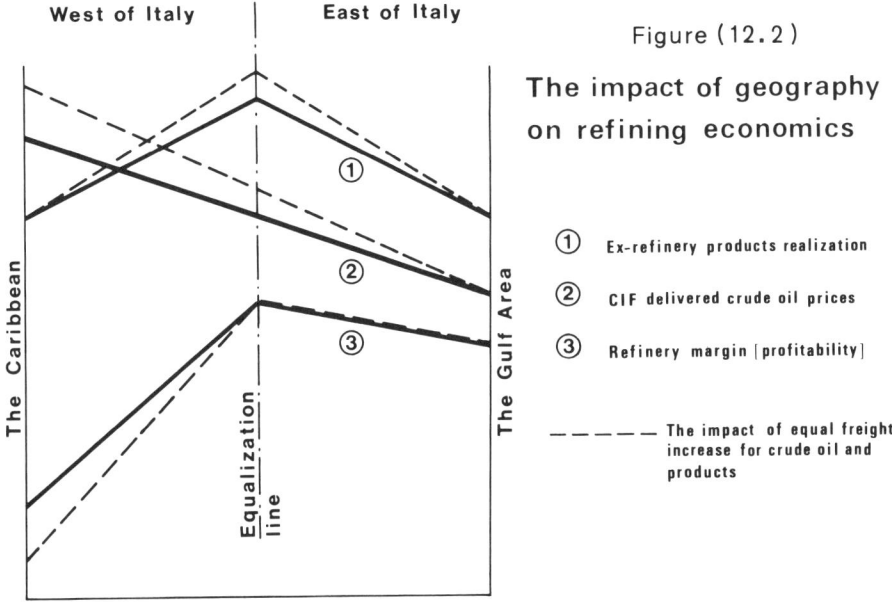

Figure (12.2)

The impact of geography on refining economics

① Ex-refinery products realization
② CIF delivered crude oil prices
③ Refinery margin (profitability)

— — — — The impact of equal freight increase for crude oil and products

These expectations were confirmed in the Report by calculating refining margins in Northwestern and Southern Europe. These calculations were very approximate because of the absence of reliable, detailed published data on ex-refinery products' realizations. A macroeconomic approach was employed, and average figures were taken for Western Europe. It is obvious that results would rarely apply to a specific refiner, since none of the results actually reflects a perfect image of the European average. The Report estimated average refining costs at about 64¢/Bbl of capacity including amortization. The actual figures were probably higher for refineries running at lower than design capacity. Under these conditions, refiners in Northwestern Europe would have suffered heavy losses in the early 1950s, when freight rates increased sharply because of the Korean War, whereas refineries in Southern Europe would have netted some 5-8 percent return on investment.

Under these conditions, one may legitimately wonder how refining capacity maintained its high pace of expansion under such distress conditions in Northwestern Europe. The answer is to be found in the extensive integration of the major oil companies, which control nearly all crude production where most of the profits are located. The apparent losses that resulted therefore were merely internal accounting exercises that did not affect over-all performance. The profitability of "integrated" refining could be easily manipulated by considering one freight rate or another, at will, for crude transportation within the normal wide range of freight discrepancies.

Nevertheless, this "explanation" would not apply to independent refiners, who can survive and prosper only if their refining activities are substantially remunerative. In fact, they have prospered, and the Report noted that about one-quarter of European refining capacity in 1953-54 was not controlled by the majors. Indeed, the existence and survival of independent refiners is politically essential to the major oil companies, which actually "subsidized" the independents' activities although they could have eliminated most of them through price wars. Major oil companies are obsessed by the threat of antitrust regulations in the United States and the oppressive accusation of cartel collusion in the rest of the world. In the nationalistic, politically maturing European countries of the postwar period, the public image of the majors should not represent an omnipresent "imperialism" commanding and manipulating the most important sector of economic life and independence. Europeans should feel effectively the master, at least, of the refining operations that take place in their countries. The existence and prosperity of independent refiners were a security premium for a harmonious political atmosphere and a barrier against the development of extreme nationalism.

The Report outlines two means for achieving the hidden subsidization. When freight rates were at their peak during the Korean War, posted prices at Eastern Mediterranean terminals were undervalued and kept below parity prices at actual freight rates (whether London award or single-voyage rates). Independent refiners were supplied from Eastern Mediterranean terminals rather than from the Gulf, thus taking advantage of privileged freight differentials.

Furthermore, most independent refiners had no direct access to the market through distribution facilities of their own. Most of them were actually operating under "processing agreements" whereby the majors supplyied them with crude oil. The specific terms of such contracts were not known; nevertheless it is safe to assume they could easily secure minimum profit margins for refiners irrespective of market fluctuations provoked and/or absorbed by the majors themselves.

The survival and prosperity of independent refiners were possible only because their independence was quite harmlessly superficial and designed to protect the public image of the majors. No independent refiner could afford to deviate from the framework of the over-all strategies designed by the majors who retained formidable means of dissuasion and retaliation by denying the independent the advantage of various hidden subsidies. However, this situation changed in the late 1950s when nonintegrated refiners were able to gain real independence by receiving low-cost crude oil from the Soviet Union or from independent producers. They were able to make their way into the highly expanding market by cutting prices, especially in those countries where state control was working to thwart the omnipresent domination by the majors of domestic markets. Although the volumes involved were rather marginal compared to total demand, the potential danger of a deterioration of the oligopolistic positions of the majors was readily and fully appraised, and this threat induced the move by the majors to cut Middle Eastern prices in the late 1950s, which in turn led to the creation of OPEC and the general freezing of prices.

In addition to the impact of the general level of products' prices outlined above, the relative structure of these prices was deemed inadequate by the Report and was suspected as a cause of the unusual attitudes and performances of European refiners. It has been indicated that products' prices were related to those posted in the Gulf and the Caribbean, which were themselves aligned with those prevailing in the Gulf of Mexico. The relationships between the prices of white products and those of residual fuels are expected to reflect supply and demand forces as a function of the market demand pattern. Such an alignment of the structure of European prices with the specific realities of the U.S. domestic market was conceivable and justifiable throughout the 1930s, since market demand patterns were quite similar in both areas and were characterized by a high demand for gasoline (more than 50 percent). The situation changed profoundly and permanently after the war; the U.S. demand pattern remained relatively unaffected, whereas European demand for gasoline dropped substantially, to the advantage of middle distillates and residual fuels. Yet, products' pricing patterns remained roughly the same. The imbalance between the market demand pattern and the products' pricing pattern would affect the stability of refining economics in Western Europe in a peculiar way: A slight modification in fuel price would have very limited consequences in the United States because of its relatively small share in the total production, whereas the impact of the same in Europe could be very great. Similarly, a slight fluctuation in the gasoline price in the United States could have dramatic consequences whereas the effect in Europe would be rather imperceptible. Thus, refining profitability and products' realization in European countries are governed by the specific conditions of the U.S. domestic market and not of their own.

From all the above, it appears that pricing patterns in the Eastern Hemisphere, which were believed to have developed in association with and under the impact of petroleum economics in Western Europe, were actually designed as functions of American interests in the U.S. domestic market and the rest of the world. The key to a genuine understanding of the real dynamic forces behind pricing policies and patterns in the Eastern Hemisphere is fully to appraise American realities and interests.

THE U.S. DOMESTIC OIL INDUSTRY

These words could easily be the title of a gigantic encyclopedia of its own, and an extensive survey of related references would cover dozens of pages. Although our interest in this subject, for the purpose of investigating worldwide pricing policies and strategies, is great, the scope of our approach in this study, confined to a subparagraph, has obvious limitations. Therefore, we have to assume that the reader is quite familiar with the facts and figures of the U.S. domestic industry in its historical background, present setup, and future prospects.[1]

The investigation of the origins and the complex evolution of pricing patterns in the United States can be made easier if one keeps in mind the specific features of the U.S. legislation on the ownership of subsoil wealth by surface owners. Since the geological conditions of a given reservoir and its behavior when producing do not recognize the arbitrary ownership divisions at the surface, the traditional "law of capture" motivated the formidable rush to the attractive oil business after Colonel Edwin Drake's lucky strike in 1859. The legal principle of the law of capture derives from the traditional fact that oil drawn from below the ground belongs to the owner of the land surface, regardless of whose land it was originally under. Rapidly, the law of capture became a devastating law of the jungle. Most oil-bearing areas were divided into small parcels, and farm owners were assailed by all kinds of businessmen, speculators, and adventurers attempting to take lease of their land. The prospects of quick fortunes were alluring, and the relatively low cost of drilling and refining in the early days helped transform some landscapes in Pennsylvania and Oklahoma into forests of derricks.

Thus, this most extraordinary waste of human energy and natural resources developed at a terrific pace. According to Harvey O'Connor,[2] only about 5 percent of all the wells drilled up to 1929 proved to be producers. The law of capture operated once oil was discovered; the lucky driller had just the time to cover his parcel of land with as many wells as he could in order to extract as much oil as possible, which otherwise would be drained in the multitude of wells that mushroomed all around in neighboring leases. Nobody seemed to care about rational exploitation or maximum recovery of reserves, and such behavior virtually "killed" many important discoveries, and possibilities of recovering invested capital vanished with the illusion of profit.

The immediate consequence of the automatic development of surplus production, irrespective of market needs or possibilities for disposal, was a dramatically sharp drop in price because of the high degree of inelasticity of demand for petroleum production in the short run.[3] Such price fluctuation was the predominant feature of the U.S.

oil industry since the early discovery of the first commercial well in 1859. This phenomenon gained particular momentum during the 1920s. The price of midcontinent crude oil, which had reached a high of $2.29 a barrel in middle and late 1926, ranged between $1.28 and $1.45 in the next four years, and dropped to $0.33 in mid-1931, when oil sold in East Texas for as little as 10 cents a barrel. Out of the first 24,000 wells drilled in the East Texas basin, with and without the permission of the Texas Railroad Commission, more than 21,000 wells were formally recognized to be absolutely useless and represented a net loss of $250 million.

The frightful waste that characterized America's exploitation of its precious oil resources and the dramatically abrupt fluctuation that followed would have remained a kind of tragic curiosity of big capitalism were it not for the strategic importance of oil and its vital role in the survival and prosperity of the nation's economic and industrial activities. Up to 1914, federal oil policy was entirely devoted to a relentless antitrust battle, especially against the gigantic Standard Oil Company; this was a time when oil was produced and refined primarily to supply illuminating kerosene. World War I emphasized the strategic importance of oil, and the industrial revolution entered a new era with the development of engines and the shift of demand to ever increasing quantities of gasoline.

In the postwar period (the early 1920s), federal doctrine was dominated by fears of the imminent exhaustion of oil reserves in the United States, with its disastrous consequences on the economic life of the nation and on U.S. military supremacy and related economic domination abroad. The up-and-down character of oil exploration, caused by the uncontrollable periodic rush of speculators and independents to the alluring mirage of easy fortune and their stubborn and irresponsible squandering of any discovered reserves, helped create serious doubts and fears about the reliability of optimistic reserve forecasts and the continuity of adequate supplies. Federal oil policy was essentially motivated by strategic considerations that ultimately took priority over antitrust and economic preoccupations. Developing domestic recoverable reserves and maintaining them at the highest level possible became primary objectives that could be achieved only by intensification of regular exploration, limitation of waste through rationalization of production, and stabilization of the market.

Strong financial incentives to encourage exploration were granted in the mid-1920s, and the percentage depletion allowance was promulgated as a federal law. Thus the producer was allowed to deduct 27.5 percent from the gross annual income of a lease or property before income tax computation. The allowance, which is tax free, may not, however, exceed 50 percent of the net income.

However, it should be pointed out that, from the viewpoints of both the federal government and the oil companies, such financial incentives could not make the industry healthy and secure unless prices were regularly maintained above a reasonable minimum, a situation highly incompatible with the absence of restrictive regulations over production. Therefore, conservation emerged as an overwhelming aim of federal policy and action. Production regulation and prorationing, initially enforced at the state level, proved to be unsuccessful in preventing the incredible sack in 1930 of the East Texas basin, where the use of the state militia to enforce restrictions could not prevent prices from plummeting. The industry was equally concerned by conservation since price collapse was a serious threat to the existence and survival of oil companies, mainly the majors. The latter were very active through the American Petroleum Institute, which drew up most of the practical aspects of conservation regulations.

Restrictive regulations at the state level were inoperative since "hot" oil produced in violation of state laws (especially in Texas and Oklahoma) was marketed in major consumer areas (Chicago, St. Louis) at dumping prices. The need for interstate regulations was deemed indispensable, and in 1935, Congress approved the Interstate Oil Compact to conserve oil and gas. This compact placed a moral obligation on member states to enact and enforce laws for preventing physical waste and to deny access to commerce for "hot" oil produced in violation of these laws. The Connally Act, passed later in 1935, transformed the latter moral obligation into a legal one, and conservation became a federal responsibility that has effectively operated ever since.

Under the new system, production was tailored to demand, and oversupply conditions were eliminated. The Federal Bureau of Mines works out monthly estimates of anticipated demand for crude oils in the various states. These figures provide the basis for calculating production allowable for each basin or well by competent state commissions. According to M. G. de Chazeau and A. E. Kahn, "The Texas Railroad Commission, which controls more than 40 percent of U.S. crude production and approximately half of estimated national proved reserves, has also found it desirable to make allowance in its estimates for anticipated production in states that do not control production or use other bases of control."[4] By thus allowing for estimated supplies beyond its jurisdiction, Texas in effect brought total available supply, including imports, within the principle of prorating to market demand. Estimated market demand minus these uncontrolled additions to supply gives the total, which is to be prorated among fields and wells in a manner designed to preserve equity among producers and to prevent and well from producing beyond its maximum efficient rate (MER).

These measures succeeded in pulling the industry out of chaos and matched federal strategic objectives. The elimination of wasteful practices was attained by relying on sound technical rules in determining the allowables, and exploration for new reserves could be expected to stand regularly at high levels since the remuneration of risk capital was ensured of high rates because of market stabilization. Because chaotic oversupply conditions were eliminated, prices were efficiently stabilized.

The direct involvement of the United States in World War II helped dramatically to emphasize the primordial strategic importance of oil. The fantastic industrial and economic expansion that followed and the irreversible move of the United States into the position of net importer underlined the necessity of securing continuity of supplies in all conditions and at any cost. This security element played a major role in shaping federal oil policy in the postwar period. Reliance on imported crude oil could not be permitted to exceed given limits, and in view of the ever increasing demand, intensification of exploration to add new domestic reserves had to be maintained and encouraged. This could be achieved only by maintaining the alluring prospects of substantial profit margins and thus high and stable crude prices. It is easy to understand that this complex system could not survive if the domestic market under the pressure of supply shortage were opened to invasion by low-cost crude oil from the Eastern Hemisphere that could be produced by U.S. companies. As long as U.S. domestic markets were not legally protected from this kind of direct threat, market conditions had to be tailored in such a way that competition from low-cost crude oils from the Eastern Hemisphere would normally be limited to a marginal share of the market. The most vital U.S. strategic interests required that the CIF price of such crude oils at the U.S. East Coast equal prevailing domestic prices so that any invasion would stop, economically, at the marketplace gate. Since pressure in the Eastern Hemisphere for lowering crude prices was irresistible, as already indicated, the level of U.S. domestic prices at the East Coast represented a limit not to be exceeded.

However, with the increasing imbalance of supply and demand in the United States, the extensive development of huge reserves in the Middle East, the irresistible and increasing pressure from European countries as outlined in the Report, the appearance of a kind of competition from independent producers, and the absence of effective resistance from producing countries, the necessity for a further reduction of Middle Eastern crude oil prices appeared inescapable. Such a move would be detrimental to U.S. strategic interests unless domestic markets were protected. This protection was ensured by enforcing control regulations through import quotas, thus giving the green light to price reductions, which followed immediately thereafter. Furthermore, domestic prices could be increased without necessitating their alignment with international prices.

Consequently, crude prices in the Eastern Hemisphere, which stand substantially below parity, are directly affected by any modification of U.S. policy regarding crude oil imports. This will be a key element to the investigation of prospects for the future evolution of crude prices.

It should be noted that the prevailing prorating system provides the U.S. Government a formidable tool for controlling the industry by influencing the level of prices. In fact, prorating strengthens the administered pricing of crude oil by elasticizing crude supply: "Surplus" is kept at ground level when demand falls short of what unregulated producers would be willing to offer at the posted price. Phillips Petroleum's unsuccessful attempt to increase crude oil prices in 1948 reaffirmed the inability of a major company to implement a price policy at variance with the general assessment of the market situation by the U.S. Government and by leaders of the industry, even though apparent economic facts and higher cost figures could justify such a move. On the other hand, the successful price increase in 1953 demonstrated the feasibility of a pure policy advance in oil prices when there is consensus among industry leaders and when the Texas Railroad Commission is sympathetic, even though supplies are adequate or more than adequate.[5]

THE ROLE OF MAJOR OIL COMPANIES

It would take another encyclopedia to investigate and analyze the historical development and role of the major oil companies. We have no choice but to assume that the reader is familiar enough with this subject to understand and appraise the full dimensions of our interest, limited here to a brief description of the role of the majors in establishing pricing patterns and strategies throughout the world. Among other references, Edith T. Penrose's The Large International Firm in Developing Countries and Michel Laudrain's Le Prix du pétrole brut should provide appropriate information.

The international oil business has been largely influenced and commanded by American interests, public and private. One of the sources of strength of the American companies operating abroad (old internationals as well as newcomers) is that they have a very remunerative business at home.[6] Therefore, the primary objective of these companies is to keep their domestic activities flourishing so as to live in harmony with powerful state and federal regulations. Domestic activities thus govern the shaping of the basic philosophy of action abroad and its limitations and motivations. Therefore, the majors' worldwide strategies should be contemplated as they are influenced by over-all U.S. strategic considerations.

Until World War I, the oil business was largely dominated by the formidable development of the Standard Oil Company, incorporated in Ohio by John D. Rockefeller in 1870, into a tentacular monster. In application of the antitrust regulations, Standard Oil was required by the Supreme Court in 1911 to divest itself of the stock of 33 constituent companies. The latter rapidly managed to integrate, and most of them regained individual postions as majors, and their common origin did not prevent conflicts of interest from developing between them. Standard Oil of New Jersey, of California (SOCAL), and of New York (SOCONY) formed the bulk of the companies known as international majors. The Texas Oil Company (Texaco) and the Gulf Oil Corporation succeeded in surviving the aggressive competition of Standard Oil, and later of its heirs, thanks to their early efforts for integration of their activities at home and for diversification of supply sources outside the United States.[7]

With Royal/Dutch Shell, British Petroleum (formerly Anglo-Iranian), and Compagnie Française des Pétroles, the above-mentioned companies form what is generally called the "eight sisters," who control the largest part of oil business outside the United States. However, the U.S. domestic market was dominated to a large extent by some 20 companies, including those of early international stature. Calling these 20 companies majors is quite significant, and their domination of the industry is better illustrated by some figures. Their collective share of crude production increased from 46.3 percent in 1926 to 52.5 percent in 1927, their refining capacity from 67.5 percent to 76.6 percent and production of refined products from 71.3 percent to 83.3 percent. Their over-all domination of the industry by 1939 is illustrated in Table 12.2.

Although these companies were fully engaged in the jungle of fierce business competition, they behaved similarly and acted as a group in the face of forces and threats originating outside the industry. Their community of thinking and interests derived naturally from their high degree of integration both horizontal and vertical. The fundamental philosophy behind the strategies and policies of the large integrated companies is one of power and domination. Their large size and high degree of market control assure them substantial profits and high return on investments on the condition that their privileged position continue and grow stronger, and that their power and independence be consolidated. The attitude and behavior of integrated majors are primarily governed by such long-term objectives, and an understanding or interpretation of any isolated action can be misleading in the short run unless it is fitted into a much larger context. A major oil company could deliberately behave in apparent contradiction with its own immediate interests, but resulting losses would be balanced by the attainment of far-reaching objectives. The conciliating

TABLE 12.2

Domination of U.S. Domestic Oil Industry
by Majors in Late 1930s
(percentage of control)

Item	Number of Companies	Percent	Year
Total investment	20	60.0	1939
Producing wells	20	23.7	1937
Crude production	20	52.5	1937
Gathering pipelines (mileage)	20	57.4	1936
Long-distance pipelines (mileage)	14	89.0	1938
Investment in pipelines	15	77.7	1938
Dwt tankers	15	87.2	1938
Pipeline revenue	15	86.4	1938
Crude stocks to be refined	20	96.5	1937
Refining capacity	20	75.6	1938
Cracking capacity	20	85.2	1938
Distilled crude	20	82.6	1937
Production of refined products	20	83.8	1937
Gasoline stocks	20	90.0	1937
Gasoline pipelines (mileage)	16	96.1	1939
Gasoline domestic sales	18	80.0	1938

Source: Michel Laudrain, Le Prix du pétrole brut (The Price of Crude Oil) (Paris: Ed. Genin, 1958), p. 159.

attitude of major oil companies in the early 1960s in Libya, mainly in refraining from discounting prices for the purpose of tax calculation and their ready acceptance of the shift to posted prices as a basis for such calculations in 1965, is a significant example. The extra payments "granted" to the Libyan Government were accepted as a reasonable price for combating and eventually eliminating cunning independents.

Most of the incomprehensible attitudes of oil companies often result from conflicts between long-term strategies and short-term plans. In the chaotic situation that characterized the domestic oil industry in the United States throughout the 1920s, the necessity of survival pushed major oil companies to participate actively in the feverish competition that contributed largely to the deterioration of the market situation. Nevertheless, they were preoccupied, at the same time, in stabilizing the market and keeping prices continuously at acceptable levels. Therefore, they welcomed and encouraged federal attempts to regulate and prorate production and supplemented them with their own efforts and action. Their contribution in this respect was made through the American Petroleum Institute, which played a primary role in designing and shaping conservation policy and regulations in "collusion" with federal authorities in the early 1930s. Their successful achievement in this respect was so beneficial to major oil companies that defense of the conservation system was established and the resulting price stabilization became a major objective, shared equally by federal authorities and domestic independents for different reasons. It was felt that in no way should this common front be threatened by the invasion of low-cost foreign oil.[8]

The fruitful coexistence of federal strategic objectives and the majors' professional interests before World War II in the U.S. domestic market was a tactical move on both sides because their interests were complementary. On the other hand, their collusion in international action was a strong and durable one since their interests were quite similar. The intimate imbrication of persons and interests between federal authorities and international majors' management is much deeper and far beyond what the average observer could imagine. French readers would better visualize the intricate situation by reading Claude Julien's L'Empire Américain. Michel Tanzer has outlined some aspects of the deep-rooted collusion between the federal government and international oil companies outside the United States. Tanzer illustrates this collusion in journalistic style by quoting Washington reporter Jack Anderson in 1967:

> The State Department has often taken its policies right out of the executive suites of the oil companies. When Big Oil can't get what it wants in foreign countries, the State Department tries to get it for them. In many countries, the American Embassies function virtually as branch offices for the oil combine. . . . The State Department can be found almost always on the side of the "seven sisters,"* as the oil giants are known inside the industry.[9]

A number of high-ranking officials in key federal positions come from the industry, and official support is not limited to diplomatic efforts. The Central Intelligence Agency (CIA) and even the armed forces

---

*This is the same as the "eight sisters" previously mentioned but excludes the Compagnie Française des Pétroles.

contribute actively and openly in crisis situations, as outlined in Chapter 24 of Tanzer's remarkable book.

The early history of the international oil industry was largely shaped by the rivalry between Standard Oil and its powerful European competitors and by the policies adopted to mitigate this rivalry to the common interest.[10] Early rivalries took place in the Far East, which witnessed the aggressive birth of Royal Dutch under Henry Deterding, who soon brought the three largest foreign rivals—Standard Oil-Royal Dutch, Shell, and the Rothschilds—into a single marketing organization. Until shortly before the outbreak of World War I, the Dutch Government and Deterding blocked Standard's efforts to obtain concessions to explore for oil in the Dutch East Indies. In Burma and India, the British Government and British interests (Burma Oil Company) successfully opposed similar efforts. Just before World War I, Standard finally succeeded in obtaining a toehold in the Dutch East Indies through its Dutch marketing affiliate, but it was not until 1922 that it discovered oil in commercial quantities in South Sumatra, and not until 1928 that the Dutch Government finally gave way to the persistent attempts of the company, assisted by the U.S. State Department, to obtain additional concessions in the territory under its control. It should be emphasized that these episodes were characterized by merciless price wars, especially in 1910-11 and 1927-28.

The European counterattack was the invasion of the U.S. domestic market by Royal Dutch/Shell, which rapidly developed into a major U.S. company. The infiltration into the American continent was thus materialized and rapidly expanded into Mexico and Venezuela.

But the real battle concerned the control of huge reserves in Mesopotamia (now part of Iraq) and Iran. The British had long dominated Middle Eastern politics before World War I. The D'Arcy concession obtained in 1901 covered almost the whole Persian territory. Its exploitation was undertaken by the Anglo-Iranian Oil Company (AIOC), in which the British Government was a major shareholder.

In the Ottoman Empire, German, British, and Dutch interests were deeply rooted in Turkish politics in their effort to obtain concessions in the lands containing the ancient oil seepages of Mesopotamia. The Turkish Petroleum Company was formed in 1911 grouping AIOC; Royal Dutch/Shell, and German interests. After the war, the German interest was handed over to the French by the San Remo Agreement of 1920 whereby the British and French not only obtained mandates over Mesopotamia and Syria but attempted to ensure that oil rights throughout the area would be exclusively in the hands of British, Dutch, and French interests.

This period witnessed a secret, fierce diplomatic war between Europeans and Americans who formally rejected British claims over "reserved territories." They succeeded in breaking through the Iraq

Petroleum Company in 1922 under the active pressure of the State Department. This agreement included the famous Red Line Clause according to which the IPC partners undertook not to conduct independent operations in a large area comprising most of the territory of the old Ottoman Empire. This restriction on collective action actually enabled other American companies—not participating in the IPC—to get a foothold in the rich Gulf Area, where they rapidly made great discoveries: Standard Oil of California in Bahrein and Saudi Arabia and Gulf Oil in Kuwait. Later, the Red Line Clause was abandoned to enable Standard Oil of New Jersey and SOCONY Mobil to get into Aramco, and AIOC to get into Kuwait along with Gulf Oil.

From the early years of this century, crude and products' prices throughout the world were based on U.S. prices prevailing in the Gulf of Mexico, which was the major source for world supplies. But the genesis of later developments in pricing strategies and patterns occurred in the famous 1928 Achnacarry Agreement, which is assumed to be the origin of the international petroleum cartel.

In the early decades, international trade was mainly based on petroleum products, and the fierce competition between American and European companies was actually over control of world markets. Price wars were not uncommon and often resulted in big financial losses to all parties. After the dismantling of the old Standard Oil, Standard Oil Company of New York got most of the Asian assets and outlets and was determined to develop these remunerative markets firmly held by Royal Dutch/Shell and Burma Oil. On the other hand, Royal Dutch and the Rothschilds held the majority of the shares of the rich Russian Baku fields when they were taken over by the Bolsheviks. However, they did not renounce recovering their property and imposed a strict boycott on "their" Russian oil. Then, in 1926, SOCONY reached an agreement with the Soviet Union for cheap supplies of products in order to eliminate Royal Dutch/Shell from Asian markets. This episode ended with government intervention on both sides and cost Royal Dutch/Shell several million pounds.

Parallel to this bitter experience, the devastating price competition in the U.S. domestic market foretold gloomy prospects for the whole industry should war prices extend to the Middle East, where the existence of fabulous reserves was considered certain. To avoid such catastrophic waste to the detriment of all, Sir Henry Deterding of Royal Dutch/Shell invited Sir John Cadman of AIOC and Walter C. Teagle of Jersey (eldest sister of SOCONY) to a "party" at his Scottish castle at Achnacarry, where a mutual understanding was reached to ensure fruitful domination of world oil. The main tenet of the agreement was to maintain the status quo among the three companies upon which their further expansion would be regulated. Seven general principles were embodied in this so-called "as is" agreement. The first formalized the "as is" sharing of world markets and virtually eliminated competition by allocating quotas on the basis of their respective 1928 positions. Two others were intended to rationalize production and distribution by joining existing facilities and by coordinating new investments. Three others were aimed at setting up new structures for international trade that would ensure price stability. In particular, crude and products prices, at a given place, would be the same, irrespective of their origin or source.

The implementation of this "as is" agreement was followed by a supervisory management association including a representative of each of the three groups. This association was in charge of fixing crude and products' prices according to a mechanism that had been mutually agreed upon. The single-point pricing pattern was formally adopted with reference to prices prevailing at the Gulf of Mexico. The association had to determine the freight rates to be applied in periods of six months. Such a pricing system is a very effective device for ensuring not only that uniform prices are quoted by all sellers but also that low-cost producers cannot use their lower costs to expand their share of the market by reducing prices.[11]

World War II, during which the strategic importance of oil emerged emphatically as a sine qua non for world domination, pushed the U.S. and British governments to reiterate the "as is" agreement and to adapt it for sharing oil-world domination after the war. On British initiative, an "oil conference," from which the French were excluded, was held in Washington in July 1944.[12] The formal document signed by Edward Riley Stettinius for the U.S. Government and Anthony Eden for the British Government was a kind of a gentlemen's agreement emphasizing the direct involvement of governments in international oil business through their majors. The main principles of the Washington Conference provided that:

1. Petroleum should be marketed in sufficient quantities and treated equally throughout the world;
2. The development of international oil business in producing countries should be based on a sound economic foundation;
3. The United States and the United Kingdom, as well as other peaceful countries, should be guaranteed access to adequate crude oil supplies even in periods of crisis;
4. Territories not yet covered by concessions should be accessible to all (open-door policy);
5. Present concessions should be respected;
6. Production, refining, transportation, and distribution should not be subject to restrictions, especially by any government that might not agree with the stated principles.

The Washington Conference thus confirmed the prewar sharing of international oil by the Americans and the British and formally emphasized the support of both governments for their oil companies to resume their expansion and hegemony after the war. In fact, the the members of the conference attempted to create a common front to resist any attempt to threaten their overwhelming positions in both producing and consuming countries. The conference could not and did not prevent conflicts of interest and tricky underground confrontations between dollar and sterling interests within the framework of their common reserved domain. This confrontation had direct impact on the pricing patterns in the Eastern Hemisphere, as will be shown in the following chapter.

NOTES

1. To be appropriately prepared for the present approach to the domestic U.S. oil economics, one should read both M. G. de Chazeau and A. E. Kahn, Integration and Competition in the Petroleum Industry (New Haven, Conn.: Yale University Press, 1959); and Claude Julien, L'Empire Américain (Paris: Ed. Grasset, 1968).

2. Harvey O'Connor, The Empire of Oil (New York, Monthly Review Press, 1955), Chapter 4.

3. De Chazeau and Kahn, op. cit., p. 66.

4. Ibid., p. 123.

5. Ibid., p. 193.

6. P. H. Frankel, Structure of World Oil Industry, 2d Management Conference, Northwestern University, March 1966.

7. De Chazeau and Kahn, op. cit., p. 84

8. Michel Laudrain, Le Prix de pétrole brut (Paris: Ed. Genin, 1958), p. 172.

9. Micheal Tanzer, The Political Economy of International Oil and the Underdeveloped Countries, (Boston: Beacon Press, 1969), p. 49

10. For a brief survey, see E. T. Penrose, The Large International Firm in Developing Countries (London: George Allen and Unwin, 1968), Chapter 3.

11. Ibid., p. 180.

12. Pierre Fontaine, Le Pétrole du Moyen-Orient et les trusts (Paris: Les Septs Couleurs, 1960), p. 70.

CHAPTER

13

A COMPREHENSIVE EXPLANATION OF PRICING GENESIS

The preceding chapter strove to clarify and explain the deep roots of the pricing strategies and policies designed and implemented by major oil companies and to outline the historical background that characterized and motivated their genesis. Dynamic forces have been outlined, as has their mechanism of action designed to achieve long-term strategic objectives, equally shared or complemented by the U.S. Government and the majors.

Except for the specific limitations of the majors' interests and the restrictions on their freedom of action within the United States, the international oil business clearly appears to be almost under the full control of the majors. In particular, the fantastic expansion of the oil industry after World War II and the extensive development of huge reserves in the Middle East actually took place within a predetermined over-all framework conceived by the majors to achieve and consolidate their supremacy, to optimize their operations, and to maximize their profits in the long run.

The rule that governed the fundamental philosophy and thinking of major oil companies and set the limitations in their international operations outside the United States should be constantly kept in mind: What is good for America is good for oil companies, and vice-versa. American oil companies not only could implement any policy or strategy anywhere in the world (assuming that it was consistent with U.S. strategic interests as outlined in the preceding chapter) but could also, in so doing, rely on the full support and unrestricted cooperation of the U.S. Government and its different departments and agencies. The fact that two of the big majors were not American but British (and Dutch) would not greatly upset the picture, and this was so for two main reasons. First the majors have many common characteristics and interests, polished and conciliated in the Achnacarry Agreement, the spirit and motivations of which regained their natural momentum in the postwar period. The cake to be shared was much bigger than in 1928, and group solidarity was felt to be more vital than ever. Second, British and U.S. interests throughout the world were either quite similar or complementary, as emphasized in the Washington Conference and in the multiple contacts made during and after World War II aimed at sharing and dominating their respective zones of influence. However, this unified front could operate only in the face of outsiders threatening the empires established before the war. It had its own centrifugal forces that were very active from within; the British Empire was declining and could no longer fulfill its international responsibilities, and the United States was eager to prepare for the takeover, starting on economic grounds. The oil industry was directly affected by conflicts between the dollar and the pound sterling, which developed once again, although in a more cunning and sophisticated manner than during the early decades of the century.

Pricing patterns in the Eastern Hemisphere are actually an extension of the American pricing system, adapted to specific regional conditions to ensure that they parallel U.S. strategic interests. Therefore, they cannot be understood within the framework and conditions prevailing in the Eastern Hemisphere itself but rather must be viewed as a modified version of the U.S. domestic pricing system. Thus, the key to a proper understanding of pricing economics in the Eastern Hemisphere must begin with a genuine understanding of the American system and of the motivations and mechanism of its adaptation to specific realities and constraints outside the United States.

POLITICS AND ECONOMICS OF PRICING IN THE U.S. DOMESTIC MARKET

The fundamental philosophy of major oil companies, implicit in their size and integration, is a philosophy of power; their policies and strategies are designed to extend and consolidate their control and domination over world markets. Their plans are optimized over long periods of time, and their force is reflected in the relative length of their strategies' duration. The more powerful a company is, the more it can resist market fluctuation and pressure, limited in time and space and induced by changing circumstances or competitors. It can afford deliberately to incur substantial losses incompatible with sound economic and commercial practices if appraised within the conditions of the moment. In fact, such an episode would only be a temporary alteration of the long-term plan, intended to regain control of the situation by eliminating competitors with shorter-term plans who are less prepared to resist and withstand immediate market pressure.

The most efficient tool for achieving long-term strategies of power is the control of pricing mechanisms. Because of the integrated nature of the industry this can be achieved in principle, at any phase of operations from the well-head to the consumers. In fact, the integration of the industry under oligopolistic conditions means that there is supply of crude oils and demand for petroleum products and that all intermediary activities can be regarded as technical services that occasionally might be achieved by outsiders. The real power of the majors is to control price formation and levels at these two key sectors in one way or another. The large size and established market position of the international majors have given rise to the view that these companies possess and exercise the power to set prices with virtually complete arbitrariness, to the detriment of the consumers or host countries or both.[1]

The first key sector is direct access to the consumer market since the prices paid by consumers actually cover all expenses and consolidated profits incurred throughout the many phases of the industry. The vital importance of the control of market outlets is emphasized by Paul H. Frankel:

> The strength of the international companies lies in the degree of their integration. That is to say, they could have been replaced because technology is no longer any secret in any activities. They could have been replaced as producers. They could have lost their professionals. They could have been bypassed as refiners because, as you know, it takes two years to build a refinery, but it takes ten years to build up a market. . . . The real power that these oil companies have is the Power of Disposal. They are the ones, and the only ones, who can move the oil from the producing companies regularly in massive quantities and can pay for it. Nobody else can, and, indeed, if the international oil companies would not provide what I like to call this "international grid," somebody else would have to find a similar structure.[2]

A control mechanism based solely on products, pricing would embody the potential danger of "putting all your eggs in one basket." The system should also be balanced by tight control of the second strategic sector of the industry: pricing and the supply conditions of crude oil. The mechanisms by which effective control of the industry is achieved may vary as to time and location, but prices always remain the mechanism through which such systems operate. In the last century, J. D. Rockerfeller established the supremacy of Standard Oil through monopolistic control of pipeline transportation and extensive concentration of refining operations. After World War II, the old Standard Oil was replaced by a large and loose oligopolistic structure with extensive fragmentation of the industry among a multitude of independents. Control of the industry could not be achieved at the production stage because of the specific U.S. conditions explained in the preceding chapter. The majors' power derived from their domination over various sectors of the industry and their almost exclusive control of market outlets, as dramatized in Table 12.2. However, alternative means of control were still based on crude oil prices, since the latter were buyers' prices determined by the few big off-takers and not by the multitude of producers.

Thus, the prices of Crude oil and petroleum products are the two levers controlling the over-all economics of the industry, in harmony with the long-term plans of the majors. Their levels are not determined by conventional market forces and the laws of supply and demand but through specific mechanisms that will be discussed below. But before proceeding with this investigation, a major question should be clarified: Since crude and products, prices are the two alternative and complementary means for achieving the same strategic objectives, is there any direct relationship between them?

Extensive investigations have shown that in the United States before World War II, changes in products' prices unmistakably led to changes in crude prices, in both directions. The relationship becomes somewhat less clear after 1940, possibly because the fluctuations are obscured by wartime price controls (during the Korean War as well as World War II) and the sharp upward trend after 1945. Crude oil prices failed to catch up with the increase in products' prices in 1948, and both rose sharply in 1953, when government controls were withdrawn. After a slight decline, products' prices recovered in 1955-56, but the increase was not matched by crude oil until early 1957.[3]

According to de Chazeau and Kahn, generally,

> the time lag of crude price changes behind products price changes is greater on the down side than on the up side. In other words, the crude price follows refined products price increases much more promptly than it follows products price declines. . . . It is not intended [here] to suggest that oil price changes are <u>caused</u> by products price changes. It is the writer's belief that the same causes are involved in both . . . for, except in a superficial sense, there is only one set of supply-demand factors in the oil industry . . . namely, the supply of crude oil and the demand for refined products.[4]

Therefore, it can be concluded that, although crude and products' prices are somewhat interrelated, they are determined rather independently according to their specific market positions and supply conditions. The interrelation is often reflected, in published statistical data, by the spread of products over crude—that is, total realization less crude prices, or what is occasionally called refinery margin. The U.S. figures are currently reported in Platt's <u>Oil Price Handbook and Oilmanac</u>.

Let us examine the conditions and the mechanism of price formation for both crude oil and products. The demand for petroleum and its products, taken as a group, is quite unresponsive to changes in price in the short run. The dominant determinant of the volume of sales is the general level of industrial activity, the standard of living of the people, the level of employment and incomes, and the weather. Some major products have virtually no alternative substitutes; this is notably true for gasoline and lubricants.

The situation is slightly different for distillates and residual fuels because of the availability of adequate substitutes, namely coal, gas, and electricity. Demand for such products is more sensitive to price fluctuation, although to different degrees. Industrial equipment is generally designed to burn a variety of fuels; competitive forces are very active, and a shift in purchases may occur quickly with a change in prices. In domestic utilization, especially for home-heating, equipment is generally designed to use a specific type of fuel, and substitution is prohibitively costly compared to the fuel's price. This makes demand for domestic fuels relatively less elastic.

These different degrees of relative inelasticity of demand have played a major role in determining the relative structure of products' pricing patterns. Profit maximization results from the higher prices of the less elastic products; gasolines are the most expensive, followed by distillates and fuel oils. This is true not only for ex-refinery prices but also for local and governmental taxation as well.

An aggregate demand that fails to respond readily to price changes exposes the industry that supplies it to wide price fluctuations. A price rise, when supplies become scarce, is not held in check by a quick contraction in the amounts buyers offer to take; a price drop in periods of surplus production is not quickly cushioned by expansion of purchases. Thus, in the oil industry, instability is a permanent potential threat, which cannot be countered unless the short-term inelasticity of demand is balanced by a flexible supply for prompt adjustment to market fluctuation.

Until the federal government began to regulate crude oil production in the United States, the chaotic conditions of the industry outlined in the previous chapter prevented the possibility of a flexible adjustment of supply. The absolute freedom previously left to producers often generated dramatic fluctuation in crude oil prices because of the compelling necessity to produce as rapidly as possible whatever was discovered due to the pressure of heavily fixed costs generally financed by borrowed money to be reimbursed and due to the law of capture. The conservation and production prorating in the mid-1930s actually tailored production to demand and thus provided the necessary basis for crude price stability.

Thus, crude oil supply has gained a large degree of controlled elasticity at the production stage. Nevertheless, the aggregate demand for crude oil remains rather inelastic with respect to price fluctuation for both crude oil and products. The major reason is the limited flexibility of refining and the very high cost of investment-related items in refining costs. Refining margins are generally in a very vulnerable position because of the necessity of operating plants very nearly to their design capacity. High investment costs and construction delays are further difficulties that limit refining flexibility to match market fluctuations.

Generally speaking, the remarkable inelasticity of demand for petroleum products with respect to prices, reinforced by the severe conditions of profitability of refining facilities, would have deep repercussions on crude oil demand, which would be globally determined by total available refining capacity rather than by price fluctuation. At any given moment, the market would be characterized by a well-defined and highly inelastic demand for crude oil as opposed to supply conditions that are equally inelastic with respect to crude price fluctuation in the short run. Crude production, inelastic with respect to actual market conditions, depends mainly on over-all economic assessment of reservoir performances and prospects of additional reserves. That is to say, it depends on prevailing prices in the immediate past. On the other hand, the general level of refining capacity is directly determined by the conditions of products' markets, not crude markets. All these factors help create a natural internal conflict between supply and demand for crude oil. Therefore, it is obvious that permanent stability of prices cannot be achieved individually by major oil companies, in spite of their large size and high degree of integration, without profound coordination and harmonization of their strategies and long-term plans. The concept of an international petroleum cartel underlies these words. Price stability has been actually achieved as long as the concerted action of the majors was not altered by political pressure from governments (ECA action and pressure from European countries) or uncontrolled professional competition in the production of low-cost oil (independents).[5]

U.S. IMPACT ON PRICE FORMATION
IN THE EASTERN HEMISPHERE

The prewar period in the Eastern Hemisphere was characterized by fierce competition between the international majors over the control of production centers and promising areas in the Middle East and by high Western European dependence on Western Hemisphere sources for the largest part of its supplies of petroleum products and crude oil. The evolution of Middle Eastern concessions, with special reference to crude pricing, and of European markets, with special reference to products' prices, took place rather independently from each other under the conjugated impact of U.S. world strategies, and the concerted, and occasionally conflicting, actions of the majors.

We have already outlined the rivalries and competition between the international majors throughout the world. Early confrontation between Standard of New Jersey and Royal Dutch/Shell materialized through fierce price-cutting competition for the control of products' markets. The Achnacarry Agreement was a salutary one that was readily accepted by other international majors. The fundamental significance of this agreement is that the majors realized that their very existence was indissolubly related to their coexistence. Oligopoly emerged and was formally adopted as the normal structural basis of the industry. This agreement could not, and actually did not, eliminate different kinds of conflicts and rivalries between the "sisters" but firmly established the basic principle upon which the industry has developed and flourished ever since. The majors should refrain from any activity that could threaten price stability throughout the world, with the corollary consequence that profit maximization would result from keeping prices at the highest possible level.

In fact, the conflicts that developed between the majors in the Eastern Hemisphere were mainly concerned with the control of promising reserve-rich new areas in the Middle East. They were characterized by the direct political involvement of the U.S., British, and French governments to reinforce their positions in this highly strategic area.[6] However, it should be noted that such competition proved to be meaningless, even for the lucky winners. The huge reserves acquired after bitter and fierce battles had to be shared voluntarily in order to protect price stability. The first concession in Saudi Arabia was obtained in 1933 by Standard Oil of California, which was the first to reap profit from the Red Line Agreement between the IPC partners. Because SOCAL lacked adequate facilities to market its crude, in 1936 it turned half of its interests in Bahrein and Saudi Arabia over to Texaco, which in return gave the new half-owned affiliate, California Texas Oil (Caltex), its marketing facilities east of Suez and $3 million in cash and undertook to pay $18 million out of the oil produced in Saudi Arabia. The first commercial field was discovered in March 1938, but production facilities were not extensively developed until after the war. Now the IPC partners, then "prisoners" of the Red Line Agreement, were anxious about the prospects of price competition that could result from the marketing of Arabian production. Instead, it was deemed vital and profitable for all to join together to control production and to ensure market stability through a concerted mechanism of price determination and internal off-take arrangements. In 1948, Aramco was formed by letting Standard of New Jersey (30 percent) and SOCONY (10 percent) in with SOCAL and Texaco (30 percent each). The features of the off-take arrangements were quite effective in avoiding price competition between partners. The counterpart of this American breaking of the Red Line Agreement was the association of AIOC with Gulf in the control of Kuwait oil.

Fabulous crude reserves in Kuwait were developed during World War II, and production prospects were far beyond the marketing possibilities of AIOC and Gulf, coowners of the concession. In the same spirit, aiming to avoid competition and to preserve price stability, two long-term off-take agreements were concluded with two other crude-lacking majors, Standard of New Jersey and Royal Dutch/Shell.7 The off-take arrangements between the members of the Iranian consortium are designed on the same model to control production and to avoid price competition.

The complex pattern thus shaped not only matched the long-term strategies and objectives of the majors but was also in harmony with the second prerequisite condition for its effective worldwide application, that is, it was in harmony with strategic U.S. domestic interests, outlined in the previous chapter. The development of Eastern Hemisphere low-cost oil had to be intensified because of the shortage of domestic resources but without endangering the price stability of high-cost domestic crudes. Because of the absence of import restrictions in U.S. markets, the delivered prices of Eastern Hemisphere oil had to remain high enough to avoid the danger of their irresistible competition. This could be achieved by setting the Eastern Hemisphere prices initially at high levels, irrespective of the actual economic conditions of their production, and by maintaining exclusive control over production and market outlets. Here we find the real roots of the price system in the late 1940s in the Eastern Hemisphere. The pricing patterns that developed in this context were designed essentially to preserve and consolidate American interests at home and abroad in spite of European opposition and the conflict of interests with the declining British Empire. This casts a new light on the evolution of pricing patterns in the Eastern Hemisphere as investigated in Part II.

Nobody discussed or challenged the validity of the pricing patterns prevailing throughout the world before the war. The "Gulf-plus" system has operated as a fact of life since the final decades of the last century and was formally recognized by the majors in the Achnacarry Agreement as a common basis for crude and products' prices throughout the world.

It is currently admitted that the formal end of the Gulf-plus system was initiated by the complaint by the British Admiralty about the abusively high price of fuel supplied to the British Fleet during the war. This was the first political pressure exerted on oil companies to lower prices and resulted in the recognition of the Gulf area as a second price-basing point at parity with actual posted prices in the Gulf of Mexico (\$1.05/Bbl for 36° API). However, the British arguments should not be exclusively confined to military considerations since the British Admiralty was a shareholder in the Anglo-Iranian Oil Company. The British claim was made at the end of the war when military considerations were already fading. Moreover, most of the dealings complained about involved the AIOC with respect to supplies of crude oils and products from the Abadan refinery—whose largest stockholder was the British Government.[8] Nevertheless, the negotiations were conducted with Aramco concerning the pricing of Arabian crude oil, which largely contributed to the supply of the Bahrein refinery, which was very active during the war. The British move certainly aimed to reduce the burden of the military budget, but it might also be contemplated as a measure to undermine the prospects for expansion of "dollar oil" in the postwar period by reducing the potential profit margins of American newcomers faced with extensive investment requirements to develop newly discovered reserves, whereas production of (British) Iranian crude could expand rapidly at much lower costs.

The formal recognition of the Gulf area as a separate price-basing point permitted the birth of specific pricing patterns for Eastern Hemisphere crude oil. The "birthday present" was to align prices with those prevailing in the Gulf of Mexico at \$1.05/Bbl for 36° API crude oils. The unknown was the future evolution of these prices as an independent entity, especially as the Middle East was soon to become a world-leading source of supply. This evolution is "explained" in the literature by the westward shift of the equalization line between Eastern and Western Hemisphere sources.

The mechanism that predetermined this evolution, and the dynamic forces that governed it, have been outlined in the preceding analysis. The real problems began in the immediate post war period when the war controls over low-level prices in the U.S. domestic market were removed because of a sharp increase in demand and especially because of the shrinking of U.S. reserves evidenced by the fact that the United States became a net importer in 1948. Due to galloping inflation and a substantial increase in exploration and development costs in the United States, American prices were pushed upwards with the blessings of the federal government so as to create and maintain strong incentives to intensify exploration activities at home. Now, how would Middle Eastern prices react to this move in U.S. prices? The outcome would be a long-standing compromise between the following forces and interests:

● Major oil companies, whose collective interest is to maximize profits by maintaining prices at the highest level possible, which could be achieved without threatening market position because or mutually accepted joint control of production;
● The economics of crude production, which would have commanded low prices for the low-cost Eastern Hemisphere crude oil in an open market governed by the conventional laws of supply and demand; and
● The pressure of consuming countries to reduce prices to a minimum.

Producing countries were absent from this confrontation. Oil companies enjoyed the historically established situation of high-level prices (in parity with U.S. prices) and had to face irresistible political pressure from the governments concerned. In the first round, the challenger was the U.S. Government, which financed European supplies under the Marshall Plan through the ECA. It is clear that the majors could by no means resist such pressure and had no choice but to comply with it, especially since this pressure had its own built-in limitations: The delivered prices of Eastern Hemisphere crude oils could not step across the barrier of the East Coast; thus, the development of domestic sources was safeguarded. On the other hand, the benefits of such a compromising position were quite impressive—namely, unprecedented profits from developing lowest-cost crude from the world's largest reserves and their marketing in the fastest-expanding markets. Due to the direct intervention of ECA, Eastern Hemisphere prices were allowed to follow U.S. increases only partially, and the reference price for 36° Arabian light crude FOB Ras Tanura was set at $1.75/Bbl as against $2.75/Bbl in the Gulf of Mexico, with the equalization line located at New York in 1950.

The challenger in the second round throughout the 1950s was the European governments, which had relatively poor bargaining positions as compared not only to major oil companies but also the U.S. Government, which could not permit further relative price cuts as long as imports into U.S. domestic markets were free of control. However, the major oil companies were not inclined to comply with

European claims for further price reductions. Integration offered a series of strong inducements for the maintenance of crude oil prices at the highest level possible,[9] as long as the oligopolistic market structure remained under tight control. Furthermore, the abundance of low-cost oil in the Middle East and the availability of large tanker tonnage at relatively low cost were viewed as a potential danger to the stability of U.S. domestic prices. Should crude oil prices in the Middle East have been reduced under the pressure of market forces, Caribbean and U.S. domestic prices would almost certainly have been forced down. Domestic producers and their representatives in Congress could hardly have been expected to permit such a development to stand unchallenged. The very real alternative to higher prices in the Middle East was the enactment of import restrictions in the United States, which the international oil companies were most anxious to avoid.[10] Such restrictions, however, proved to be indispensable in the late 1950s under the combined effect of the increasing import needs of the United States, the temptation of the majors to market as much low-cost oil as possible from the Middle East in the lucrative American market, and the deterioration of the oligopolistic control of crude oil production due to the increasing availability of low-cost oil produced by independent producers with no integrated downstream outlets of their own. Oil from the Soviet Union equally participated in the price-cutting competition, which was particularly beneficial to the European countries, which were striving to induce substantial price reductions in the Middle East.

Under these conditions, a new assessment of the situation was deemed necessary. It became obvious that yielding to European pressure by lowering crude oil prices in the Eastern Hemisphere had become advisable in order to ensure the harmonious continuation of market control under new political and professional circumstances. This led to a new strategy aimed at maximizing profits through production maximization. Price reduction was conceived as an efficient way to encourage market development and expansion along with the consolidation of harmonious coexistence with the mounting nationalistic feelings in Western Europe.

This new philosophy could not be implemented without prior protection of U.S. domestic high-cost oil from the irresistible invasion of much lower-cost oil from the Eastern Hemisphere. This was achieved by imposing strict control regulations on imports to the United States in March 1959 by a mandatory quota system. Shortly thereafter, posted prices in the Eastern Hemisphere were cut twice, as already detailed in Chapter 6.

The new strategy completely ignored the growing nationalistic feelings in producing countries perhaps because it was assumed that they could be kept under tight control. The popular explosion in Arab countries after the 1956 conflagration in the Middle East was not considered serious, although it resulted in the interruption of oil flow across the Suez Canal and Syria. It was only with the creation of OPEC, symbolizing a high degree of political maturity and determination, that the majors were convinced of the emergence of a new potential danger that could undermine easy access to huge low-cost reserves. The mounting fury of producing countries had to be neutralized, especially since their behavior and reaction could be irrational and not easily foreseen. No one can assert that posted prices would have been cut further had OPEC not been created. Nevertheless, by allowing OPEC to achieve "price stabilization" in spite of the aggravation of real competition in the market from independent producers, the deterioration of products prices', the ever increasing pressure from consuming countries, and the active resistance of European national oil corporations, major oil companies succeeded in diverting feelings of frustration in producing countries into harmless routine work in the OPEC offices in Geneva and later Vienna.

INTERCARTEL CONFLICTS: DOLLAR OIL VERSUS STERLING OIL

Our analysis of price evolution throughout the 1950s has considered major oil companies as a homogeneous entity with individual companies having common interests governed by production control arrangements. This is almost true for American companies, whose interests and strategic objectives were in harmony with those of the U.S. Government. European majors, namely British Petroleum, Royal Dutch/Shell, and to a lesser extent Compagnie Française des Pétroles, faced the situation with a mitigated outlook. They indeed acted in solidarity with the American majors against any force that could threaten their long-established and concerted domination of world markets. Nevertheless, they were very active in opposing the supremacy of dollar oil within the common framework thus elaborated. Secret wars between European and American interests have actually never stopped; they have directly affected petroleum economics and pricing patterns in the Eastern Hemisphere, as will be outlined below.

CEP has never been a strong opponent of American strategies because of its relatively small size and its limited and poorly diversified crude oil sources (mainly from Iraq). These factors contribute to make CEP's strategies more sensitive to short-term fluctuations in market conditions. "Its optimum strategic plans extend over relatively shorter periods than other Major companies, thus weakening its position vis a vis its big sisters."[11] These conditions induced a specific CFP attitude within the IPC that contrasted with the strategies of other partners. It has always strived to push for higher rates of production from its exclusive major source of supply. It chose to maintain an understanding and conciliatory attitude toward Iraqi authorities, even during the dramatic conflicts with General Kassem before and after the promulgation of Law 80 expropriating 99.5 percent of the IPC concession in Iraq. Its cooperative policy was demonstrated in setting the new posting of Basrah crude oil FOB Khor el Amaya at its full value, thus refusing to align with other member companies in the retaliatory move, as already noted in Chapter 6.

British Petroleum (formerly AIOC) and Royal Dutch/Shell were tough opponents of comparable magnitude backed by a long-established empire protecting it from the assaults of Americans. The era of devastating direct price war, as in the Far East during the 1920s, had been abandoned in the Achnacarry Agreement, which recognized the disastrous impact on all parties of such competition. In fact, the conflict survived in a more discreet way; it regained momentum in the postwar period, when the British were striving to minimize the dollar element in the production cost of sterling crude and to expand market outlets for the latter as much as possible—that is to say, to engage in direct confrontation with dollar crude.

In the United Kingdom, the problem of the dollar element was rapidly overcome in spite of violent conflicts. British and American oil interests throughout the world were inextricably interlaced; competition, if ever it existed, was more effective between American companies themselves than between British and American Companies, and it was relatively easy for a country disposing of powerful companies to reach a mutually acceptable compromise with dollar companies when problems arose acutely in 1949.

Petroleum activities occasioned great needs in dollars for the United Kingdom. In countries belonging to the dollar zone, the expenses of British companies were important; investments, concession bonuses, royalties, and so on had to be paid in dollars; the same was necessary for the purchase of petroleum from American companies, for the freight of dollar tankers, and for the purchase of American drilling and refining equipment. The same was true, although to a more limited extent, regarding operations in countries belonging to the sterling zone itself. In the face of these great needs of dollars, British companies realized large dollar profits from their operations in the United States, Canada, and Latin America. The objectives of the governmental policy was to achieve balance, at least.

In 1949, British companies succeeded in improving their position vis-à-vis their dollar needs (then largely covered by the Marshall Plan—that is, by American taxpayers). Under the impulse of ECA, they largely increased their crude production. On the other hand, the British equipment manufacturing industry broke through the American monopoly, and its exports doubled between 1946 and 1950, and negotiations over royalty payments to Iran reduced the dollar component of the payments. At the same time, exchange control regulations limited American companies' allocations for crude imports to the United Kingdom. This move was the starting point of the direct confrontation.

Petroleum sales of American companies account for the major part of dollar expenses of the United Kingdom. They represented about half of the total dollar deficit of the sterling zone in 1949. However, British companies that agreed to a given international market-sharing with their American partners of the cartel (the Achnacarry Agreement in 1928 and the Washington Conference in 1944) have now developed enough sterling crude reserves to substitute for dollar crude so as to shift the latter out of sterling zone markets. The British Government took the initiative of imposing discriminatory measures on dollar crudes.

The British Government was aiming to substitute sterling crude for up to 42 percent of crude and fuel imports. British expenses for dollar crude were limited to 2 million tons. In Denmark, they were reduced by 30 percent. Furthermore, in 1949, it was announced that, from January 1950, British companies would be able to offer 4 million tons more to the sterling zone. Stopping dollar purchase of fuel and refined products was recommended. Several measures were taken to implement this policy. Exchange control authorities reduced allocations to American companies, and the latter could not accept non-freely-convertible currencies, as British companies could. The British Treasury refused to convert sterling credits of other countries to cover sales of American companies. Further measures were taken by signing direct bilateral agreements with some countries outside the sterling zone such as Egypt, Sudan, Brazil, and Argentina. Even France was granted privileged facilities for the purchase of sterling crude.

In the face of this spreading danger, two American companies, Caltex and Standard of New Jersey, started advocating some kind of over-all sterling policy for oil trade. Caltex proposed to sell its crude against sterling pounds, which would enable it to finance refinery expansion in Europe and the Middle East. The company envisaged marketing all its Aramco and Bahrein production in the sterling zone by retroceding all dollars from its sales to the sterling pool, which, in return, would provide it with the dollars necessary for the purchase of its shares in Aramco and Bahrein production. Similarly, Jersey envisaged selling its Venezuelan, Iraqi, and Arabian production at special rates, covered in dollars to the extent of 50 percent only, and the balance would enable them to finance expansion of refining capacity in the United Kingdom.

The British Government refused such an arrangement, fearing that extensive accumulation of pounds sterling in the hands of American companies would generate dangerous pressure on the convertibility parity of the currency. The use of sterling profits by American companies for further investments in the sterling zone was unacceptable because of the alienation of the industry and the necessity of transferring future dividends. Moreover, it was almost impossible to control the movement of merchandise. The British Government could not encourage the selling of sterling zone merchandise by American companies in the United States that would have been marketed there anyhow. Therefore, it suggested a program listing sales and purchases that could be done in this way.

American companies reacted vigorously to these objections and emphasized that the sterling issue was only a pretext to cover unfair competition to the detriment of American interests. The State Department adopted similar views and suggested that 30 percent of British crude imports be in dollar oil. A vehement press campaign was encouraged, and in June 1950 Texas Senator Tom Connally called on the government to address an ultimatum to the British Government demanding that it accept dollar oil or renounce the support of the ECA. Effectively, the latter suspended its financial assistance in expanding British refining industry for a time.

Finally an agreement was reached between the two governments to loosen disputed restrictions. This move was helped by the sterling devaluation, the surging demand in the United States and the "free world," the impact of the Korean War, and the necessity of collective Western solidarity in the face of the danger of Losing Iranian oil in a much-feared precedent-setting nationalization.[12]

Conscious of this peculiar atmosphere, one can better understand the real motivation for the price reduction of Kuwait crude alone by 15¢/Bbl, achieved in April 1954 by Bp with the participation of Gulf Oil. The British devoted great effort to securing maximum outlets for their sterling crude and undertook specific market commitments in that respect. Then suddenly they were deprived of their main source of supply by the nationalization of Iranian oil in 1951. Indeed, Western solidarity and the unified front of the majors collaborated closely and unanimously to combat the devil and to eliminate, once and for all, the seeds of revolt. Nationalization had to prove impossible and disastrous to the country itself so as to provide a severe lesson to any other nationalistic campaign that might develop in the area.

Nevertheless, temptation for American companies to retaliate legitimately against the aggressive British was irresistible. And they did not hesitate to take the best advantage of the situation. In Iran itself, their direct intervention was of decisive importance in the elimination of Dr. Mohammed Mossadegh.[13] The reward for this "solidarity" was the handing over of 40 percent of Iranian oil to American companies in the consortium agreement reached in September 1954.

More harmful to British interests throughout the world were the active attempts and maneuvering of American companies to take over established British markets, then facing serious difficulty in ensuring continuity of supplies. Although production was pushed up in other producing areas (mainly Kuwait, Iraq, and Saudi Arabia), incremental production was no longer wholly British owned as in Iran, and the British lost the initiative of production control in their aggressive strategy against dollar crude. The weakness of their commercial and political positions could not be overcome by the settlement of the Iranian crisis, prepared secretly in early 1953, since they would be indebted to the Americans for its successful outcome.

Should it be put in economic and monetary terms, the British oil situation in early 1953 would be similar to a recession period resulting from an acute imbalance between the possibilities of the national economy and international commitments of the nation. The conventional remedy to such a situation would be a carefully balanced devaluation, which would stimulate production and exports without threatening the economy with the danger of uncontrolled inflation. In the present case, sterling crude was the exchange currency, and its devaluation had to be achieved without undermining the over-all price structure of the industry.

Kuwait crude oil was the most suitable for such a strategic move. Although it was not wholly British-owned, the coowner, Gulf, could be an accommodating partner because of its long-term supply contract with Shell and because its own market outlets were less than its production capacity. Taking into consideration the huge Kuwaiti reserves (20 percent or more of total world reserves) and the very low cost of production, the temptation to increase market penetration through price reduction was strong enough to induce Gulf's adherence to the British move.

Over and above these political considerations, the choice of 31° Kuwait crude oil could be happily justified on technical and economic grounds because of its lower gravity and higher sulfur content. In fact, the argument is no more valid than if the expansion of Kuwait crude production have been conducted at lower rates in which case no surplus production of heavy fuel would generate in the market place.[14] But, the objective was, precisely, to push Kuwait production at the expense of such other alternative sources. This would result in surplus heavy products, which could require upgrading in order to fit into market demand patterns in Western Europe. This could be achieved by adding cracking facilities to refineries—that is to say, further processing expenses to be covered by price reduction. In other words, although the technical and economic arguments forwarded by AIOC (later BP) to justify the price cut in Kuwait could be accurate, their validity should be viewed within the political framework outlined above. Normal professional considerations could dispense with such price reduction, if one were to accept a moderate expansion of Kuwait production. This move developed later into a fundamental break in the over-all pricing pattern in the Gulf. In fact, American companies had no choice but to align, for the same reasons of market penetration, with the price cut of heavy crudes. The investment discontinuity penalty has been a basic fact of life ever since.

It is most difficult to visualize how such a fundamental move germinated in the thinking of the British. It is too easy to imagine afterwards the complex evolution of the situation during the Iranian nationalization and to sum it up in the analysis given above. In fact, it is highly doubtful that British strategists could have thought of such far-reaching strategic objectives in the feverish early days of nationalization. It is most probable also that the Americans were motivated by solidarity considerations and were preoccupied by filling the gap resulting from the Iranian defection. Allocations for incremental expansion in the Middle East could hardly be made outside the 1950 arrangement just reached for settling the dispute between sterling and dollar crudes. The defecting Iranian crude was sterling, and it had to be replaced to a great extent by another sterling crude. Since prospects for recovering Iranian production in the near future were very gloomy, extensive cracking facilities were installed in European refineries to accommodate an excess proportion of heavy Kuwait crude over the average demand pattern in Western Europe. Table 13.1 outlines the relative importance of cracking capacity in North Western Europe, where the demand for light products was no more than 30 percent. These extra investments created new facts of life that could not be abandoned once the situation became normal. Cracking facilities were there to be used, that is, to continue processing heavy crudes. The extra investments involved had to be repaid, that is, to be recovered by an appropriate increase of the price of the crude oil responsible.

However, effective price reduction was not implemented by the time the shift to Kuwait crude was at its height in 1951-52. It took place in 1953, when it became clear that Iranian crude was to be shared by the Americans, who were very active in reinforcing their penetration in European markets as well as east of Suez to the detriment of British political and economic interests. The far-reaching strategy outlined above crystallized and emerged through a normal process of evolution imposed by the political conditions and arrangements prevailing in the early 1950s.

TABLE 13.1

Refining and Cracking Capacity, July 1, 1956

Area	Number of Refineries	1 Topping (1,000 barrels per day)	2 Cracking	2 as Percentage of 1
United Kingdom	16	599.15	190.50	31.7
France	15	503.47	163.65	32.5
Italy	34	522.45	64.43	12.3
W. Germany	29	309.17	89.00	28.8
Netherlands	2	261.00	62.00	23.7
Belgium	8	135.00	29.70	22.0
Spain	3	80.70	3.50	4.3
Others	24	161.65	22.10	13.6
Total Western Europe	131	2,572.59	624.88	24.3
U.S.A.	318	8,720.00	5,100.00	58.5
Canada	43	644.85	341.87	53.0
W. Indies	2	650.00	487.00	74.9
Venezuela	11	519.19	78.00	15.0
Middle East	19	1,336.57	217.70	16.3
Total World	627	18,058.52	7,853.26	43.5

Source: Michel Laudrain, Le Prix du pétrole brut (The Price of Crude Oil) (Paris: Ed Genin, 1958), pp 68-69.

THE MAJORS' STRATEGIC OBJECTIVES AND PRODUCTS' PRICING PATTERNS IN WESTERN EUROPE

The structure of products' pricing patterns in Western Europe was extensively discussed and analyzed in the previous chapter, and the matter was thoroughly investigated in the Report prepared and published by the United Nations. To sum up, the situation that prevailed in Western Europe in the early 1950s can be characterized as follows: (1) ex-refinery products' prices were generally aligned with parity values of products' posted prices in major exporting centers (the Caribbean or the Gulf area) plus actual freight cost; and (2) posted prices were aligned with those prevailing in the Gulf of Mexico (lows of Platt's) without any substantial adjustment for geographical location or the specific pattern of production.

The price stability at each exporting center was remarkable, and fluctuations were generally limited except in crisis periods. It should be noted, however, that price differentiation started to materialize in the late 1950s subsequent to the enforcement of mandatory import quotas in the United States. Posted prices of petroleum products in both the Gulf area and the Caribbean remained roughly the same but at a lower level than those prevailing in the Gulf of Mexico (lows of Platt's) by some 10-20 percent. Further differentiation was introduced between posted prices of heavy fuel oil from the Caribbean and the Gulf area.

A preliminary analysis of products' pricing patterns prevailing in Western Europe in the early 1950s is highly confusing since the pricing mechanism appears to be at odds with the realities of the situation; its concept and consequences are rather irrational when checked against the conventional economic concepts. Some of the largest disparities in the system will be outlined below.

Before World War II, the largest part of international oil trade was in the form of petroleum products, mainly from the Gulf of Mexico and the Caribbean; and as with crude oil, it was quite normal that the Gulf of Mexico enjoyed a leading role in worldwide products' pricing. Although the Abadan refinery first came into operation in 1913 with an initial design capacity of 120,000 tons per year,[15] refining capacity in the Gulf Area actually developed in the postwar period along with the extensive expansion of crude production.

Total refining capacity in the early 1950s in the Gulf almost amounted to 1 million barrels per day, half of it being located at Abadan, which did not operate during the nationalization period. Nevertheless, most refinery production was for fuel bunkering of tankers loading in the Gulf, local consumption, and exports to markets located east of Suez. Very little was marketed in Western Europe, as indicated in Figure 12.1. Refining capacity was further increased in the Gulf, mainly in Kuwait and Bahrein, and new plants were added in Qatar and the Neutral Zone. Total capacity in 1968 reached about 1.5 million barrels per day. However, the export marketing pattern did not change, and European imports of petroleum products from the Gulf area have always been insignificant.

Under these conditions, products' pricing economics in the Eastern Hemisphere would appear to be marked by two apparent contradictions. The structure of posted prices in the Gulf area, which was almost identical to the one prevailing in the Gulf of Mexico, was not well-adapted to the production pattern made up of about 60 percent of heavy fuel oil, whereas the gasoline pool was dominant in the Gulf of Mexico. Furthermore, the Gulf area cannot be expected to play a major role in setting products' prices in Western Europe since it represents a very minor source of actual supply and a very poor alternative to other sources in crisis periods. These two factors also apply to the economics of Venezuelan products. However, products' pricing patterns in Western Europe as well as in producing areas were merely aligned with pricing patterns prevailing in the Gulf of Mexico. An attempt to explain this will be made below.

The structure of products' pricing patterns in the Gulf of Mexico reflects the specific characteristics of the American domestic market, the mechanism and dynamic forces of which were outlined earlier in this study. Competition in products' marketing is to a large extent decisive in the United States, and the majors do not have a monopoly on the drawing up of over-all pricing strategies, as they do abroad. The second significant factor is the overwhelming share of gasolines, which accounts for more than half of the U.S. market demand. By contrast, export refineries in producing areas are mainly designed to produce heavy fuel oil destined to be exported to the U.S. East Coast as far as Caribbean refineries are concerned, and to bunkering activities in the Gulf area. Working from a sound economic basis and applying the mechanism of supply and demand in the determination of products' pricing patterns would have caused more stress to be placed on the price of heavy fuel and would have reduced, in relative terms, the large gap with the price of white products displayed by the American pricing pattern. Enjoying the ability exclusively to determine prices, the majors merely applied American prices everywhere else, irrespective of specific local conditions.

There were two motives for this pricing strategy. The first was to ensure the stability and continuation of the American domestic price system by extending it on a worldwide basis so that no other precedent could be established. Otherwise, allowing different systems to prevail elsewhere, although they may appear harmless, would entail the potential danger of repercussions on the American market. In particular, should gasoline prices be relatively lower in other areas, the temptation would be great to import large quantities into the lucrative U.S. market and thus to threaten the stability of domestic prices. Second, the high degree of inelasticity in the demand for petroleum products with respect to prices largely limited the impact of supply and demand forces that would have shaped the products' pricing pattern differently in Western Europe. Furthermore, there was no significant opposition to the majors' pricing from European governments. The economic theory would suggest a relatively smaller difference between the prices of heavy fuel and gasoline. The price of heavy fuel oil, as set by the majors in Western Europe, happened to be quite low indeed, since it was cheaper than alternative sources of primary energy, mainly coal. The European Industrial Revolution was based on coal, which still accounted for some 74 percent of total primary energy consumption in 1950, as shown in Table 12.1. However, coal's relative share dropped to 53 percent in 1960 and to 29.3 percent in 1968, despite substantial government subsidies. It appears as if the low price set by the majors for heavy fuel oil was to ensure market penetration and control. This process was further reinforced be political considerations. Since European reconstruction was financed by American dollars, it would be advisable to invest in enlarging the outlets for fuel oil largely derived from dollar crude rather than in reinforcing and reviving local coal industries.

The European governments could only welcome and support such a pricing basis, which matched their basic economic policy and aimed to ensure the production of primary energy at lowest cost. Thus, the reference price of heavy fuel oil set by the majors in Western Europe was naturally accepted by the European governments as an established fact of life. A divorce between American and European interests could have developed in the case of white products, mainly gasoline, whose price could be expected to be lower than the values set by the majors. However, the latter contemplated maximizing their profits by keeping these prices as high as possible because the demand for gasoline was highly inelastic with respect to prices. In fact, the European governments not only did not offer any serious oppostion to this strategy but indeed relied on it to impose substantial taxes on the unlucky consumer.

NOTES

1. H. J. Frank, Crude Oil Prices in the Middle East: A Study in Oligopolistic Price Behavior (New York: Praeger Publishers, 1966), p. 125.

2. P. H. Frankel, Structure of World Oil Industry, 2d Management Conference, Northwestern University, March 1966.

3. M. G. de Chazeau and A. E. Kahn, Integration and Competition in the Petroleum Industry (New Haven, Conn.: Yale University Press, 1959), p. 414.

4. Ibid.

5. Ibid., pp. 64-74; and Michel Laudrain, Le Prix de pétrole brut (Paris: Ed. Genin, 1958), pp 206-228.

6. A brief description of some of the episodes of these open and secret wars between big powers through their big companies was offered in the previous chapter. The reader is referred to the specialized literature for further details (with special reference to

Benjamin Shwadran, The Middle East, Oil, and the Great Powers 1959 (New York: Council for Middle Eastern Affairs Press, 1959); and Z. Mikdashi, A Financial Analysis of Middle Eastern Oil Concession, 1901-65 (New York: Praeger Publishers, 1966).

7. Laudrain, op. cit., p. 117.
8. Frank, op. cit., p. 21.
9. Ibid.
10. Ibid, p. 70.
11. Laudrain, op. cit., p. 239.
12. Ibid, pp. 264-266.
13. The coup Fazlollah by Zahedi that forced Mossadegh out of office and restored the Shah to power is generally credited to the direct action of the CIA. See in particular, Claude Julien, L'Empire Américain (Paris: Ed. Grasset, 1968), p. 315; Michael Tanzer, The Political Economy of International Oil and the Underdeveloped Countries (Boston: Beacon Press, 1969), p. 325; and G. W. Stocking, Middle East Oil (Nashville, Tenn.: Vanderbilt University Press, 1970), p. 156.
14. This will be demonstrated in the model described in Chapter 14.
15. The technical information and data are derived from S. H. Longrigg, Oil in the Middle East (London: Oxford University Press, 1968).

CHAPTER

14

**GRAVITY DIFFERENTIAL
AND INTEGRATED ECONOMICS**

Crude oil prices in the Eastern Hemisphere are known to be related to each other in terms of a gravity differential of 2 cents per each full degree (°API). This value appears as an established historical fact with no clear significance or justification. Furthermore, it is not altered by price variations, which affect prices as a group without altering their internal structure. This permanence in time and space outlines gravity differential as the strategic component of the majors' pricing philosophy, whereas the actual level of prices reflects their tactical adaptation to market evolution. However, the concept of gravity differential is as confused and obscure as the majors' pricing philosophy and practices. Not only is its value of 2 cents per degree almost arbitrary, but the concept of retaining gravity as a differential measure of value seems doubtful since gravity is no more than an empirical measurement of a physical characteristic of the crude oil, which, consequently, has no value as such. The clarification of these questions would enable us to have a better understanding of the pricing issue itself.

The consideration of gravity differential can be confusing since this concept actually has two distinct significances. The conventional one, publicly stated and commonly utilized in the profession, concerns the adjustment of the posted price of each crude oil <u>individually</u>, should its actual gravity fluctuate by more than one full degree as compared to the announced quality. The second one outlined in this study concerns the basic pricing structure and relates posted prices of <u>different</u> crude oils with respect to their gravities (G), which may extend over a large range. The two concepts embodied in the same wording are then quite different. In order to avoid any possible confusion, the first will be called gravity fluctuation differential (GFD) and the second basic gravity differential (BGD).

GFD figures are commonly published in association with posted prices, whereas BGD significance and value have never been publicly announced. It so happens that both GFD and BGD have the same value—2¢/Bbl/ degree—in the Gulf and other producing areas in the Eastern Hemisphere, but this is not a general rule and an area's figures often can be quite different, as was demonstrated for Venezuelan crude oils and CIF pricing patterns. However, the two concepts of gravity differential reflect one basic idea: Lighter crude oils are credited with a higher value than heavier ones, in recognition of the higher value of products derived from them.

For the purpose of investigating pricing patterns, attention will be focused on basic gravity differential, which constitutes their backbone, whereas the gravity fluctuation differential has only marginal commercial significance. Relating gravity differentials to the structure of posted prices would mean recognizing higher valorization for products derived from higher-gravity crude oils. This cannot be achieved consistently unless two prerequisite conditions have been satisfactorily investigated: (1) to verify that higher-gravity crude oils actually yield larger quantities of light products congenial to market specifications; and (2) to appraise the actual market acceptance of such light products and to weigh their relative contribution to the over-all realization.

THE REAL SIGNIFICANCE OF GRAVITY

Gravity, as expressed in terms of API degrees, is empirically related to specific gravity (sp. gr.; 60° F) by the following formula:

$$G, °API = \frac{141.5}{sp. gr.} - 131.5$$

Thus, API degrees provides a measurement of the average gravity of a crude oil, with higher API degree values corresponding to lighter crude oils. However, this could be the consequence of one or both of the following: (1) the relative preponderance of lighter products in terms of percentage-of-yields pattern derived by distillation, or (2) the special chemical nature of crude oil with an exceptional predominance of paraffinic hydrocarbons, which makes the gravities of products derived therefrom relatively lighter than in the average case.

In the first case, gravity reflects the physical properties of crude oils, which are directly related to products' realization through the physical operation of distillation. This current situation is upset by specific chemical features of the second case, where the higher gravity of crude oils does not mean larger yields of lighter products as in the first case, but reflects the fact that the gravity of each product is relatively lighter than expected.

It is obvious that gravity cannot have the same economic value in both cases. Conventional economic valuation has been most concerned with the physical significance of gravity, and the first case is, by far, the dominant feature of the industry. Paraffinic crude oils were mostly unknown in the Middle East. They have been available for a long time in the Far East (Minas crude oil in Indonesia) and more recently in Africa (Libyan high-pour crude oils).

Therefore, before commencing an economic assessment of the basic gravity differentials of a set of crude oils, it is essential to see in which category each of them would fit. To achieve this accurately, it is necessary to work out the material and economic balances of processing schemes designed to enable each crude to meet market requirements at specifications. Such calculations are too complex and sophisticated to be of any practical use to the nonspecialist. Therefore, a simple, practical, and reasonably accurate criterion is needed to compare different crude oils readily and to determine qualitatively to which category they belong. A first indication is provided by comparing their true boiling point (TBP) curves plotted on the same figure. These curves are perfect indicators of the products' yields, which can be obtained from a given crude oil in a simple hydro-skimming refinery, irrespective of its chemical nature. This comparison, although qualitative and approximate, is quite satisfactory for initial classification. Furthermore, it is highly representative since it covers continuously the full range of distillation.

The comparison of TBP curves is quite easy even for the nonspecialist who may not even know the significance of the specific terminology of the oil industry. When the curves are plotted together, as in Figure 14.1, a curve lying farther to the right would yield relatively larger quantities of lighter products than one to its left, irrespective of the actual gravity values of the corresponding crude oils. Should two curves cut at one point or more, their relative positions would interchange from one set of products to another. Thus, a simple glance at TBP curves plotted together would make it possible to compare and classify corresponding crude oils.

TBP curves of Middle Eastern and North African crude oils are regular in form except for low boiling range cuts. They gradually and progressively shift to the left along with the lower gravities of corresponding crude oils. The situation is completely different for paraffinic crude oils. Libyan high-pour—35.5° Amna (Amal Nafoora), 36.5° Serir, and 44.0° Agip—crude oils perform similarly, irrespective of their different gravities. Their TBP curves lie to the left of 31.7° Kuwait crude, and thus they would yield comparatively less white

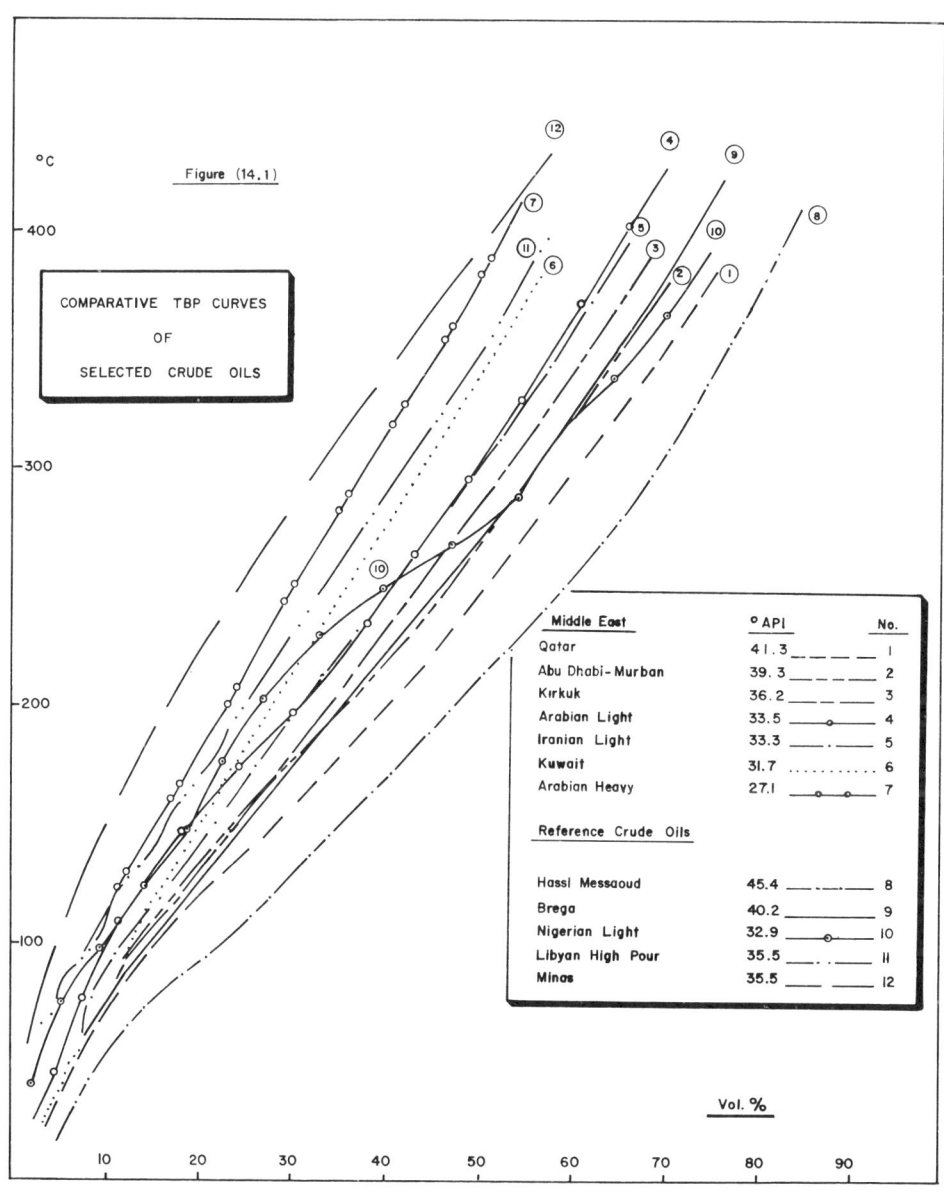

products and the over-all valorization of products derived therefrom would be less than what would be indicated by their actual gravities.

The TBP curve of 32.9° Nigerian light crude oil has a quite peculiar form with respect to Middle Eastern crude oils. It cuts through the different TBP curves, and its gas oil yield exceeds the expected value by some 11-12 percent at the expense of fuel oil yield. This results in higher realization per barrel and might be suspected to be partly responsible for its price premium.

A more accurate quantitative appraisal can be achieved by investigating possible correlations between crude gravity and products' yields derived therefrom. Petroleum products obtained by distillation are defined by their cut points, irrespective of the crude oil concerned. They are more or less close to the specifications of ultimate finished products, and they generally undergo further processing specific to each crude oil, the cost of which would affect differently its relative value.

There is a very large variety of finished products obtainable from a given crude oil. However, the refining operations an oil goes through after distillation can be grouped in four major categories: (1) gas processing C_5; (2) gasoline and naphtha processing (sweetening, re-forming, and so on), for which the average cut range will be taken at C_5 - 180° C; (3) distillates' processing, for which cut ranges will be 180-240° C for kerosene and 240-371° C for gas oil; and (4) atmospheric residuum at 371° C+.

TBP cut points, in volume percent, at C_5, 180, 240, and 371° C are given in Table 14.1, along with corresponding yields for the four bulk products as defined above. Products' yields are plotted against the gravity of corresponding crude oils in Figure 14.2.

Bearing in mind possible fluctuation in both gravities and yields, for a given crude oil, we see that Figure 14.2 shows that products' yields of Middle Eastern crude oils fit remarkably into a linear correlation with respect to gravity. This lends more meaning to the statement that they belong to the same family of comparable chemical nature for which gravity has the same economic significance. Similarly, Algerian and most Libyan crude oils seem to belong to a different family characterized by a distinct set of linear correlations.

Libyan pariffinic crude oils have specific correlations that are completely different from the two other categories. The difference is so great in both form and magnitude that gravity obviously does not have the same economic significance. The paraffinic nature of these crude oils is clearly illustrated by the much lower yields of lighter products. The relative excess of heavy products is counterbalanced, in terms of gravity, by the lower density of waxy cuts. From this point of view, Indonesian Minas crude oil seems much more paraffinic than Libyan crudes.

TABLE 14.1

Eastern Hemisphere Crude Oils

No.	Crude Oil	Gravity (°API)	K_{uop}*	C_5	°C TBP Cut Points (Volume percent)			Products' Yields (Volume percent)			
					180	240	371	C_5-180	180-240	240-371	371*
● Middle East											
1	Qatar	41.3	11.70	3.2	35.5	48.5	73.0	32.0	13.0	24.5	27.0
2	Murban	39.3	11.71	2.4	31.0	44.7	69.0	28.6	13.7	28.3	31.0
3	Kirkuk	36.2	11.61	2.1	30.2	41.6	65.0	28.1	11.4	23.4	35.0
4	Arabian lt.	35.5	11.70	1.6	26.0	39.0	61.2	24.4	13.0	21.8	38.8
5	Iranian lt.	35.3	11.58	2.6	27.5	38.7	62.0	24.9	11.2	23.3	38.0
6	Kuwait	31.7	11.63	3.1	24.8	34.4	54.2	21.7	9.6	19.8	45.8
7	Arabian h.	27.1	11.57	2.4	20.0	21.0	49.0	18.0	9.0	20.0	51.0
∆ Algeria											
8	H. Messaoud	45.4	11.70	4.9	42.6	56.8	79.5	37.7	14.2	22.7	20.5
9	Arzew	43.1	11.78	3.2	33.9	47.8	74.0	30.7	13.9	26.2	26.0
10	Zarzaitine	41.4	11.94	2.5	3.04	42.2	71.8	27.9	11.8	29.6	26.2
□ Libya											
11	Zustina	42.1	11.90	2.3	31.3	42.7	72.3	29.0	11.4	29.6	27.7
12	Brega	40.2	11.78	2.8	31.5	44.0	68.5	28.7	12.5	24.5	31.5
13	Es Sider	39.6	11.72	1.9	33.8	44.7	68.0	31.9	10.9	23.3	32.0
14	Sirtica	37.6	11.80	3.2	27.5	39.0	66.4	24.3	11.5	27.4	34.6
■ Paraffinic											
15	Agip†	44.0	12.37	0.2	19.7	30.0	56.7	19.5	10.3	26.7	43.3
16	Serir	36.5	12.07	1.3	22.2	31.3	55.2	20.9	9.1	23.9	44.8
17	High-pour	35.5	12.00	0.4	21.7	30.5	52.5	21.3	8.8	22.0	47.5
18	Minas‡	35.5	12.29	1.0	14.0	23.0	45.7	13.0	9.0	22.7	54.3
19	Nigerian Lt.	32.9	11.57	0.7	26.1	36.5	70.2	25.4	10.4	33.7	29.8

*Kuop, characterization factor considered as an indicator of the chemical nature of the crude oil.
†Agip field still under development with no production facilities for the moment.
‡Indonesian.

Source: Compiled by the author.

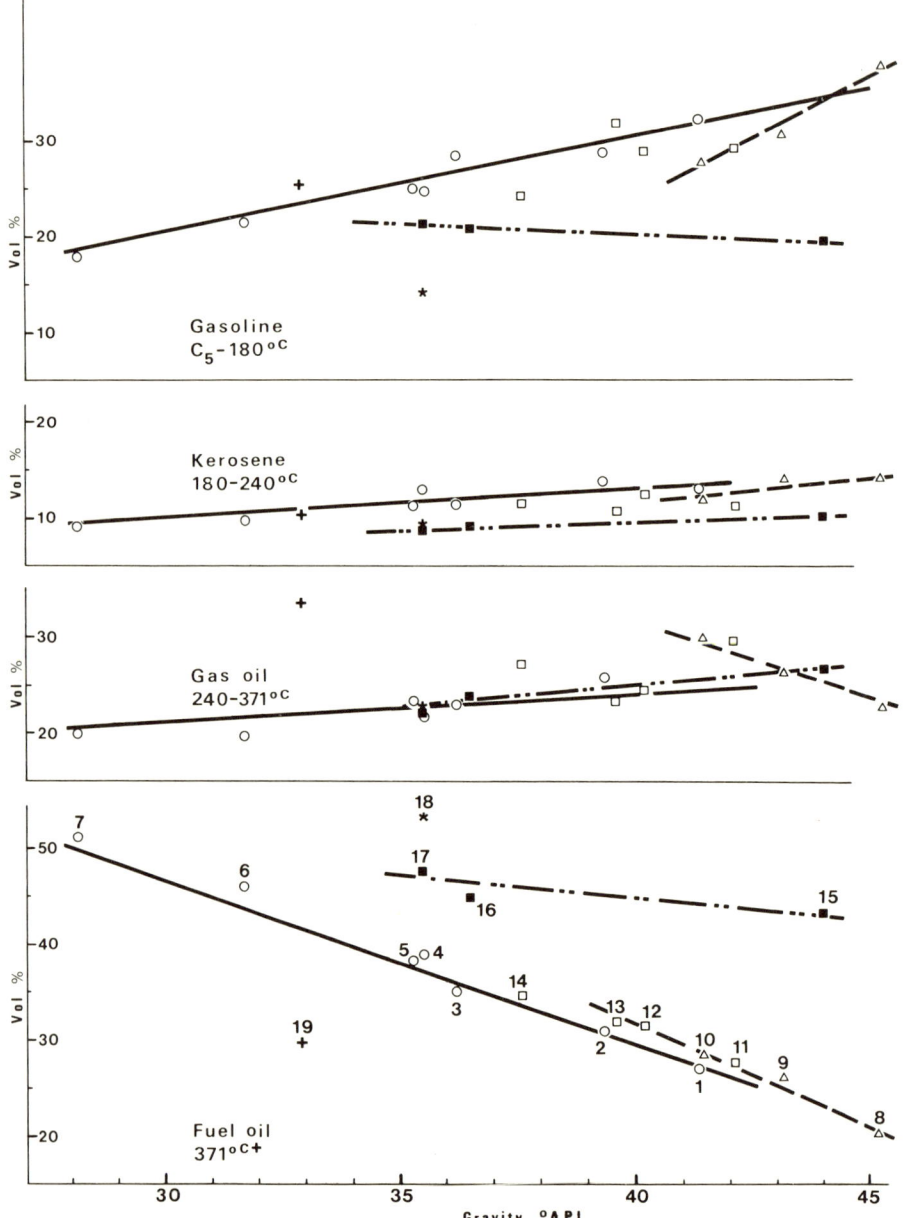

Figure (14.2)

TBP volume yields of major bulk products versus gravity of crude oils

(see Table (14-1) for key numbers)

The peculiar performance of Nigerian light crude oil is illustrated by the fact that it would fit within the Middle Eastern category, except for the fuel cut point, which shifts by some 11-12 percent in favor of gas oil.

To conclude, it might be stated that for a given family of crude oils of comparable chemical nature, yields of straight-run products are linear functions of crude gravity. For any given product (i), the yield (α_i) can be expressed as

$$\alpha_i = a_i G + b_i$$

Thus, each crude family can be characterized by a series of values (a_i, b_i) for the different products obtained by distillation. These values can be determined once and for all so that the yields that can be obtained from any specific crude oil would be automatically derived from its gravity without having to refer to its assay. It should be noted that the total yields of all products derived from any crude oil is equal to 100 percent. It follows that

$$\sum_i \alpha_i = \sum_i a_i G + b_i = 1$$

where

$$\sum_i a_i = 0 \text{ and } \sum_i b_i = 1$$

THE ECONOMICS OF INTEGRATED OIL

The linearity of both crude prices and products' yields with respect to gravity suggests that they might be related and that such a presumed relationship represents the mechanism of price formulation. This mechanism should reflect the process by which the cash flow generated at the consumers' end by the sales of petroleum products can be allocated to the different crude oils that contribute to produce them. A model simulating such integrated operations will be tentatively elaborated.

Now, let us assume that a given integrated major oil company has access to a large variety of crude oils with such large reserves and actual and potential production capacity as to meet all its market requirements in the long run. This is the case in the Gulf, where crude ownership is largely shared between the majors, which hold direct concession rights as well as long-term sales contracts. It is also the case in Venezuela, where each major has a large variety of crude oils extending over a wide range of gravity, and in the United

States, where the majors are producers or major buyers of most of the independent production. These crude oils are transferred from one affiliate to another to be transported, refined, distributed, and marketed to the ultimate consumers (practically valorized at the refinery gate) who pay a given price x_i(\$/Bbl) for each product (i). The total proceeds cover all expenses incurred in the different phases of the industry and provide the bulk of the consolidated profits that can be distributed, by an internal accounting system, to the different sectors of the integrated operations through the mechanism of "internal transfer prices." For our purposes x_i is assumed to be tax-free.

Each major oil company has to meet a given market demand both in terms of total volume and products' patterns. It disposes, at a given moment, of an adequate network of processing facilities designed to provide the necessary operational flexibility and to optimize both investment and operating costs. Market requirements are generally met by processing adequate blends of different crude oils, some of them being individually unadapted to demands when processed alone. Crude oils compensate each other with respect to yield and quality, so as to bring over-all total expenses to a minimum. That is to say, within the limits of compatibility, all the products from each crude oil are fully valorized even if they are occasionally in excess of the relative market demand for the product considered. Within the integrated framework of majors' operations, the latter have no reason to favor one crude oil over another on its own specific merits; each crude is actually valorized up to its relative contribution to the collective enterprise.

The basic principle of integrated pricing strategies was formulated in the Achnacarry Agreement, which marked the birth of the international petroleum cartel. Crude oil prices at a given place and moment would be the same, irrespective of the source or company of origin—that is to say, any pricing pattern would necessarily be a universal one. Furthermore, the principle embodies the concept of the net-back formula by which pricing patterns in different areas are related by the equalization of CIF prices of all crudes at the place of their consumption.

Another consequence of this principle is that our model need no longer be restricted to one major company or another but can apply to all of them collectively. Therefore, global market demand volumes and patterns in a given area (Western Europe, for example) can be safely considered since the majors have almost all the markets outside the United States under full control. This eliminates the difficult task of working out specific market share patterns for one major company or another, since none of them has a specific pattern of its own. There is only one universal pattern, collectively designed and implemented by the majors . . . and the others.

A significant pricing model should reflect the over-all performance of the majors' integrated operations by which market demand is met as closely as possible by refining adequate blends of crude oils. If we keep in mind that products' demand patterns are fixed at any given moment, it follows that the composition of crude supply blends depends primarily on the structure and the flexibility of existing refining facilities and on the availability of different grades of crude oils diversified enough to compensate for the shortcomings of each other. Consequently, crude supply patterns, and the lifting programs related to them, are not directly influenced by crude oil prices; they are controlled by the flexibility of existing logistics under prevailing price conditions. Should different supply patterns be possible, an optimum blend would be selected according to a given criterion of optimization such as minimizing over-all expenses or maximizing consolidated net profits. In other words, the pattern of crude oil prices is independent of the demand pattern for petroleum products, which, in turn, affects the composition of supply blends at prevailing prices. This conclusion is consistent with the fact that crude oil prices are stable in nature and universal in scope while market demand is continuously subject to fluctuation.

Crude and products' prices are the commanding factors of the over-all economic balance of the majors' integrated operations under the conditions indicated above. The mathematical formulation of our pricing model is detailed in the Notes at the end of this chapter.[1] The formulation reflects the fact that products' valorization at the refinery gate pays for the cost of crude oil supply and for the refining margin that comprises operating costs and profits. Keep in mind that

- crude oil prices are linearly related to gravity: $X = aG + b$;
- products' yields are linearly related to crude gravity: $\alpha_i = a_i G + b_i$;
- the valorization of product (i) results from the sale of (α_i) at a fixed price (x_i), so that total valorization is equally a linear function of crude gravity: $V = \sum_i x_i \alpha_i = \sum_i x_i a_i G + x_i b_i$;
- the refining margin in dollar per barrel of crude oil (r) has a more or less fixed average value independent of the crude gravity.[2]

Keeping also in mind that the combination of different linear relations with respect to gravity results in an equally linear relation, it follows that the pricing pattern of crude oils determined by the values of the basic gravity differential (a) and of the pricing factor (b) is related to the physical characteristics of the crude oil family defined by the values of (a_i, b_i), to products' prices (x_i), and to the refining margin (r) through the following equations:

$$a = \sum_i x_i a_i \qquad (\sum_i a_i = 0)$$

$$b = \sum_i x_i b_i - r \qquad (\sum_i b_i = 1)$$

These equations confirm our previous conclusion that crude oil pricing pattern does not depend on the demand pattern for petroleum products but solely on their prices. In other words, market demand patterns affect lifting programs of the majors in conjunction with existing refining facilities—that is to say, supply and demand are adjusted at prevailing prices. It should be noted that the crude oil prices considered above correspond, in fact, to CIF prices at the refinery gate. The assumption that CIF prices are governed by a linear pattern with respect to gravity is consistent with the findings of Chapter 7, where it was demonstrated that the freight cost does not alter the original linearity with respect to gravity, which characterizes the pattern of FOB prices.

THE SIGNIFICANCE OF GRAVITY DIFFERENTIAL

As indicated earlier, price variations that took place in the Middle East did not affect the over-all pattern governing them but resulted in collective moves of the same amplitude. That is to say, the value of the basic gravity differential (a) was not altered and the price moves were reflected in the variation of the value of the pricing factor (b). This permanance and continuity of the structure of crude pricing patterns contrast with the ever fluctuating factors and parameters making up petroleum economics. In particular, (a) as defined above results from the combination of two factors that are normally subject to fluctuation, that is, (a_i), which results from the statistical correlation between products' yields and crude oil gravity, and (x_i) products' prices. Now, how can these fluctuations and approximations be adapted to the necessity of keeping (a) constant under all conditions?

In fact, crude pricing patterns account for these fluctuations, although indirectly, by keeping the price constant over each full range of one degree gravity. This means that the economics of the model elaborated above embodies an intrinsic margin of indetermination, the magnitude of which is governed by appropriate relationships between the fluctuation of the different parameters governing the model. The mathematical calculations are presented in the Notes section of this chapter,[3] and the main findings will be summarized below.

Let us begin with the investigation of the impact of the fluctuation of products' yields ($\alpha_i = a_i G + b_i$) which, by the way, will serve to clarify the difference between the two concepts of basic gravity differential and gravity fluctuation differential.

Gravity fluctuation (ΔG) has no real meaning in the model developed above since the gravities of crude oils making up a given supply blend can be exactly measured, so that ($\Delta G = 0$). It would be more convenient to talk about gravity variation when considering different, and fixed, values of gravity (G) corresponding to different crude oils making up the blend. Gravity variation actually means fluctuation in the average gravity of the blend resulting from the fluctuation of its composition, whereas the gravity of each specific crude oil making up the blend is constant. Under these conditions, the pricing pattern relating the prices of each crude would be governed by the basic gravity differential (a).

Gravity fluctuation would have true significance when one is considering each crude oil individually, not at a given moment as a constituent of a blend intended to meet a given market demand in given economic and professional conditions but through its production life during which it can participate differently in various supply patterns. Gravity can actually fluctuate along with the evolution of reservoir performances, the relative contribution of different producing horizons, the modification of gathering and stabilization systems, and so forth. Under these conditions, it would be improper to talk about the economic performance of the model as such, and price "fluctuation," resulting from gravity fluctuation, should not be taken as an economic measure of the deformation thus caused in the model. Price "fluctuation" is measured in terms of the gravity fluctuation differential (a'), which can have a different value with respect to (a) since it derives from a quite different concept.

To be more specific, we will attempt to define the two concepts in more quantitative terms. Let us consider one of the correlation lines relating products' yields (α_i) to crude gravity (G) for a given family of crude oils, as illustrated in Figure 14.3=A. The equation

$$\alpha_i = a_i G + b_i$$

was developed as the best statistical correlation between (α_i) and (G) and does not represent any natural phenomenon as such. Its linearity has provided the basis for linearizing crude prices (X) with respect to gravity (G).

Let us consider a blend of crude oils eligible for the model. Each crude oil is defined by its effective gravity as measured physically (G_1, G_2, G_3, and so on), and products' yields derived therefrom (α_i) have fixed values given by its crude assay. The actual values of those yields do not fit exactly into the correlation line ($\alpha_i = a_i G + b_i$), which

Figure (14.3)

Illustration of the pricing model

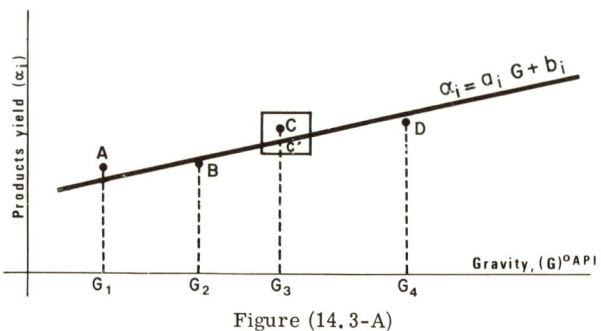

Figure (14.3-A)

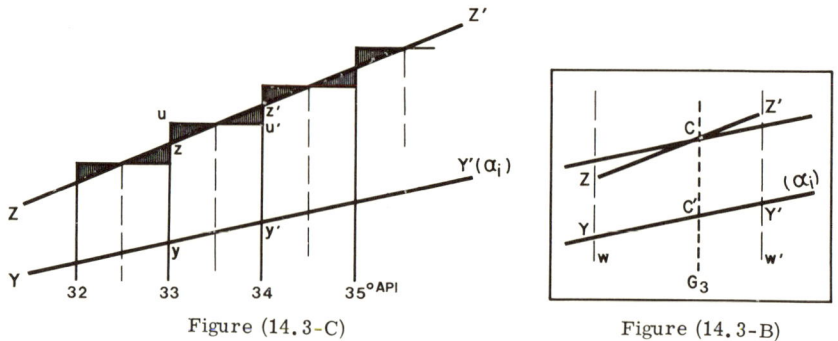

Figure (14.3-C)   Figure (14.3-B)

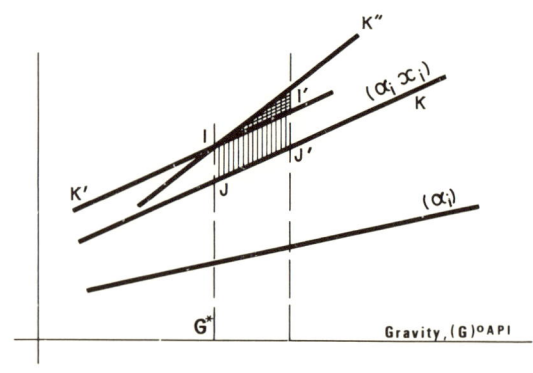

Figure (14.3D)

is only a statistical approximation. Considering a given product (i) from a given crude, for example (G_3 in Figure 14.3-B), the actual yield would be located at C, whereas the average value defined by the correlation would be at C'. When we consider the collective contribution of crude oils to meet market demand, the elaboration of the crude pricing pattern implicitly assumes that crude performances are governed by the correlation ($\alpha_i = a_i G + b_i$) for each fixed value of gravity. In other words, crude oils are represented by C' instead of C. The difference in the economic performance of each crude oil, induced by this approximation, should have been taken into consideration, as indicated below.

Now, if we consider each crude oil separately, its gravity could fluctuate with time, so that the effective yield of the product considered would be represented by the line ZZ', which is not necessarily parallel to the correlation line. If we consider a given range of gravity fluctuation ww', a variation in crude price (ΔX) would be induced by a variation in products' valorization, at the prices of the moment, resulting from a change in products' yield. In other words, price deviation is the economic measurement of the surface (ZZ' Y'Y) of Figure 14.3-B.

This price indetermination reflects the approximation embodied in the model, and the pricing pattern derived therefrom cannot determine crude prices with a degree of accuracy exceeding these limitations. However, it is inconvenient to talk about the deviation induced by the economic value of yield fluctuation, and it is most difficult to express this complex notion in simple quantitative terms, whereas the pricing pattern should be as simple as possible. In practice, price approximation is achieved by keeping the price constant over the full range of one degree gravity, which is the equivalent of taking into consideration the intrinsic indetermination of the whole approach. Limiting the interval to the full range of one degree is arbitrary, but it is highly practical and very easy to remember and apply. Figure 14.3-C shows that the surface (ZZ'Y'Y) can be subdivided into successive ranges of full degrees of gravity (32, 33, 34, for example). The surface of each range (zz', y'y') is equal the surface of an equivalent range (uu', y'y), (uu') being horizontal and cutting (zz') at its middle corresponding to the midvalue of gravity. Thus, the total continuous surface (ZZ', Y'Y) is replaced by the equivalent total of discontinuous surfaces (uu', y'y). In this manner, in each full range of one degree, product yield (α_i) is made constant (since uu' is horizontal)—that is to say, it corresponds to a constant value of gravity corresponding to the midvalue of the range in question. The same would apply to the following range and so on.

In other words, gravity fluctuation differential (a') accounts for individual price deviation related to the physical properties and performances of each crude oil as they compare with the properties and performances of the correlation representing the crude family. Price deviation is the economic measure of the deviation thus induced in the valorization of products derived therefrom.

The concept of gravity differential is so popular that very few realize that the expression "the price varies in terms of 2 cents per each degree of gravity" has no real meaning since gravity is only an empirical way of determining a physical characteristic of the crude oil and, consequently, has no value as such. On the contrary, gravity fluctuation might induce a modification in the valorization of the products derived therefrom, which is of real economic significance. Keeping prices constant over each full range of one degree gravity actually substitutes the abstract concept of gravity fluctuation differential (a') for an economic entity in such a way that they would simultaneously cancel each other out so as to make $\Delta X = 0$.

Now, if we consider once again the blend of crude oils, as represented in Figure 14.3-A, assumed to be eligible for the model, the elaboration of the pricing pattern assumes that products' yields (α_i) are actually located on the correlation line $(\alpha_i = a_i G + b_i)$. In other words, it assumes that the value of each crude oil is determined within a range of approximation defined by (a'), as indicated above. Under these conditions, the collective performance of crude oils $(\sum_i x_i \alpha_i)$ could vary as a consequence of the fluctuation of the products' prices (x_i). It can be easily demonstrated that if products' prices move collectively without altering their relative positions with respect to each other, the structure of crude oil prices would not be affected since (a) remains constant, but the result would be a collective move of crude oil prices by an appropriate change in the value of the pricing factor (b). On the contrary, if market evolution induces a modification in the prevailing structure of products' prices, then the basic gravity differential governing crude oil prices would be affected by an equivalent indetermination over and above the impact on the pricing factor. The stability of crude oil pricing patterns is governed by the stability of products' pricing patterns.

These findings clearly illustrate the intimate interrelationship between the stability of crude and products' pricing patterns, without prejudging the structure of the latter. This casts a new and revealing light on the majors' worldwide pricing philosophy and strategies by which products' pricing patterns throughout the world were uniformly elaborated; products' prices in Western Europe have been based on the parity values of posted prices in the Gulf and the Caribbean, which were identical to those prevailing in the Gulf of Mexico. In this manner, long-term strategic stability of crude pricing patterns was intimately related to the stability of the U.S. domestic market kept under close control by federal authorities and major oil companies. Necessary deviations and adjustments that could be induced by actual market fluctuation or evolution (in Western Europe) can be reflected through the pricing factor (b), without altering the basic structure of crude prices. In other words, basic gravity differential (a) can be taken as the strategic component of the majors' pricing philosophy, whereas the pricing factor (b) would be its tactical one.

Figure 14.3-D illustrates the mechanism of price formation, where the line (K) representing $(x_i o_i)$ is derived from the correlation line (o_i) for each product. Should products' prices move collectively without changing their structure, the impact on crude prices would affect only the pricing factor (b), and the corresponding line (K') would be parallel to (K). If the move would alter internal price distribution, the impact would affect both (a) and (b), and the corresponding line (K") would have a different slope. However, any variation in products' prices can be broken down into a collective move, which would induce a corresponding move of the pricing factor so that the crude pricing pattern would not be affected, and a relative deformation, which would induce a deformation in the structure of crude prices (Δa), so that crude prices would be determined with corresponding approximation.

The above would apply irrespective of the actual composition of the blend or its average gravity. Should gravity vary (and not fluctuate), then the indetermination of crude prices would be represented by the surface (II' J'J), for example, as indicated in Figure 14.3-D. As in the previous case, this notion is difficult to formulate in the simple, practical terms required by the pricing pattern. The difficulty can be overcome by subdividing the surface into intervals of one full degree gravity and by substituting the total continuous surface by the equivalent total of discontinuous surfaces so that in each interval, crude prices remain constant and correspond to midway gravity. Under these conditions the basic gravity differential (a) would add the collective indetermination resulting from the fluctuation of products' prices to the individual gravity fluctuation of each and all crude oils.

Further mathematical calculations[4] would serve to clarify these concepts in quantitative terms. It would be easy to demonstrate that the fluctuation of crude pricing patterns, induced by the fluctuation of the parameters governing the pricing model, is made up of two items: The first accounts for the fluctuation of products' yields, which is represented by the gravity fluctuation differential (a'), and the second accounts for the fluctuation of products' prices, which can similarly be represented by a market fluctuation differential (a"). Bearing in mind that a collective variation in products prices does not affect the basic gravity differential (a) but is accounted for by a corresponding variation in the pricing factor (b), it follows that (a") is related only to the indetermination induced by a deformation of the structure of products' prices—that is, to the degree of stability and control of products' pricing patterns by the majors, irrespective of actual price levels.

The mathematical formulation of these concepts shows that the basic gravity differential (a) governing crude pricing patterns is equal to the sum of the gravity fluctuation differential (a') and market fluctuation differential (a"),

$$a = a' + a''$$

In practice, only the value of (a') is publicly announced in the industry; the value of (a) can be calculated by statistical correlations, as mentioned in Chapter 6. Their difference would theoretically yield the value of (a"), the existence of which has not been mentioned before.

NOTES

1. (Q) Bbl, being the total market demand volume to be met, the over-all material balance would be expressed as

$$Q = \Sigma_i Q_i = \Sigma_j Q_j$$

(Q_i), Bbl = total volume of each product (i)
(Q_j), Bbl = total volume of each crude oil (j) making up the total supply.

This material balance assumes that refinery fuel, bunkering, and so on, are included in the total market demand pattern (Q_i). Furthermore, it assumes that all market needs are met by processing adequate blends of crude oils (Q_j) so that there is no need for importing refined products.

(Q_i) and (Q_j) are related to each other through the products' yield (α_{ij}), which was demonstrated to be a linear function of crude gravity (G_j):

$$Q_i = \Sigma_j \alpha_{ij} Q_j = a_i \Sigma_j Q_j G_j + b_i$$

or

$$\frac{Q_i}{Q} = a_i \Sigma_j \frac{Q_j}{Q} G_j + b_i \qquad (14.1)$$

Market demand pattern $(\frac{Q_i}{Q})$ is predetermined at a given moment; it commands the relative composition of the crude supply blend through

Equation 14.1, which is the prerequisite for the validity of the model. In fact, $\left(\dfrac{Q_j}{Q}\right)$ represents the relative share of each crude oil (G_j) in the total supply, so that $\Sigma \dfrac{Q_j}{Q} G_j = G^*$ can be taken as the mean average gravity (G^*) of the blend. This approximation is valid in the range of 20-45° API according to the same reasons outlined in Figure 7.1.

The over-all economic balance of integrated refining operations can be written as follows:

$$\sum_j X_j Q_j + R = \sum_i V_i \qquad (14.2)$$

Crude oil price (X_j), whether FOB export terminals or CIF at the refinery gate, is a linear function of gravity (G_j):

$$X_j = aG_j + b$$

The refinery margin (R), or spread of products over crude, comprises operating costs and profits. Its value per barrel ($\dfrac{R}{Q} = r$) is reported to be independent of crude gravity.

Products' valorization (V_i) results from the sale of products (Q_i) at ex-refinery prices (x_i), so that

$$V_i = x_i Q_i = x_i a_i \sum_j Q_j G_j + x_i b_i$$

Under these conditions, Equation 14.2 would be

$$a \sum_j \dfrac{Q_j}{Q} G_j + b + r = \sum_i X_i a_i \sum_j \dfrac{Q_j}{Q} G_j + \sum_i X_i b_i$$

The average price (X^*) of the blend (G^*) would be given by the following equation:

$$X^* = aG^* + b = \sum_i X_i a_i G^* + \sum_i X_i b_i - r \qquad (14.3)$$

Equation 14.3 should apply to all markets and for any and all values of (G^*) compatible with Equation 14.1. It follows then that

$$a = \sum_i x_i a_i$$

and

$$b = \sum_i x_i b_i - r$$

2. W. L. Nelson "Crude Gravity Little Affects Operating Cost," Oil and Gas Journal 61, 33 (1963): 100.

3. Let us consider a given blend of crude oils (Q_j, G_j) satisfactorily meeting market demand, that is to say, governed by the following equations:

$$\frac{Q_i}{Q} = a_i \sum_j \frac{Q_j}{Q} G_j + b_i \qquad (1)$$

or

$$\frac{Q_i}{Q} = a_i G^* + b_i = \alpha_i^* \qquad (2)$$

Crude oil prices are governed by the following equations:

$$X = aG + b \qquad (3)$$

where
$$a = \sum_i a_i x_i \qquad \sum_i a_i = 0 \qquad (4)$$
$$b = \sum_i b_i x_i - r \qquad \sum_i b_i = 1 \qquad (5)$$

Crude oil prices remain constant over the range of each full degree gravity. In the present calculations, G will represent a mid-degree value so that gravity fluctuation is represented by $\Delta G_j = \pm 0.5$, or $\Delta G_j \leq 1$ (in absolute terms). Each crude oil making up the blend is subject to fluctuation in terms of gravity (G_j), price (X_j), and products' yields (a_i, b_i, or o_{ij}). It is easy to demonstrate that the fluctuation affecting the different crude oils can be reflected in and represented by the fluctuations in the blend taken as an average crude oil and represented by (G^*, X^*, o_i^*, etc.). That is to say,

$$\Delta G^* \leq 1 \quad \text{if} \quad \Delta G_j \leq 1$$
$$\Delta X^* \leq \Delta X \quad \text{if} \quad \Delta X_j \leq \Delta X$$
$$\Delta o_i^* \leq \Delta o \quad \text{if} \quad \Delta o_{ij} \leq \Delta o$$

The crude pricing pattern is represented by

$$X^* = a G^* + b = \sum_i x_i a_i G^* + \sum_i x_i b_i - r \qquad (6)$$

which can also be written as follows:

$$X^* = aG^* + b = \sum_i x_i o_i^* - r \qquad (7)$$

Because of the integrated structure of the model, the refining margin (r) can be taken as constant under given economic conditions, that is, $\Delta r = 0$. It then follows that

$$\Delta X^* = a\, \Delta G^* + G^* \Delta a + \Delta b \tag{8}$$

or

$$\Delta X^* = \sum_i x_i \Delta o_i^* + \sum_j \Delta x_i o_i^* \tag{9}$$

or

$$\Delta X^* = \sum_i (x_i a_i \Delta G^* + x_i \Delta a_i G^* + \Delta x_i\, a_i\, G^*)$$

$$+ \sum_i (x_i \Delta b_i + \Delta x_i b_i) \tag{10}$$

All these equations are equivalent to each other, and we will attempt to clarify their significance and evaluate their magnitude in order to appraise the actual inaccuracy affecting price determination (ΔX^*).

The first term of Equation 9—that is, $\sum_i x_i \Delta_i^*$—actually accounts for the specific inaccuracy resulting from the approximate statistical correlation by which products' yields (o_{ij}) were linearized in terms of crude gravity (G_j). Random fluctuation of gravity (ΔG^*) can account for random fluctuation of the correlation (Δa_i, Δb_i), and the crude price is affected by an amount equal to the economic valorization of this deviation from the model ($\Delta X^* = \sum_i x_i \Delta o_i^*$), assuming that products' prices remain constant ($\Delta x_i = 0$). This term reflects the relative indetermination of the model with respect to the physical characteristics of crude oils. Other indetermination can be induced by market fluctuations, as will be noted below.

Let us consider a given market demand pattern (o_i), of (n) products (i = 1, 2 ... n). One of these (i = n) will be taken as a reference product (gasoline, fuel oil, and so on) so that the others would be represented by (i, 2 ... n - 1). The price structure of these products at a given moment, whatever it is, can be represented by the relative distribution ($x_i - x_n$) of the (n - 1) prices with respect to the reference price (x_n); it then follows that

$$\sum_i a_i = 0 \qquad a_n = -\sum_i^{n-1} a_i \tag{11}$$

$$\sum_i b_i = 1 \qquad b_n = 1 - \sum_i^{n-1} b_i \tag{12}$$

229

Now,
$$a = \sum_i x_i a_i = \sum_i^{n-1} x_i a_i + x_n a_n \quad (13)$$

or
$$a = \sum_i^{n-1} (x_i - x_n) a_i \quad (14)$$

and
$$\Delta a = \sum_i^{n-1} (x_i - x_n) \Delta a_i + \sum_i \Delta(x_i - x_n) a_i \quad (15)$$

Similarly, we have
$$b = \sum_i x_i b_i = \sum_i^{n-1} x_i b_i + x_n b_n$$

or
$$b = x_n + \sum_i^{n-1} (x_i - x_n) b_i \quad (16)$$

and
$$\Delta b = \Delta x_n + \sum_i^{n-1} (x_i - x_n) \Delta b_i + \sum_i^{n-1} \Delta(x_i - x_n) b_i \quad (17)$$

It is obvious from Equations 15 and 17 that if products' prices remain constant ($\Delta x_i = \Delta x_n = 0$), then variation in the pricing pattern ($\Delta a, \Delta b$) would be related solely to variations in the physical performances of crude oils ($\Delta a_i, \Delta b_i$). On the other hand, assuming that the physical characteristics of crude oils (G_j, α_{ij}) are perfectly known and stable, then the indetermination of crude prices (ΔX^*) would be due to market instability (Δx_i) as reflected in the model. Under these latter conditions we have

$$\Delta a = \sum_i^{n-1} \Delta(x_i - x_n) a_i \quad (\Delta a_i = 0) \quad (18)$$

$$\Delta b = \Delta x_n + \sum_i^{n-1} \Delta(x_i - x_n) b_i \quad (\Delta b_i = 0) \quad (19)$$

and price indetermination (ΔX^*), limited to the second term of Equation 9 would be expressed follows:

$$\Delta X^* = \sum_i^{n-1} \Delta(x_i - x_n) a_i G^* + \sum_i^{n-1} \Delta(x_i - x_n) b_i + \Delta x_n \quad (20)$$

If market fluctuation affects products' prices collectively without altering their specific structure $(x_i - x_n)$, then we have $\Delta(x_i - x_n) = 0$, that is to say, $\Delta a = 0$ and $\Delta b = \Delta x_n$. Such collective market evolution does not affect the structure of the crude pricing pattern since (a) remains constant but is reflected in the value of the pricing factor (b). On the other hand, if market evolution induces a deviation in the prevailing structure of products' prices, then the basic gravity differential would be affected by an equivalent indetermination given by Equation 18. The same deviation would affect the pricing factor (b) over and above the collective variation (Δx_n). The stability of crude pricing patterns is commanded by the stability of products' pricing patterns.

4. The collective performance of crude oils in the pricing model is represented by the mean average gravity of the blend (G^*). We thus have

$$G^* = \sum_j \frac{Q_j}{Q} G_j$$

$$\Delta G^* = \sum_j \frac{Q_j}{Q} \Delta G_j + \sum_j \Delta\left(\frac{Q_j}{Q}\right) G_j \tag{21}$$

In Equation 21, the first term represents the modification of the average gravity resulting from the gravity fluctuation of each crude oil taken individually, whereas the second term represents the modification of the average gravity resulting from the relative modification of the blend composition, with each crude oil keeping its effective and constant gravity. The first term will be called gravity fluctuation (ΔG^*_F), and the second will be called gravity variation (ΔG^*_V), so that

$$\Delta G^*_F = \sum_j \frac{Q_j}{Q} \Delta G_j \tag{22}$$

$$\Delta G^*_V = \sum_j \Delta\left(\frac{Q_j}{Q}\right) G_j \tag{23}$$

and
$$\Delta G^* = \Delta G^*_F + \Delta G^*_V \tag{24}$$

Now, combining Equations 10 and 24 we get

$$\Delta X^* = a' \; \Delta G_F^* + \Sigma_i \; x_i \; \Delta \alpha_i^*$$

$$+ \; a'' \; \Delta G_V^* + \Sigma_i \; \Delta x_i \; \alpha_i^* \qquad (25)$$

If we consider each crude oil individually for a determined blend composition and under given market conditions, that is to say (ΔG_V^* = o and Δx_i = o), then the price indetermination (ΔX^*) will be caused by gravity fluctuation (ΔG_F^*) and measured by the economic impact induced by the shift from the average statistical correlation ($\Delta \alpha_i^*$), or

$$\Delta X^* = a' \; \Delta G_F^* + \Sigma_i \; x_i \; \Delta \alpha_i^* \qquad (26)$$

The assumption that crude oil price remains constant over the full range of one degree gravity in spite of actual gravity fluctuation is equivalent to giving a nil value to (ΔX^*) over the same range; that is to say,

$$\Delta X^* = a' \; \Delta G_F^* + \Sigma_i \; x_i \; \Delta \alpha_i^* = o \qquad (27)$$

$$\Delta G_F^* = 1$$

As explained in the text, Equation 27 shows that, by maintaining the price constant over the full range of one degree of gravity fluctuation, the actual economic distortion ($\Sigma_i x_i \Delta \alpha_i^*$) is replaced by an abstract concept (gravity differential a'), which has an identical absolute value, or

$$\Delta X^* = o \text{ if } a' = - \Sigma_i \; x_i \; \Delta \alpha_i^* \qquad (28)$$

$$\Delta G_F^* = 1$$

Similarly, if we consider a given blend of crude oils, characterized by the average performances of the crude family, that is,

(ΔG_F^* = o and $\Delta \alpha_i^*$ = o), then we have

$$\Delta X^* = a'' \; \Delta G_F^* + \Sigma_i \; \Delta x_i \; \alpha_i^* \qquad (29)$$

As above, the assumption that the price remains constant over the full range of one degree of gravity variation (ΔX^* = o) means that the economic distortion ($\Sigma_i \Delta x_i \alpha_i^*$) induced by the fluctuation of products' prices is replaced by another abstract concept (gravity differential a"), which has an identical absolute value, or

$$\Delta X^* = o \text{ if } a" = - \sum_i \Delta x_i \alpha_i^* \qquad (30)$$

$$\Delta G_V^* = 1$$

(a') is the gravity fluctuation differential substituted, mainly for practical purposes, for the indetermination related to the deviation of crude physical performances ($\sum_i x_i \Delta \alpha_i^*$). Similarly, (a") could be called the market fluctuation differential since it substitutes the indetermination resulting from the deformation of market conditions ($\sum_i \Delta x_i \alpha_i^*$) for a practical notion related to a theoretical variation in gravity. Comparing Equation 25 with Equation 8, and bearing in mind that a collective variation in products' prices is accounted for by a corresponding variation in the pricing factor (Δb), it follows that (a") is related only to the indetermination induced by a deformation of the structure of products prices [$\Delta (x_i - x_n)$], that is to say, to the degree of stability and control by the majors of products' pricing patterns, irrespective of actual price levels.

Under these conditions, we would have the symbolic relation:

$$a \Delta G^* = a' \Delta G_F^* + a" \Delta G_V^* \qquad (31)$$

Substitution of the two above-mentioned aspects of price indetermination in terms of (a') and (a") was made in equal intervals of gravity indetermination ($\Delta G^* = \Delta G_F^* = \Delta G_V^* = 1°$ API); it then follows that

$$a = a' + a" \qquad (32)$$

The basic gravity differential (a) governing crude pricing patterns is equal to the sum of gravity fluctuation differential (a') and market fluctuation differential (a").

CHAPTER
15

**A TENTATIVE EXPLANATION
OF PRICING PATTERNS**

INTEGRATED ECONOMICS AT THE SERVICE OF
THE MAJORS

The pricing model elaborated in the preceding chapter provides a fairly realistic approach to the pricing philosophy of the majors before World War II in the United States and just after the war in the Eastern Hemisphere. The situation prevailing at that time could be quite easily accommodated to the apparent oversimplification of the model when compared with the extraordinary complexity of the international oil business of today. The key to an over-all and genuine understanding of today's pricing patterns and practices would be to begin with the model as representative of the early 1950s and then to try to conceive how the evolution of the industry could have affected the initial model and/or have been influenced by it.

All the above might appear exciting and satisfactory on intellectual grounds, but one can legitimately wonder how well these new concepts would agree with the actual facts and figures of the industry. In fact, checking the validity or accuracy of our approach, or of any other pricing pattern, is highly difficult, if not impossible, because the necessary information and data are kept secret and confidential by oil companies. The majors and the independents have always refrained from publishing any significant or comprehensive data. Furthermore, their contribution to public debate about pricing, if any, is generally intended to induce further confusion and complications. Among other things, their tactics could be designed to guide the curiosity of observers and critics of the industry to investigate the pricing issue on a partial scale both in time and space, that is to say, to concentrate only on the tactical aspects of pricing policies concerning the adaptation of long-term strategies to local short-term plans and to ignore the far-reaching political and strategic essence of the subject. The majors would be most delighted to have prominent figures of the international intelligentsia devote their efforts to, and focus world attention on, the relative values of crude oils with the intention of taking such an approach as a basis for setting or giving uniformity to crude pricing patterns. In the light of the above, and as already discussed in Chapter 11, the relative values of crude oils bear no direct relation to crude prices, and the two concepts are completely different in essence and scope.

This negative attitude of the oil companies was illustrated in 1970 in the discussion that followed P. H. Frankel and W. L. Newton's public conference concerning the "Comparative Evaluation of Crude Oils." Suffice it to quote one of the distinguished representatives of a major oil company: "I might suggest that representatives of oil companies are probably attending this meeting first for a rather selfish reason, i.e., to see whether an outside consultant of repute could provide us with a more sophisticated system of crude oil evaluation than we already employ. Sheer self-preservation requires us to attend this kind of meeting." Another representative partly unveiled the majors' philosophy, when he stated, "We should not forget that, even today, the countries from which some of the oil supplies are being drawn are chosen for political or politico-economic reasons rather than solely on the basis of strict commercial considerations."[1]

This attitude of oil companies is dictated by many factors. One of the most important is the frightful, and feared, specter of antitrust proceedings, which looms over any discussion about price formation in the United States, and its shadow extends wherever American companies operate. This explains why oil companies candidly invoke "competition" whenever they are forced to take a position on the pricing issue. On the other hand, the mechanism of price formation as elaborated above clearly shows that it reflects the real center of power and embodies the vital levers of command that make major oil companies what they really are, the unchallenged masters of world oil. The mystification they strive to maintain over the pricing issue is intended to achieve at least two prime objectives. First, to keep their supremacy intact and invulnerable by protecting their main source of power from the curiosity of outsiders and opponents and by diverting attention or action to harmless areas by astute camouflage. Secondly, to be able to implement any policy or plan of action intended to optimize their activities or to maximize their profits, in the short or long run, even though they could be in contradiction with the facts and figures of the industry or in opposition to the interests of the countries in which the companies operate. Thus, secrecy and diversionary measures became specific virtues of the oil industry whenever attention happened to focus on prices. The corollary of this philosophy was to enjoy exclusive and complete freedom in setting prices, an established privilege that oil companies have always vigorously defended.

As indicated earlier in this study, our approach to the pricing issue, and the model that enabled us to elaborate the crude pricing pattern within the framework of the integrated operations of the majors, should be considered only as providing the basic grounds for further thinking and refinements. The model has obvious limitations because of the necessity of simplifying and averaging so that the whole industry could be included in simple formulas. The reality is far more complex, and complexity has prodigiously intensified since the relatively simple situation of the late 1940s that provided the direct framework for our approach. Therefore, the model cannot be expected to explain in detail such and such an action or event in the recent history in the industry.

Apart from the fundamental difficulties outlined above, checking the validity for accuracy of the pricing model is hampered by a lack of adequate information and data. "Even inside the American industry, which has the most elaborate price reporting and statistical series of any oil business in the world, it is not too easy to get rational discussion of the way in which these prices are in fact formed."[2] On technical grounds, the limitation of the representativity of the model comes from the fact that the linearization of products' yields with respect to crude gravity applies to straight-run streams, which means that the refining facilities are assumed to be of the hydro-skimming refinery type.

Despite these shortcomings, the model, and the crude pricing pattern derived therefrom, might be taken as fairly representative for Eastern Hemisphere oil, especially in the late 1940s and early 1950s, when crude prices were shaped. The linearity of products' yields of the Gulf crude oils was demonstrated in Figure 14.2. The gravity of average demand in Western Europe was in the magnitude of 32-33° API and could easily be met by the variety of crude oils available in the early 1950s, ranging from 31° Kuwait to 40° Qatar. Under these conditions, the hydro-skimming-type refinery was well-adapted to market requirements, and the development of some cracking facilities was probably due to an oversupply of Kuwait crude oil after the nationalization of the oil industry in Iran. The preponderance of such a relatively heavy crude oil in the European supply pattern was largely responsible for the implementation of the investment discontinuity penalty, as already indicated in Chapter 13.

Now, let us assume that the crude pricing pattern in the Gulf area has been elaborated according to the model outlined in Chapter 14. The value of the basic gravity differential ($a = \sum_i a_i x_i$) can be calculated on the basis of the value of (a_i) as derived from Figure 14.2 and the value of (x_i) as indicated in the U.N. Report.[3] Such a numerical exercise would lead to a differential of a = 3.5 cents per degree API. This value is consistent with the findings of W. L. Nelson, who investigated the subject in an empirical way. In his remarkable book Petroleum Refinery Engineering, a paragraph[4] of Chapter 23 is devoted to the "realization of crude oils," where it is stated that "the proper differential per degree API is a function primarily of the relative prices of gasoline and residual fuel, and it may be estimated from Eq. (23.1) in which G, D and B are the prices respectively of gasoline, distillate and Bunker C fuel oil, and C is the operating cost, on a barrel basis:

$$\text{Differential} = \frac{2G + D - 3B - 3C}{150} \quad (23.1)"$$

It is obvious that in Nelson's thinking, gravity differential is related only to products' prices irrespective of market demand pattern. His calculations are illustrated in Figure 15.1, which shows that the basic gravity differential governing CIF values of Middle Eastern crude oils has an average value of 3.6 cents per degree.

The value computed by the application of the formula derived by our model (a = 3.5¢/degree) represents the basic gravity differential of CIF prices as governed by the economics of integrated operations. If we consider the impact of gravity on the transportation cost as investigated in Chapter 7 and if we assume that the value (a = 3.5¢/degree) applies to CIF prices in Northwestern Europe (Rotterdam), the corresponding differential governing FOB prices in the Gulf area would be in the range (a = 3.9-4.1¢/degree).

This substantial deviation from the established value of 2 cents per degree might be tentatively explained as follows. Up to October 1949, all crude oils in the Gulf area were quoted at flat prices equivalent to the price of 36° light Arabian FOB Ras Tanura, in spite of their different gravities. Later on, a specific pricing pattern was elaborated by displaying a basic gravity differential of 2¢/degree with respect to the reference gravity of 36°. Under these conditions and in comparison with the theoretical differential of about 4¢/degree, it is obvious that prices with gravities higher than 36° are underpriced and crude oils with gravities below 36° are overpriced. However, the actual and prospective demand patterns in Western Europe were such that the relative share of light crude oils (over 36°) was most likely very limited so that the majors would not suffer substantial losses because of their underpricing. On the contrary, most of the crude supplies were less than 36°, thus enabling the companies to recover a part of price reductions they were forced to accept under the pressure of the U.S. Government through the ECA. It is reasonable to expect that the companies would not have full freedom to benefit from

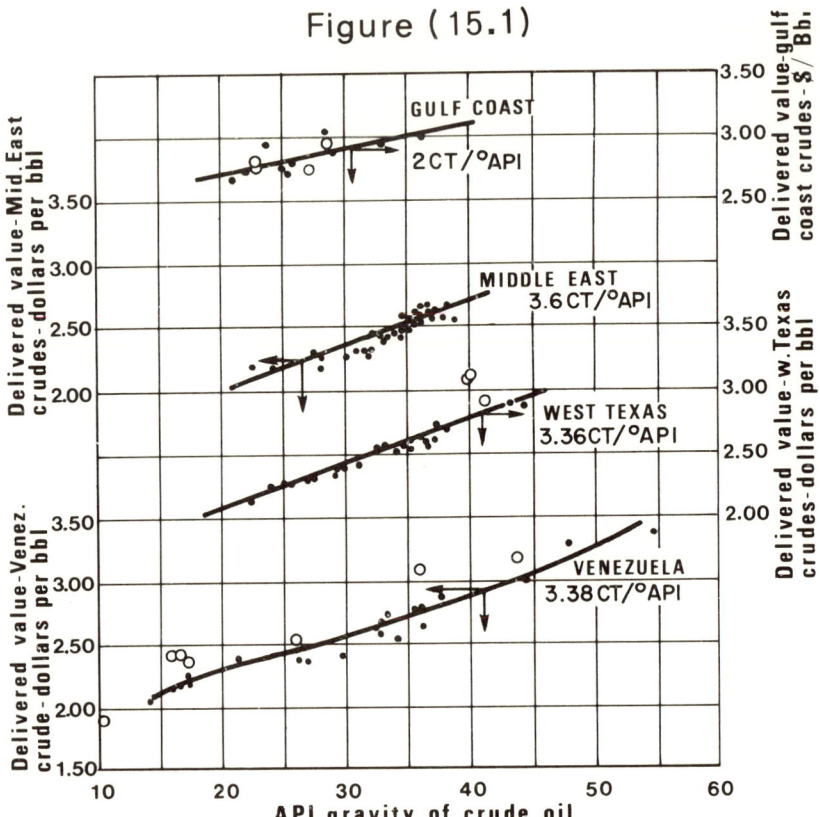

Crude-oil value vs. gravity for various types of crude oils, based on Gulf Coast cargo product prices of June 23, 1954. *(Oil Gas J.)*
SOURCE: *W. L. Nelson.*

such manipulation of the pricing structure should the mechanism of price formation within the framework of integrated economics have been publicly known and accepted as such by the industry.

PRICING OF LIBYAN HIGH-POUR AND NIGERIAN CRUDE OILS

The approach to crude oil pricing within the framework of integrated economics as elaborated in the preceding chapter might be applicable to Eastern Hemisphere oil in the early 1950s, when the structures of the industry were rather simple. It is obvious that such a simplified model cannot account for the ever increasing complexity characterizing the industry in the 1950s and 1960s. Apart from the profound changes in the political environment both in Western Europe and the Middle East, which resulted in the erosion of the tight oligopolistic control of the industry by the majors, the professional and technical conditions became more and more complex and sophisticated. Parallel to soaring demands in Western Europe and Japan, the number of refined products available in the market increased substantially and their specifications were more stringent. Average processing schemes became more complex because of increasing size and technological advances. The construction of extensive networks of pipelines in Western Europe and the development of inland refining created regional distortions in the market, which could not be accounted for by a simple economic model, even for the same oil company.

As for crude supplies, more and more varieties of crude oils were made available in increasing quantities from nonintegrated sources. Although volumes were rather marginal, they had a far-reaching political impact disproportionate to their relative importance. In fact, they provided a good pretext to exert and justify all kinds of pressure for price reduction. Such campaigns, which were very active in the 1960s, have actually helped make price reduction, both for crude and products, appear as a natural and inevitable process outside the United States.

Another disturbing factor related to crude supplies started to play an increasing role in the early 1960s. More and more new crude oils were discovered and brought into the market. Crude oils discovered in the Gulf fitted naturally into the prevailing pricing pattern, since they proved to belong to the same family (in terms of products' yields and qualities) and were located in the same area. Thus, they did not introduce any specific distortion in the pricing pattern as it stood, although the latter has not evolved to compensate for the new conditions of the industry.

The matter was completely different for crude oils discovered in new producing areas outside the Gulf. The diversification of geographical location of supply sources introduced major distortions into the pricing model, as noted in Chapter 7. Another distortion factor was the fact that new crudes were mainly of high gravity and fitted very poorly into the family of the Gulf crude oils. The first new family of crude oils to shift substantially from the Gulf pattern were Algerian oils brought into production in 1958 under French jurisdiction. On technical grounds, they occasioned the start of a small "revolution" in European supply patterns because of their high gravity ($40°$ and above) and because of their very low sulfur content. On economic and political grounds, they did not induce substantial modification in the prevailing pricing patterns because they were isolated from the majors' circuits and were rather captively disposed of within the framework of French interests and markets.

The first significant change in crude supply patterns in the Eastern Hemisphere took place in 1961 with the start of exports of Libyan 39° Zelten crude oil by Esso FOB Mersa el Brega at $2.21/Bbl. The grounds and circumstances of this posting were fully discussed and analyzed in Chapter 8 from technical, economic, and political points of view. The importance of this posting derives from the fact that it set a precedent and a reference point for further Libyan postings at a time when such crude oils had not been discovered or when they were not yet being produced. In other words, in considering the pricing model, one cannot consider, a priori, a Libyan family of crudes that did not yet exist in 1961, and Zelten crude had to be priced according to its individual merits as compared with the performances of the Gulf crude oils. Fortunately, the main characteristics of Zelten crude oil proved to be almost equivalent to the average performances of the Gulf crude oils (See Chapter 14), so that its price could be safely determined by the same mechanism.

Serious trouble began in 1966 with the pricing of 36° Libyan high-pour crude oils at $2.10/Bbl FOB Ras Lanuf, first by Mobil and Gelsenberg (Amal) and then by Amoseas (Nafoora). The Following year BP posted its 37° Serir crude oil at $2.10/Bbl FOB Mersa el Hariga (this price embodies a 2¢/Bbl penalty for the freight differential between Hariga terminal near Tobruk and other Libyan terminals in the Gulf of Sirtica). Another 44° high-pour crude oil was discovered by Agip, but it is not yet in production, and thus no price has been posted for it.

If we consider, a posteriori, the whole range of Libyan high-pour crude oils, they would appear to belong to a specific family of their own, as outlined in Figure 14.2. If we assume that they would fit into the over-all economic model, then the mechanism of price formation would set their pricing pattern with a basic gravity differential of about 0.5-0.6¢/Bbl/degree. These calculations are not given here in order to avoid any error of interpretation because the assumption formulated above is not valid. A more realistic approach to the pricing of Libyan high-pour crude oils would be based on the grounds presented below.

The problem was first raised by Mobil/Gelsenberg in 1965 with the development of the Amal field. This crude oil had to be priced individually and not as a representative of a specific family of crude oils, the existence of which could not even be suspected at that time. In other words, Amal crude oil had to be priced on its own merits as compared with the prevailing pricing pattern in Libya, which in turn, was aligned with the Gulf pricing patterns. From this point of view, the new price would be determined by the gravity of the crude evaluated in terms of a 2¢/Bbl/degree basic gravity differential. Compared with the Libyan "reference" price of $2.21/Bbl for 39° Brega crude, the posted price of 36° Amal crude should have been $2.15/Bbl unless further adjustments were to be introduced in terms of specific quality differentials.

Amal crude oil has a very low sulfur content, like all other Libyan crude oils, in marked contrast to the relatively high sulfur content of the Gulf crude oils. However, the advantage of this quality was denied to Libyan postings in general, and it was not taken into consideration in the pricing of Libyan high-pour crude oils in particular.

From technical and economic points of view, the handling and processing of these crude oils encounter specific problems that are generally deemed to induce price penalties. These problems will be enumerated below before we discuss their eventual impact on the mechanisms of price formation.

If we consider 36° Libyan high-pour individually, as indicated above, its performance, in terms of products' yields, would have to be compared with the Gulf and other Libyan crude oils. Under these conditions, Figure 14.2 obviously shows that it performs rather as a 30° API crude oil; its apparently high gravity is mainly due to its paraffinic nature, as demonstrated in Table 14.1. The relatively lower yields of lighter products are balanced by higher yields of residual fuel oil of high pour point. Yields of middle distillates are comparable to those obtained from the Gulf crude oils; they are generally of excellent quality (high diesel index) but are off specifications with respect to wax content and pour point.

As indicated by its name, the pour point of this type of crude is relatively high (+18° C or almost 65° F), occasioning special problems in storage, handling, and transportation. These problems can be easily overcome in Libyan weather conditions during production and exportation operations. They can become really serious in European weather conditions. They are generally overcome by blending with other low-pour crude oils or by continuous heating up to delivery to the refinery storage tanks. Problems encountered during refining operations can be overcome by adequate blending with streams from other crude oils, by changing the straight-run yields, so that products would be at specifications, or by adding further processing units (viscosity or pour breaking, and so on).

All these problems are summarized in Table 15.1, which compares the performance of Libyan high-pour crude oil with some other crude oils. However, their impact on prices can be questionable. Each of the shortcomings listed above could certainly lead to substantial extra expenses, which would affect the relative value of the crude. However, crude pricing is basically a matter of philosophy and strategy, and the impact of such shortcomings can be more or less effective

TABLE 15.1

Main Characteristics of Selected Crude Oils

Characteristic		Qatar	Brega	Murban	Kirkuk	Light Arabian	Lt. Nigerian	Libyan H.P.	Serir
Gravity	°API	41.3	40.2	39.3	36.2	34.5	33.5	35.5	36.5
Sulfur	Wt. percent	1.07	0.23	0.75	2.0	1.63	0.26	0.18	0.14
Pour point	°C	-16	+2	-16	-34	-26	+12	+18	+27
Gasoline C_5 - 100									
Yields	Wt./vol.	10.0/12.1	10.0/11.9	9.2/11.1	9.4/11.7	6.7/8.9	8.9/9.1	5.2/6.1	6.5/7.9
Mercaptans	Wt. percent	0.043	0.0082	0.007	0.02	0.015	—	0.0003	0.0002
RON	(R)	55	67	60	64	55	82	72	65
Naphtha 100-180° C									
Yields	Wt. vol.	17.6/19.2	15.7/16.8	15.7/17.5	14.7/16.2	14.0/15.5	11.8/13.1	11.7/13.8	11.3/12.5
Sulfur	Wt percent	0.04	0.028	0.12	0.10	0.04	0.08	0.0006	0.0017
RON	(R)	36	43	40	40	30	67		37
Kerosene 180-240									
Yield	Wt./vol.	13.5/13.4	12.0/12.4	13.5/13.7	11.0/11.6	12.0/13.0	13.5/13.9	9.2/9.3	9.0/9.6
Sulfur	Wt percent	0.09	0.055	0.06	0.30	0.24	0.04	0.008	0.008
Smoke pt.	mm	26	24	23	23	24	22	29	33
Freezing pt.	°C	-40	-48	-51	—	—	—	-25	-39
Diesel oil 240-371 C									
Yield	Wt./vol.	25.3/24.8	25.0/23.3	25.3/24.3	23.5/23.5	22.7/22.3	35.2/34.3	20.7/22.0	23.4/23.9
Sulfur	Wt percent	1.0	0.15	0.65	1.3	1.15	0.16	0.07	0.08
Pour point	°C	-9	-5	—	-7	-9	-7	+7	-1
Diesel index		61	63	60	61	56	50	73	68
Fuel Oil 371°									
Yield	Wt./vol.	31.2/27.3	35.5/32.8	34.7/31.0	40.0/34.8	43.3/38.7	31.0/28.8	52.3/47.5	48.8/44.8
Sulfur	Wt. percent	2.46	0.4	1.67	3.9	3.05	0.31	0.18	0.25
Pour point	°C	—	+35	+35	+29	+16	+42	+40	+46

Source: Compiled by the author.

depending on the selection of one or another approach to crude prices. To be more specific in the present context, it depends on whether the pricing of high-pour crude oils is to be considered within the over-all framework of the integrated international oil business or from the viewpoint of independent operators. The present example is most significant since the pricing of 36° Libyan high pour in 1966, which set a precedent for this kind of crude, was achieved jointly by two co-owners, the big integrated Mobil Oil and the German independent Gelsenberg.

Within the framework of the integrated operations that have actually governed pricing patterns in the Gulf, the impact would be rather limited. As indicated throughout the elaboration of the overall economic model and the pricing patterns derived therefrom, crude prices bear no direct relation to relative quantities of products' yields by different crude oils contributing collectively to meet a given market demand. Shortcomings and extra advantages compensate each other with respect to quantity and quality and command the relative share of each type of crude oil in the supply pattern. It is only the extra cost occasioned by the necessity of processing more of a given crude oil that should be charged as a penalty against the value of that crude oil. The average production of Libyan high-pour crude amounted to about 7 percent of total world production of Mobil Oil in the period 1967-69, and this limited percentage could have been easily balanced by the large variety of crude oils from other sources. Handling problems and their extra costs could have been largely absorbed by consolidated expenses of the group, since the eventual special facilities costs at discharging ports cannot be exclusively attributable to a given crude oil; they are generally included in operating and handling costs within the framework of operations optimization schemes. To support this view, the following item should be kept in mind. It is well-known that the mammoth tankers of 200,000 dwt and above, which have been growing at high rates in recent years, require special port facilities that are not always available. They have to be built at very high costs; the Bantray Bay facilities constructed by Gulf Oil Corporation is an example. Now, nobody has ever claimed that crude oils transported by super-tankers should be penalized because of their utilization of these facilities and that such penalty should be reflected in crude prices. On the other hand, the pour points of both Nigerian light crude oil and the fuel oil derived therefrom are of comparable magnitude, if not higher than the corresponding values for Libyan high pour, and yet no such penalty was introduced when the posted price of Nigerian crude was announced at almost the same time. It is noteworthy that this posting was made by BP and Shell, both large integrated major oil companies.

The situation would be completely different for independent Gelsenberg, which has a very limited ability to adjust its supply patterns so as to balance the specific problems related to the high-pour crude listed above. In fact, most of the production of this crude oil was destined for the German market, and Gelsenberg had to provide extra facilities for the handling and processing of this crude—which accounted for a large part of its supplies—in order to meet its market requirements. Under these conditions, penalties are equally attributable to poor qualities of crude oil and to the limited, nondiversified structure within which it is disposed of. A rough estimate of the economics of handling and processing 36° Libyan high-pour crude oil in the German market in the late 1960s is given in Table 15.2.

Under these conditions, the relative value of 36° Libyan high-pour crude to Gelsenberg would be about $1.94-1.99/Bbl, in comparison with the parity price of $2.15/Bbl FOB Ras Lanuf. The actual posting was set at $2.10/Bbl, and it is almost impossible to visualize how this figure was determined without the active contribution of the companies concerned. However, it is obvious that the impact of the big integrated Mobil was more telling than the impact of the smaller independent Gelsenberg.

The circumstances that governed the elaboration of posted prices in 1966-67 for Nigerian crude oils were detailed in Chapter 8. Although BP and Shell were obliged to align with Libyan postings, 34° Nigerian

TABLE 15.2

Additional Costs of Disposing of High-Pour
Crude Oils

Costs	Cents a barrel
Penalty for:	
additional cost of tanker transportation	1-2
additional cost for scheduling at dischange ports	3-4
additional cost for pipe-transportation in Europe	1-2
Total for handling and transportation	5-8
Penalty for higher refining cost and lower valorization	11-13

Source: Unpublished study undertaken by an oil company involved in the production of paraffinic oils.

light crude oil was overpriced by about 17¢/Bbl with respect to the average parity value based on and derived from the Gulf and Libyan postings. Furthermore, 27° Nigerian medium crude oil was not subject to the Investment Discontinuity penalty introduced by B. P. itself in 1953.

These spontaneous and unprovoked shifts from established patterns clearly emphasize the profound changes in the overall economic model governing international integrated operations and the evolution of the pricing philosophy and strategies of the majors. From technical and economic points of view, the relative value of Nigerian crude oils, with respect to their parity positions in prevailing pricing patterns, would be commanded by the balance of the penalty for high pour point, the premium for high products valorization, and the premium for low sulfur content

Higher products valorization results from the fact that Nigerian Light crude oil yields some 11-13 percent of higher value middle distillates over and above what could be expected from crude oil of the same gravity in the Gulf. The yields of lighter products are of the same magnitude and the balance is actually made to the detriment of residual fuel. This would enhance the relative value of the crude proportionally to the difference between prices of Diesel oils and fuel oils. Considering the average European prices for the period 1967-68 the impact on products valorization would show a net premium of about 15-18¢/Bbl for Nigerian Light crude oil. This extra valorization is of the same magnitude as the premium granted to Nigerian postings. Therefore, one might be tempted to explain and justify the overpricing in this way, and to admit, implicitly, that the penalty for high pour point and the premium for low sulfur content counterbalanced each other.

However, this hasty conclusion is not significant and does not necessarily reflect the deep motivations that pushed BP/Shell to retain the posted price of $2.17/Bbl for 34° Nigerian Light FOB Bonny. The high degree of integration of these two companies and their access to a very large variety of crude oils would have amply justified the integrated approach to crude pricing, that is to say, higher yields of distillates would balance the deficit of the same from other crude oils in an optimized supply pattern. In other terms, Nigerian crude would subsidize in the overall consolidated accountings of the group, the deficits of other crude oils so that the higher valorization of the crude oil taken individually would not be totally reflected in its price. Assuming that the contrary would mean that the low-sulfur premium (with a much higher value than the High Pour penalty) would have been denied to the same crude, that is to say, low sulfur Nigerian Light would subsidize higher sulfur crude oils making up the bulk of supplies from the Gulf at a time when sulfur pollution considerations became serious. The debate about sulfur pollution is essentially political as already outlined in chapter 9. Another political move was the elimination of the Investment Discontinuity Penalty which reflected the divorce from the conditions which prevailed in the early fifties and which were dominated by outmoded competition between dollar and sterling crudes. This move was the prelude to the general elimination of this penalty in the Gulf four years later.

    The pricing philosophy and strategies of the late sixties have become quite different from those which command the elaboration of crude pricing patterns which have since remained basically unchanged. The implicit contradictions of the international oil business have been emphasized and reinforced by profound changes in the forces present and the interests in conflict, and by a major crisis in 1967 symbolizing the far more dangerous Palestinian tragedy. The situation became potentially explosive, and the detonator was the Libyan revolution. It would be erroneous to investigate crude pricing in the late sixties now that the then prevailing pricing patterns have become a kind of historical anomaly. Pricing philosophy and strategies should instead be considered in the new and original framework which resulted from the recent international oil crisis and from the Teheran and Tripoli agreements.

NOTES

1. Journal of the Institute of Petroleum 56, 547 (January 1970): 9.

2. J. E. Hartshorn, Oil Companies and Governments, 2d ed. (London: Faber and Faber, 1967), p. 28.

3. Products' prices for Northwestern Europe in the early 1950s are given in United Nations, The Price of Oil in Western Europe (Geneva, 1955), Table 9.

4. W. L. Nelson, Petroleum Refinery Engineering, 4th ed. (New York: McGraw-Hill, 1958).

PART

IV

LOOKING INTO THE FUTURE

CHAPTER

16

CRISIS IN MOTION

Teheran, February 15, and Tripoli, March 20, 1971 will certainly appear to historians as two of the most important places and dates in the long and eventful recent history of the international oil business and as the starting point of a new era. Within a few months, a major crisis roared unexpectedly over the industry and rapidly developed into a direct confrontation between a cartel of producing countries and an enlarged cartel of 23 international oil companies. The atmosphere was dramatic in many respects, and for a while, a happy ending seemed highly doubtful since the battle was rather unequal and extravagant. On one side there was a solid front representing the all-powerful and invulnerable major oil companies, joined for the occasion by some independents and fully backed by the U.S. Government, which had suspended antitrust procedure regulations so as to allow oil companies to unite over the pricing issue. On the other side, there was an assembly of major oil producers, which had been vainly crying out for higher prices for 10 years through OPEC and which had failed to agree on effective procedures to regulate and/or to limit the relentless competitive rush for increased production, which had over the years developed an oversupply capacity that was exerting a continuous downward pressure on prices.

The outcome came rapidly with an unprecedented, total victory for the producing countries on all fronts. For the first time, they had played the leading role in the process of price formation, a strategic privilege that had always been denied to them by the oil companies. And for the first time, they enjoyed this new capacity and proved to be capable of masterful achievements by inducing a substantial price hike that would have appeared unbelievable only a few months earlier. The political victory was further reinforced by the fact that negotiations were completed within a few weeks and in accordance with the agenda set by the producing countries. It was the minimum time necessary to formulate and draft the agreements, and the oil companies seemed to have renounced any opposition to the producing countries' claims even before beginning the negotiations. The victory was in marked contrast to what had taken place in Teheran some 20 years earlier during the Mossadegh episode and to the cautious attitude of oil companies whenever they had previously been presented with modest claims for price readjustments by such countries as Iraq, Kuwait, and Libya.

In fact, this dramatic confrontation was the culminating episode of a major crisis born in late 1969 and in 1970. The winds of change started to blow on September 1, 1969 with the Libyan revolution, which replaced a monarchic regime with a young nationalistic government. The Libyan revolutionaries selected the oil issue as one of their main battlefronts and engaged in a bold solitary confrontation with the all-powerful companies. Their action and achievements acted as a catalyst in bringing about the subsequent upheavals, and it would not be exaggerating to say that the producing countries should be grateful to Libya for their new fortunes. One could fairly call 1970 the "Libyan Year," and the subsequent agreements cannot be fully understood and appreciated without a probing analysis of the apparent facts and their underlying motivations, which came into play during the genesis period.

Paralleling the Libyan events, other significant changes took place on different fronts and contributed, in one way or another, to a hastening of the process of change. The Algerians continued their hesitation game with the French while making a sustained drive to gain a foothold in U.S. domestic gas markets. The imbalance in American supplies developed into a real crisis for the first time on the East Coast, where the shortage of fuel oil and gas had to be met in distress conditions. This episode was in fact a symptom of the basic energy problems in the United States, which were outlined in a report submitted to President Richard M. Nixon by a special task force recommending the gradual abolition of the import quota system. The future prospects of the U.S. domestic oil industry had been highly confused and confusing, and their uncertainties indirectly affected international oil economics. The mini fuel shortage, enhanced by the enforcement of stringent regulations designed to reduce sulfur pollution, diverted large quantities of fuel oil from Western Europe to the lucrative American market. Prices went up for the first time in many years and contributed to a tightening of the tanker market, which could not withstand the discontinuation of oil flow from Saudi Arabia through the Tapline in the early days of May 1970 and the reduction of Libyan production, which in combination cut Mediterranean supplies by some 50 million tons. Rapidly, freight rates soared to all-time highs, a phenomenon that, in turn, supported the Libyan case for higher prices.

1970: THE LIBYAN YEAR

It should be recalled that successive Libyan governments have never been satisfied with the level of Libyan postings since the first postings by Esso in 1961. Official protests were regularly presented on the occasion of each new posting, and the conflict became more complex with specific cuts in the posted prices of Libyan high-pour paraffinic crude oils. However, the governmental action under the monarchy never attempted to challenge seriously the companies' negative attitude towards pricing claims, and the occasional debates were stopped at the legal barrier of endless and helpless wrangling about, on the one hand, the exclusive privilege of oil companies to set prices as stipulated by Regulation No. 6 and, on the other, the alleged contradiction between that regulation and the provisions of the Petroleum Law. It was obvious that the old regime was not prepared to initiate or enter into any direct conflict with the oil companies over such thorny and complex issues.

The situation changed radically with the Libyan revolution of September 1, 1969, and it was clear that oil would be one of the immediate targets of the young nationalists then in command. The unduly low level of posted prices was regarded as proof of the collusion of the former regime with "Western Imperialism and big monopolistic companies which had been long allowed to exploit natural resources at the expense of national interests."[1] And, in fact, technical and economic studies of Libyan prices were undertaken by specialized departments of the Ministry of Petroleum in the early days of October. An official Price Commission was created by a ministerial decree of December 22, 1969, to review these studies, to make recommendations about "fair and just" level of Libyan crude prices, and to conduct negotiations with oil companies in order to attain these objectives. The commission set actively to work and completed its investigations within a few weeks and official negotiations were solemnly launched on January 29, 1970 by Colonel Muammar el-Qaddafi, president of Libya's Revolutionary Command Council and prime minister. Addressing the opening session of the talks, he set the tone of the Libyan tactics by calling upon the assembled representatives of 21 multinational oil-producing companies in Libya to accept his government's demand "gracefully." While expressing his confidence in the good faith of both sides in the forthcoming price talks, Qaddafi nevertheless struck a note of warning by declaring that "the Libyan people have lived without oil for five thousand years, and they can live without it again if their legitimate rights are not attained."

The Libyan approach to crude prices was based on professional considerations and arguments; emotional and political considerations were only stimulating accessories that set the general atmosphere and limitations of the over-all framework for the negotiations. It was sincerely hoped that oil companies would be responsive and cooperative, unlike their past negative attitude. And, indeed, the Libyans intended to undertake the negotiations on professional grounds. Their claims were stated on the following grounds, without any commitment to specific figures:

- Libyan crude oils are fully eligible for a gravity differential of 2¢/Bbl/degree (paraffinic crudes might be an exception);
- Libyan crude oils have been denied a freight advantage of about 8-10¢/Bbl when compared to Eastern Mediterranean postings;
- Libyan oil should be entitled to a Suez premium greater than what has been indirectly involved in the temporary elimination of allowances related to the royalty-expensing agreement;
- A sulfur premium for Libyan crude oils should be recognized in two respects: (1) savings in refinery maintenance costs, and (2) savings in sulfur pollution abatement;
- A price increase should be implemented retroactively.

Working sessions were held with all companies operating in Libya; these meetings were requested to present proposals for price increases of their own. However, the companies' attitude was invariably as negative as in the past. They declared that, in their opinion, the level of Libyan postings was fair and satisfactory, and that there was no valid justification to warrant a price increase. The only positive aspect of their attitude was their acceptance of debate, in principle, without questioning the legal support and framework of governmental demands.

Having thus demonstrated vainly their good will, the Libyans became convinced that price negotiations could not be conducted on rational professional grounds. The negative attitude of the companies in their unanimous refusal to recognize the possibility of any price increase or to appraise the strong nationalistic feelings of the young Libyan revolutionaries helped revive and stir up profound feelings of injustice and frustration. The failure of this first technocratic round clearly demonstrated the political and strategic nature of the forces commanding the level of crude prices, and Libyan officials were left with the impression that any satisfactory outcome of the confrontation could be achieved only if it was imposed on the oil companies.

Therefore, the technocratic price commission appointed by the ministry was replaced on April 4, by a new negotiating team headed by former Prime Minister Mahmoud Mughrabi—a hard-line nationalist—who had to report directly to the Revolutionary Command Council. Further technical investigations and negotiations were undertaken, but it became clear that as high a price increase as possible should be induced, or imposed, without regard to its justification, a priori, on technical or economic grounds. Its magnitude would actually depend on the balance of bargaining positions, and once a satisfactory settlement was reached, then it could be checked against professional arguments in order to explain its breakdown or justification, a posteriori.

The new commission set out to determine adequate tactics for the confrontation. The technocratic approach of company-by-company negotiations was abandoned, and pressure was concentrated on two companies representative of the oil industry in Libya: Esso, which was the leader of the majors and had set the first reference for Libyan postings, and Occidental, which was the largest single producer in Libya and the most vulnerable of the independents. The scope and nature of the arguments that could be brought against one or the other were different and reflected the relative vulnerability of each. The worldwide integrated activities and responsibilities of Esso made it sensitive and responsive to political arguments so that pressure was firm but cautious and diplomatic. On the contrary, Occidental could not survive any drastic move limiting its Libyan production, so that nerve-breaking insinuations and threats were propagated suggesting that Occidental assets in Libya might be partially or totally taken over. The most telling result of these campaigns was a dramatic drop in the value of Oxy's shares on the stock markets, a situation the company could not stand for a long time.

The Libyans were successful in creating an atmosphere of uneasiness and uncertainty, and the confrontation was given different colorations at different levels. The Price Commission displayed a hard-line nationalistic attitude charged with potential threats. Its claims for price increases were set at extremely high levels occasionally reaching 80¢/Bbl. Under these conditions, the companies found their contacts with Oil Minister Izz el-Dine Mabruk rather comforting since he was reportedly ready to welcome a 25¢/Bbl increase. In the meantime, oil companies continued to be harassed by general political statements by higher officials of the Libyan Revolutionary Command Council. In a speech delivered in early April, Qaddafi exhorted the Libyan people to mobilize for the "coming fight with oil companies," which he linked with "world Zionism and local forces of reaction."[2]

The ambiguity of Libyan demands was matched by uncertainty about the position of the companies, which were reportedly prepared to offer a 5-10¢/Bbl price increase in April, an increase that was readily rejected as insufficient. As indicated earlier in this chapter, it is beyond the scope of this brief recapitulation to give an exhaustive account of the price confrontation in Libya. Notwithstanding, we will present below some of the main factors in the situation:

- Indirect pressure and intimidation continued to be exerted to demonstrate Libyan determination to achieve a full-scale victory in the oil battle. The potential threat of partial or total nationalization was hinted at as the inevitable ultimate measure, should negotiations fail to satisfy the Libyan revolutionaries. This heavy, nerve-rending atmosphere had a highly depressing effect on the independents' mood and on their position on the stock exchange. By mid-May, Occidental's chairman, Dr. Armand Hammer, felt obliged to deny persisting rumors of the nationalization of "any" company operating in Libya.[3] Similarly, Esso was faced with an interdiction of liquefied natural gas (LNG) exports from its Brega plant to Italy and Spain. In fact, this measure was due to a specific conflict over the gas price that had been hanging over the project for years, but the timing was such that it was inevitably associated with the conflict over crude oil prices.[4]

- Within the same spirit, it so happened that the Technical Department of the Ministry of Petroleum started to order production cutbacks during that critical and confused period. In fact, this move was no more than the implementation of the OPEC-based conservation regulation that Libya had previously promulgated in 1968 as Regulation No. 8. These measures had become technically advisable for some fields where reservoir performances were alarming. Allowable production levels assigned to different fields were calculated in accordance with conservation practices applied in Canada and the United States. The first move affected Occidental, with a production cutback of about 300,000 Bbl/day, from an average of about 800,000 Bbl/day to 500,000. Most other producing companies were affected shortly thereafter. By early September, Libya's aggregate oil production had been reduced by some 800,000 Bbl/day, from about 3.6 million in May down to 2.8 million. Although production cutbacks were motivated essentially by technical considerations, as demonstrated later by the fact that restrictions were not lifted with the satisfactory settlement over crude prices, they were felt by the industry as a direct pressure and intimidation measure, which helped further undermine its resistance. The effect of this reduction in Libyan production was enhanced by the fact that Libyan oil had been supplying about one-third of European needs so that a "true gap" was created and widened between the current needs of oil companies in Europe and what they could lay a hand on quickly. Contracted tankers started to be routed out of the Mediterranean to the Gulf.

- The sharp drop in the availability of Mediterranean oil by cumulative Libyan reductions of about 800,000 Bbl/day was dramatically reinforced by stoppage of the flow of some 475,000 Bbl/day of oil from Aramco fields in Saudi Arabia to the Mediterranean terminal at Sidon through the Tapline. The line had been "accidentally" punctured by a bulldozer in Syrian territory on May 3, and Syrian authorities systematically refused to give the necessary authorization for repairing the line. The Mediterranean gap thus created rapidly induced an acute freight crisis, which saw spot rates shoot up within few weeks from Worldscale 129 in April 1970 to nearly Worldscale 300 in September-October, and the upward move also affected AFRA rates for all categories of vessels.

• The situation of Mediterranean oil further deteriorated when Algeria broke off its negotiating approach to standing problems with France by deciding unilaterally to raise tax reference prices from $2.08 to $2.85/Bbl. However, it was obvious that the Algerian move was a strong challenge to the traditional structure of the international oil industry, and that such a challenge was unlikely to remain a purely Algerian affair. In fact, Libya and Algeria had been in close and continuous contact to coordinate their policies and unify their efforts. The Libyan Government was reportedly very close to a final decision to emulate the Algerian example by imposing a new tax price on the companies, following the rejection of the latest price offers by Esso and Occidental. And, indeed, the atmosphere became more tense and gloomy in Libya with the nationalization of local marketing and distribution on July 4, 1970.

• The crisis atmosphere took on unexpected momentum with the energy shortage on the East Coast of the United States, which will be discussed in the next chapter. The U.S. Government's appreciation of the political implications of the price confrontation was stated by an official of the Interior Department in the following way:

> Bright as our hopes may be for discoveries in the North Sea and the North American Arctic, the fact remains that the bulk of the oil going to the Free World outside our own country is now being supplied by the Arab Nations, and will be for as far into the future as we can reliably foresee. Like it or not, we must deal with reality, and the reality is that 15 million b/d, including 80 per cent of Europe's supply, comes from the nations of the Arab world.[5]

The warning to oil companies was clear enough. The United States was not prepared to face a major oil crisis and would not support oil companies in any move that could lead to such a crisis.

As a consequence of the above-mentioned factors, the confrontation seemed to outside observers in midsummer to be sliding progressively in favor of the Libyans, and the companies were reportedly on the defensive, especially Occidental, which was almost panic-stricken by nationalization psychosis. The atmosphere became more intense as the whole pricing issue was moved upwards into the Revolutionary Command Council, with the negotiations then being conducted personally by Major Abd al-Salam Jallud, deputy prime minister and Interior minister, who had proved to be a tough and smart negotiator during the discussions over the evacuation of military bases. A final ultimatum was addressed to Esso and Occidental to present "realistic" price offers within one week. Otherwise, most observers guessed that the approaching September 1, the first anniversary of the Libyan revolution, would provide an excellent opportunity for a spectacular move against "monopolies and imperialism." In practical terms, it had become obvious that the time for give-and-take discussion on the economic and technical levels was over; the decision-making had passed to the highest political level and what mattered then was simply the size of the companies' offers in terms of hard cash and whether these could meet the government's minimum demands. In such circumstances, it was to be expected that the matter would be resolved in one of two ways—either by a negotiated settlement or by unilateral action on the part of the government—in the fairly near future.

And, indeed, the settlement came along as rapidly as expected, but its magnitude was a bit surprising even to the best-informed observers. Occidental Petroleum had no choice but to surrender to Libyan demands and conditions, but the terms of the arrangement and the level of the new price announced by Colonel Qaddafi on September 4 were higher than what had been generally expected. "The settlement caught other international firms by surprise and the size of the increase jolted them."[6]

The terms of the deal, with which all other companies operating in Libya had no choice but to align shortly after, were as follows:

- The posted price of Libyan 40° API crude oil was increased by 30¢/Bbl, from $2.23 to $2.53/Bbl effective September 1, 1970; to rise by a further 2¢/Bbl on January 1 of each year from 1971 through 1975 to reach $2.63/Bbl in 1975;
- Henceforth the gravity differential would be fixed at a 2¢/Bbl increase for each full degree of gravity above 40° API and 1.5¢/Bbl decrease for each full degree of gravity below 40.9° API. Previously, the gravity differential had been subject to a ceiling of 40.9° API with flat prices for higher-gravity crudes, and with 2¢/Bbl decrease for each full degree of gravity below 40.9° API;
- The basic tax rate would be maintained at 50 percent, but Occidental would pay a further 5 percent because of its contractual commitment for the Kufra agricultural project, which was taken over by the government; and a further 3 percent as a package settlement for the retroactive effect of the price increase back to January 1, 1965. The total tax obligation amounts to 58 percent.

- Along with this "mutually acceptable settlement," Occidental's allowable production was increased to 700,000 Bbl/day, in comparison with the drastic cut in output down to 440,000 Bbl/day.

It took almost six weeks for all other companies to reach formal agreements along the Occidental pattern. The next to fall were three independent partners of the Oasis Group—Continental, Marathon, and Amerada-Hess. The fourth partner having refused the deal, the Libyans reacted promptly by closing Shell's exports on September 22, and this move was intended to be a warning to all others, mainly the majors, that might feel they were in a better position to resist Libyan pressure than the independents. Finally, everyone came into line, with particular adjustments in specific cases, including
- The elimination of the 5¢/Bbl high-wax penalty previously implemented by Mobil-Gelsenberg, Texaco-Chevron, and BP-Hunt on their high-pour production;
- The elimination of the 2¢/Bbl freight differential penalty affecting BP-Hunt exports from the Hariga terminal.

Combined with the modification of the gravity differential above and below 40° API, actual price hikes were different from one case to another. For example, the reference price for BP's 37° Serir crude oil rose by 38.5¢/Bbl to $2.485/Bbl; Mobil-Gelsenberg's and Texaco-Chevron's 36° Amna crude rose by 37¢/Bbl to $2.55/Bbl; and Esso's 40° Brega crude rose by 30¢/Bbl to $2.53/Bbl.

However, with the exception of Occidental Petroleum, none of the companies were granted restoration of their cut-back production, although, in some cases, companies got minor approvals to produce new development wells, enabling them to maintain their production allowables. This aspect of the new situation was intended to demonstrate the technical grounds of conservation measures and to deny alleged connection with the price negotiations.

Retroactive payments were the most controversial issue of the price settlement. Libyan officials, and particularly Jallud, have been always keen on presenting their demands as a fair and just move to correct and compensate for past undue underpricing of Libyan crude oils with reference to their geographical, gravity and low-sulfur advantages. The new posted prices agreed upon as "fair and mutually acceptable" were to apply retroactively as of January 1, 1965, and incremental payments for the past period up to August 31, 1970 could be made, at company discretion, in one of the three following ways: total payment in cash; future cash payments in installments with interest; and payment in terms of fixed percentage of gross profits calculated in such a way that total retroactive dues would be paid in about five or six years. In fact, this amounted to an increase in the income tax rate which would be in effect in perpetuity, thus resulting eventually in total payments exceeding retroactive dues throughout the lifetime of the concessions.

Most companies elected to apply the third alternative, following the precedent set by Occidental, which had to bear a further 5 percent increase because of its Kufra project. The reasons for this choice were many: to avoid cash payments of huge sums, to take advantage of U.S. income tax credit for American companies, and to pass corresponding "costs" along to customers under cost-escalation clauses in supply contracts that would not operate if they were recognized as "retroactive charges" for the producer. Also to be considered was the fact that higher tax rates running "to perpetuity" would not have a very heavy financial effect when one considered the discounted value of payments that would be made after the time when the present value of incremental payments would be equal to the retroactive dues to be paid eventually in cash. The five to six years considered by the Libyans to recover total sums without interest, would actually correspond to about 15 years from the companies' financial standpoint, when discounting future payments at 8-10 percent.

In light of the above, it was natural to expect incremental tax rates to be different from one company to another due to differences in production and in total retroactive payments. The tax rates were finally set as follows: Occidental 53 percent (plus 5 percent for the Kufra project); Oasis group 54 percent; Esso, BP-Hunt, Texaco-Chevron, and Libyan American 55 percent; and Mobil-Gelsenberg 55.5 percent. However, Standard of Indiana, Murphy, Aquitaine, and Hispanoil elected to keep their tax rate at 50 percent and to pay total retroactivity (very small amounts indeed) in cash. Phillips Petroleum preferred to relinquish its concessions, including the 3,500 Bbl/day Umm Farud field, rather than accede to Libya's high new tax-price terms.

ASSESSMENT AND CONSEQUENCES OF THE LIBYAN SETTLEMENT

The Libyan settlement was not only a "happy ending" to a serious confrontation but also and primarily the starting point of a new era for the international oil industry. Even before final settlements were concluded with all companies operating in Libya, the ramifications and far-reaching impact of this defeat of the international majors in the face of the determination and credibility of governmental action in producing countries, began to be felt throughout the oil industry. The precedent-shattering hikes in prices and tax rates were fascinating not only because of their magnitude but also and primarily because of their political significance since they demonstrated that the major oil companies were no longer invulnerable and that they no longer formed a power center independent of the governments of the countries in which they operate. "Finally," according to the Middle East Economic Survey, "it might as well be admitted that those oil economists who have been so skilfully and persuasively predicting a steady (and perhaps precipitous) downward spiral in crude oil prices over the next decade, are likely, in practice, to be proved dead wrong."[7]

In fact, the thunder of the Libyan settlement started its ball-rolling effect throughout the oil-producing countries before economists and observers of the industry had enough time to evaluate the scope and consequences of the arrangement and to foresee its far-reaching impact. Very rapidly, the dramatic confrontation between the Libyan Government and the oil companies appeared almost as a harmless skirmish in comparison to the whirling, unprecedented battle between the community of producing countries and a hastily reconstituted cartel of oil companies, which resulted in profound changes in international oil economics. Therefore, there is no need to conduct an extensive evaluation of the Libyan settlement since it proved to be a temporary one and its consequences materialized only during a few months up to the Tripoli Agreement of March 20, 1971.

The collective flat increase of 30¢/Bbl (plus 2¢/Bbl yearly) was presented, and mutually accepted, as a package deal for the satisfactory settlement of the thorny and complex pricing issue. The government was eager to generalize the model agreement reached with Occidental, and the companies had no time or possibility to challenge the figures or to discuss them. It was a take-it-or-leave-it deal, with heavy retaliation measures at hand, just in case. Therefore, there was no formal explanation or justification of the figures, and nobody seemed really interested in their breakdown or in checking them against the different items on which governmental demands were officially articulated. This involuntary (or deliberate?) ambiguity later proved to be somewhat troublesome in the drafting of the Tripoli Agreement of March 1971. The freight-adjustment component of about 10¢/Bbl (aligning Libyan prices with Eastern Mediterranean postings with respect to the Gulf area) could be easily agreed on as an objective fact. The balance, say 20¢/Bbl plus 10¢/Bbl over five years, was considered by Libyan officials as fair remuneration for what they considered an unaccepted sulfur premium; the progressive increase of 2¢/Bbl would account for the expected aggravation of pollution and the implementation of more stringent regulations in industrialized countries. Oil companies claim that part of the balance should account for the Suez and/or short-haul premium, a position that the government rejected on the grounds that it had never raised the question of the closure of the Suez Canal during the negotiations, as evidenced by the fact that retroactivity goes back to 1965 (the year in which posted prices acquired their full financial significance) and not to 1967 (the year in which the Canal was closed to navigation).

This profound misunderstanding was readily emphasized on September 28, when Esso and BP announced not only "voluntary" increases, effective September 1, 1970, of 30¢/Bbl in the posted prices of their Libyan crude oils but also and at the same time increases of 20¢/Bbl in the posted prices of Iraqi crude oils at Eastern Mediterranean terminals in Banias (Syria) and Tripoli (Lebanon) from $2.21/Bbl to $2.41/Bbl for 36° grade "A" export and from $2.07/Bbl to 2.27/Bbl for grade "B" export (Bai Hassan/Jambur blend). Esso announced similar hikes in the posted price of 34° Saudi crude from $2.17/Bbl to $2.37/Bbl FOB Sidon. Other major oil companies followed suit shortly afterwards. Since these crude oils have a high sulfur content in contrast to Libyan crudes, the price increases could only be justified on the ground of a freight premium due to the closure of the Suez Canal. The 10¢/Bbl balance in Libyan increase being the normal freight differential readjustment, it was obvious that the major oil companies did not recognize any sulfur premium for Libyan crude oils. Confirmation of this interpretation is to be found in BP's statement that "the increases are a reflection of the transportation advantages of Mediterranean crudes"[8] at a time when the freight crisis was nearing its culmination.

The move by the majors was really a "voluntary" one, although it merely marked an alignment with the Occidental settlement reached four weeks earlier. It occurred without their entering into formal negotiations with the Libyan Government and without their being faced with any specific demand from the Iraqi or the Saudi governments. The reason for this may be found in the survival of, and the adherence of the majors to, the mirage of the long-cherished principle of the legal right to change posted prices without consulting the governments concerned. However, by so doing, they committed a psychological error by bluntly defying the Libyans at a time when nationalism was at its height. The settlement thus "conceded" by the majors and intended to terminate the confrontation would be only emphemeral since the Libyans could not afford not to have full recognition of their sulfur premium. It is safe to assume that the majors were quite aware of the different consequences and implications of their move, which appeared quite mysterious at that time. The genie was out of the bottle and nobody could predict the extent of the ravages it would cause. Once the turmoil was ended after the final settlements resulting from the Teheran and Tripoli Agreements, one may wonder whether the profound and unexpected changes in the over-all pricing structure were really feared or were, rather, secretly desired by the major oil companies, a question among many others that will be raised in the next chapter.

Another feature of the Libyan settlement was most likely to prove to be a source of trouble instead of peace in the area. The oil companies, aligning with the precedent set by Occidental, elected to make their retroactive payments in terms of specific increases in tax rates, in perpetuity, from among the three alternatives offered to them by Libyan officials. The oil companies were fully aware that such a fundamental break in the profit-sharing pattern would fatally shake the foundations of concession terms throughout the Middle East. And indeed,

> To nobody's surprise, Iran promptly became the first Middle East country . . . to ask for a higher tax rate in line with the new terms agreed to by most major oil companies in Libya. Iran made its point to the Consortium of international oil companies even before most of the majors had put pen to paper on their new accords with Libya . . . much less before the ink was dry.[9]

In the following weeks, oil companies agreed to increase their income tax liability from 50 to 55 percent in the major oil-producing countries in the Middle East. It was only natural, and predictable, to expect the Libyans to ask for a similar hike in their tax rate, since previous increases at different rates were only for retroactive payments. One may wonder whether these contradictory and uncoordinated moves by the majors were the result of unconsidered hasty action or rather a deliberate move within the framework of an over-all new strategy. Expectedly, Nigerian posted prices were increased in line with the new Libyan and Eastern Mediterranean prices. An agreement was reached in late November to increase Nigerian postings by 25¢/Bbl, retroactive to September 1. Once again,

> the companies—notably Shell and BP, the principal producers—officially attributed the increase in postings "exclusively" to the geographical advantage of short-haul crudes for Europe. This advantage has risen markedly, as a result, in the rise of freight cost over the past few months, the companies pointed out. The 25¢ hike was designed to bring Nigerian postings into line with those for other short-haul crudes, notably Libyan and East Mediterranean, Shell noted.[10]

Less expectedly, the major oil companies took the initiative of eliminating the investment discontinuity penalty, which had affected the posted prices of crude oils of $31°$ and below. The posted prices of $31°$ Iranian heavy and Kuwait crude oils were increased by 9¢/Bbl retroactive to November 14, and the oil companies obtained a quit claim up to that date against any retroactive demands. This price hike was coupled with an increase in the tax rate from 50 percent to 55 percent, and the governments concerned agreed that "the said arrangements constitute a fair, appropriate and final settlement of all matters related to the applicable bases of taxation and the level of posted prices up to the present date."[11] The investment discontinuity penalty, which we have estimated at 10¢/Bbl, was completely eliminated by a further increase of 1¢/Bbl, which brought back the posted prices of heavy crude oils to their parity positions in the linear price scales.

In the following weeks, all other distortions and differences affecting the pricing pattern in the Gulf area were smoothed over, especially the "political" penalty that had affected the posted price of Basrah crude oil FOB Khor el Amaya throughout the 1960s. It seems as though the major oil companies were seeking absolution for all past price anomalies while trying to avoid any potential source of conflict or discord in the new era of the industry they were in the process of forging. This spirit of conciliation largely helped resolve some of the many long-standing issues between Iraq and the IPC group and prepared favorable grounds for further arrangements. In an official statement on October 27, the Iraqi Government announced that, along with the increase of 20¢/Bbl in Mediterranean posted prices effective September 1, IPC would pay an extra 6¢/Bbl on exports from the northern fields and 7¢/Bbl on exports from the southern fields, effective January 1, 1971, and because of royalty expensing (which had not been implemented in Iraq) production would be increased by 4.5 million tons, and the companies agreed to extend a no-interest loan of £20 million pounds ($48 million).

During this feverish period, the lights of world attention focused on the Middle East with such intensity that the stage of Venezuelan oil was left in the shadows. However, in early December, Venezuela "stunned the oil industry with a sudden tax-reform move" that would have raised the prevailing "70/30 profit-sharing ratio to almost 80/20 in the government's favor."[12] The income tax rate for oil companies was raised to a flat 60 percent, from the prevailing graduated scale with its maximum of 52 percent, effective retroactively to January 1, 1970. Furthermore, the chief executive of the government cold "unilaterally" set tax reference values for crude and products' exports for "up to three years"; this prerogative was exercised to increase Venezuelan prices a few weeks later, just after the Teheran Agreement. Although the change in the taxation structure and its retroactive effect looked "pretty catastrophic" to the oil companies, the spotlights of attention shifted to the Venezuelan scene just to report briefly what was going on, and promptly returned to the Middle East.

FEBRUARY 15, 1971: THE TEHERAN AGREEMENT

The confrontation over crude oil prices had been developing throughout the 1960s, both through individual initiatives from producing countries and through the collective action of OPEC, but it had invariably failed to burst forth because it was confined to technical and economic grounds. By contrast, it took only a few months for the entire industry to flare up once the confrontation had been put on political grounds by the Libyan initiatives, and the response of consuming countries and oil companies to the challenge of producing countries was conceived, formulated, and assessed on political and strategic grounds. Technical and economic arguments came in, just at the last moment, to present and justify mutually agreed-upon arrangements.

Up to the last moment, even the best-informed observers of the industry did not suspect such a large increase in the cost of oil supplies. The oil community did not fear any moves by OPEC, which had displayed only limited capabilities throughout the decade. Professional and political rivalries between producing countries were as acute as ever, and oversupply capacity was assured at the well-head. The predominant characteristic of the industry was, and had always been, the unchallenged domination of the majors. Their doctrine and philosophy were largely influenced by their past experiences with producing countries: "Dealing with weak, quarreling, and often ignorant regimes, they frequently drove onesided bargains. If one country bucked the system, more oil could be pumped up somewhere else."[13] Yet, OPEC member countries did unite, unexpectedly, into a formidable political challenging power. Certainly, spirits were mature and circumstances were favorable. This situation notwithstanding, the rapid psychological evolution was stirred up, hastened, and directed by skillful maneuvers by certain officials in producing countries. The leading role in this respect was played by Iran, and its actions were highly publicized in the international media. It was Iran that took the lead in asking for higher tax rates immediately after the early Libyan agreements, and Iran was the first to get a price increase for heavy crude oils. Early information that leaked out about possible collective moves by OPEC countries presented the emerging cartel of oil producers as the fruit of Iranian initiative.

The Iranian leadership is quite significant, because the prestige and credibility it represented for Western public opinion aided the achievement of an acceptable compromise. Still more significant was the fact that Iran was the most unprepared country to take any initiative for a price increase of any magnitude, not to speak of the incredible demands then formulated in Teheran. In fact, during recent years, Iran had not been interested in a price increase but had engaged, on the contrary, in active campaigns to make the consortium increase its production by a staggering 20 percent a year. In "routine" annual meetings. Iran had always stressed increasing revenues by increasing production and had never mentioned prices. Such a preferential increase in production, it was publicly stated by the Shah himself, would be achieved at the expense of Arab producing countries. In an interview in the London <u>Financial Times</u> of May 6, 1969, the Shah was quoted as stating, "We can say that this is our oil—pump it. If not, we pump it ourselves." Apparently referring to some U.S. members of the consortium, deeply involved in Arab oil operations, the Shah told the <u>Financial Times,</u> "There are people in the consortium whose first love is somewhere else. We have proposed they sell out."[14] This declaration provoked angry reactions from Arab nationalists, symbolized by an open letter to the Shah of Iran by Skeikh Abd Allah al-Tariki.

Since the 1967 crisis, Iran had been conducting a worldwide campaign to demonstrate to Westerners that only its oil was sure and reliable in contrast to Arab oil. The argument of security of supplies from a stable pro-Western country, at reasonable prices, was advanced to support claims for higher production, along with humanitarian justifications in connection with Iran's large population and its meritorious efforts for economic and social development. In support of these campaigns, the Iranian Government had engaged the services of international public relation firms, and one of them published a booklet entitled "Iran-Oil in the Service of the Nation," which was discussed by the <u>Middle East Economic Survey</u>: "Unfortunately, however, the presentation of this case is marred by an almost total disregard for considerations and interests other than those of Iran. Even a ritual nod in the direction of cooperation or at least coordination with the other oil producing states is nowhere to be found. The slogan might almost be 'Growth in One Country'."[15]

Again quoting the <u>Survey</u>:

> Arab-Iranian rivalry in the matter of oil production growth is nothing new. But until recently it has been played down, in the public arena at least, in the interest of producer-country solidarity and the pursuit of common aims through OPEC. Now that the quarrel has broken out to open polemics, the great danger is that it could poison the entire climate of oil cooperation at the producing end of the business.[16]

Although producing countries displayed a very strongly united front in the confrontation over crude oil prices, the question remains as to whether this solidarity would prove permanent and resist the pressure of potential internal conflicts over production growth at higher, more profitable levels. The future will depend, to a large extent, on the ability of OPEC member countries to conciliate their individual interests with their collective responsibilities deriving from their new role as a producers' cartel. Practical indications in this respect would be provided by the outcome of OPEC attempts to set up a joint production program, an objective that OPEC has failed to achieve up to now and that would be an indispensable instrument in the support of price stability at present high levels. Without prejudging the future, we may note that all indications seem to conform that producing countries did not deviate from their race to higher and higher individual production. Iranian claims were emphatically reiterated in the heat of the battle for price increases without any apparent contradiction between the two objectives. In early September 1970, "under heavy pressure from the Shah for more revenue, the international oil Consortium in Iran . . . programmed additional crude oil production of 35 million barrels from now through December."[17] While Iran was pushing for a substantial price increase through OPEC, the Iranian Prime Minister declared in an address to the Iranian Parliament on December 8:

> The government has also made it absolutely clear to member companies of the Oil Consortium that Iran is determined to maintain her historical role as the biggest exporter of oil in the Middle East through exports of oil from the Agreement Area. Besides Iran's historical role, this position is regarded by the Government of Iran as an undeniable and a conclusive fact in view of the country's population, its size, the needs of its development plans, as well as the regional commitments of Iran, which serves as a factor of peace and stability of this part of the world. This principle forms the basis of the country's oil policy, and the Oil Consortium has recognized it and assured us that it will be reflected in Iran's oil revenues.[18]

At the 21st OPEC Conference held in Caracas, Venezuela, from December 9 to 12, Iranian proposals for the across-the-board hike in postings were endorsed by other member countries, and the conference decided on a unified strategy vis-à-vis oil prices and planned to put it into operation with the utmost vigor. It was generally believed that OPEC member countries had also agreed to enact their demands by legislation in the event that no agreement could be reached with the companies.[19]

The tactics envisaged to achieve these objectives were to adopt a hard-line attitude as a united front, with the clear determination to resort to extreme action, if necessary, including unilateral legislation and over-all stoppage of oil flow. A "regionalization" approach was adopted, under which the price problems of each particular region within the OPEC area as a whole would be treated as a separate question. The first front to bear the brunt of the confrontation would be comprised of the Gulf area producers. The second front would group the Mediterranean exporters.

Resolution XII-120 established a committee consisting of the representatives of Iran, Iraq, and Saudi Arabia to conduct negotiations for the Gulf area producers in Teheran. Furthermore, OPEC set a very tight schedule for these negotiations, which would begin no later than January 12, 1971. Then, the committee was given only one week to report back to the OPEC secretary general on the results of the Teheran talks, following which, within 15 days—that is, by February 3—an extraordinary OPEC conference would be convened to give its final verdict. Oil companies would have no loophole for escape, and in just one month there would be a momentous dénouement one way or another for the future of the oil industry—either agreement on a new and higher level of oil prices worldwide or confrontations with the producers with all its consequences.

In the face of this united and determined front of oil-producing countries and their unprecedented demands for fantastic increases, the oil companies reacted promptly and regrouped to undertake collective bargaining in the name of the industry in a coordinated and cohesive way. In fact, this was the lesson derived from the Libyan settlement, when the majors had been taken by surprise and left with no choice but to align with the terms resulting from the defection of Occidental Petroleum. The latter was not to blame, and any other independent producer would have reacted similarly if faced with immediate prospects of losing more than 90 percent of its producing assets. But the attitude would certainly have been different, and the bargaining power much stronger, if this same independent had been actively supported by the majors and the entire oil community as well.

With the formal benediction and encouragement of the U.S. Government and having obtained antitrust clearance to negotiate crude oil prices collectively, the major oil companies working together elaborated a common strategy to which the independents were made to adhere. The "companies' cartel" came into formal existence on January 16, 1971; when a collective memorandum was submitted to the Vienna headquarters of OPEC and to the governments of its 10 member countries. The message came just in time to ease the general atmosphere of acute crisis that resulted from the abortive first-round talks held on January 12, and, according to the Middle East Economic Survey, "it was understood that the responsibility for the consequences of all decisions which the Extraordinary Conference [convened in Teheran on January 19] may take with a view to attaining the rights of member countries will fall upon the oil companies which have even refused to set a date for negotiations."[20]

This move by oil companies had profound implications for the industry and would directly affect future strategy and plans. First, it pointed out emphatically the political nature of the whole pricing issue, and the profound and direct involvement of the U.S. Government and the White House in the confrontation, as will be indicated below. Secondly, OPEC obtained its "titles of nobility" and achieved its long-awaited political objective of being officially recognized as a "valid negotiator" representing and speaking for the community of oil producers. Furthermore, a moral responsibility was conferred on OPEC to guarantee the stability and application of any agreement for a period of five years. The companies proposed that they and OPEC member countries conduct global negotiations on the general lines indicated in the memorandum, which recognized the necessity of a general price increase and harmonization and accepted the principle of a temporary freight advantage premium for Libyan and other short-haul crudes, all in exchange for formal stabilization of international oil economics.

Although OPEC member countries were greatly pleased with this fundamental and unexpected capitulation by the oil companies, serious fears arose in reaction to the collective move, which was thought to be a maneuver intended to undermine the united OPEC front, which had agreed on a "regionalization strategy." During a week of intricate diplomacy in and around the Teheran price talks, the term "regionalization" emerged as a key factor in the success or failure of the negotiations. In other words, would the "January 16 Group" of oil companies agree to OPEC's demand that the Teheran talks should aim at an agreement only with the Gulf member countries of OPEC—Iran, Iraq, Saudi Arabia, Kuwait, Qatar, and Abu Dhabi—or would they continue to insist on an "over-all" settlement with all the OPEC member countries collectively?

For the obvious reasons that the Shah himself had outlined at his press conference, it was made clear that OPEC formally rejected the collective approach, which would prove not to be in the favor of the oil companies themselves. "The companies," he said,

> must come to terms with the Gulf countries as a region; otherwise even the "moderate" OPEC countries would have to align with extremist demands of the hard-liners [Libya or Venezuela]. On the companies' side, the question of abandoning the insistence on an overall settlement with all the OPEC countries at once, embodied in the January 16 message to OPEC, has been fraught with problems. It seems that this approach was adopted in the first place upon the instigation of the White House and

the U.S. Independents who felt that only the global umbrella would protect their highly vulnerable position in Libya.[21]

And, in fact, Libya formally rejected not only the collective approach on the governments' side, but also on the companies' side, and insisted on company-by-company individual talks.

Finally, the oil companies agreed on January 28 to negotiate with the Gulf countries on a regional basis. The initial antitrust clearance, which applied only to an over-all agreement with all OPEC member countries, was modified and extended accordingly.

The final aspect of the strategic significance of the Teheran price negotiations, which we would like to outline in this brief analysis, concerns the direct and profound involvement of the U.S. Government and President Nixon himself. The motivations and strategic objectives behind the open intervention of U.S. officials in what could have been thought to be a professional deal of commercial nature will be discussed and analyzed in the following chapter. The most significant and least publicized aspect was the dispatching of a high-ranking official—Undersecretary of State John Irwin II—to prepare for "professional talks and commercial deals" in which the U.S. Government was not a partner, and in which the countries directly concerned, namely Western Europe and Japan, were significantly absent.

The time that elapsed between the companies' acceptance of OPEC's "regionalization approach" on January 28 and the formal signature of the Teheran Agreement on February 14, was replete with exciting events, dramatic breaks, and moments of revived hopes. An extraordinary OPEC conference was held in Teheran on February 3-4 and a decision was made to set a firm deadline, namely February 15, for oil companies to comply with minimum terms for higher prices as decided upon in the same extraordinary conference; if there were no compliance, these minimum terms would be enforced by legislation.

Whether producing countries would have actually imposed a total and firm embargo on world oil supplies should an acceptable agreement have failed to materialize is the kind of question that will never by answered. However, such an eventuality was taken seriously in some circles, especially since the "minimum demands" of OPEC were not published, and most of the information available at that time was rather speculative.

An agreement was finally reached at the last moment, just in time. Its main terms and features are briefly outlined below:

- An immediate base increase of 35¢/Bbl in the posted price of all crude oils (of which 33¢/Bbl as a uniform increase in postings and 2¢/Bbl in satisfaction of claims related to freight disparities in the Gulf area).

- The application of a gravity differential of 1.5 instead of 2¢/Bbl/degree for crude oils of less than 40°, while keeping the conventional 2¢/Bbl/degree above 40°. The case of heavy crude oils of less than 30° to be agreed upon later.
- Gravity escalation to apply on a one-tenth degree API basis.
- Elimination of marketing and "OPEC" allowances.
- An annual increase of 5¢/Bbl plus a 2.5 percent increase in posted prices, to apply on June 1, 1971, and again on the first of January 1973, 1974, and 1975, to compensate for worldwide price inflation and the fall in the purchasing power of the dollar.
- Specific adjustments for certain crudes including an extra 6¢/Bbl for Basrah crude oil and an extra 1¢/Bbl for Iranian heavy, Arabian medium, and Kuwait crude oils.
- The income tax rate to be 55 percent along with full royalty expensing at the rate of 12.5 percent of the posted prices.

Against these fantastic across-the-board price increases, the magnitude, or even the very idea of which would have seemed almost inconceivable only a few months earlier, oil companies were satisfied by obtaining the following:

- Over-all stability for a period of five years.
- Guarantees against "leapfrogging" in financial terms.
- Guarantees against any embargo action in support of demands outside OPEC Resolution 120.
- Limitation of the extra short-haul premium of Mediterranean oil to 21.5¢/Bbl.
- Guarantees that the Gulf countries would not be entitled to demand higher prices to align with Libyan prices in case the latter would not adjust downwards as might be required by a "balancing amount" system with an ultimate ceiling set at 21.5¢/Bbl.

In light of this agreement, the price increase on February 15, 1971 ranged between 35¢/Bbl (Qatar 40°) and 49.5¢/Bbl (31° API crude oils, including a specific adjustment of 10 cents—the investment discontinuity penalty). The total increase for the period November 14, 1970-January 1, 1975 would range between 77.4¢/Bbl (Qatar 40°) and 92.3¢/Bbl (Iranian heavy 31°). Resulting additional payments to the six producing countries concerned would be approximately $1.2 billion, rising to about $3 billion for 1975 (assuming a 10 percent increase in production) and netting an estimated total revenue increase of over $10 billion over the five-year period.[22]

MARCH 20, 1971: THE TRIPOLI AGREEMENT

The hopes raised by the first Libyan settlement of September-October 1970, were known, from the very beginning, to be ephemeral. The ambiguities and misunderstandings that characterized the general atmosphere within which agreements were reached with oil companies were most likely to provoke the Libyans and to stir up their ardor for another round of price increases rather than fill them with pride and satisfaction for having achieved their national objectives. As outlined earlier in this chapter, the improvements gained in posted prices and tax rates were, from the Libyan viewpoint, no more than a "fair correction of past errors," which would bring Libyan prices into parity with the Gulf and Eastern Mediterranean postings. In particular, one could note the following:

- The increase in tax rates (by different values varying from one company to another) was the option selected voluntarily by the companies to pay for retroactivity back to January 1, 1965.
- The increase in posted prices was to pay for normal readjustment of freight (10¢/Bbl, restoring parity with Eastern Mediterranean) and for the long-overdue sulfur differential (20¢/Bbl + 2¢/Bbl yearly over five years).
- No advantage was included for the Suez and short-haul premiums.

Now, when the major oil companies "voluntarily" increased the income tax rate by 5 percent in the Gulf and prices at the Eastern Mediterranean by 20¢/Bbl to reflect the Suez and short-haul premiums, it was natural to expect the Libyans to consider these increases as specific new advantages with which they would have to align. Their frustration and distrust vis-à-vis the oil companies were intensified by what they believed to be a systematic negative attitude of oil companies operating in Libya, which was reflected in a general slowdown of exploration and development activities as evidenced by the drop in the number of rigs in operation from about 50 to some 12-15 in the last quarter of 1970, accompanied by a general freezing in expenditures, which caused general stagnation in the economic life of the country. The motivations of this attitude, as explained by the companies concerned, were due to production restrictions under Regulation No. 8, which left a large existing production capacity unexploited, so that any further drilling under such conditions would be a waste. This argument was only a pretext to conceal the general apprehension of the industry towards possible unilateral steps by the government that might make any investment highly risky.

Once the OPEC member countries had collectively agreed in Caracas to a hard-line strategy to induce a substantial price increase along with the generalization of the 55/45 profit-sharing pattern, Libya wanted to take the lead, once again. On January 3, representatives of oil companies operating in Libya were informed of the government's demands by Major Jallud, deputy prime minister for Production. These demands have never been published, and oil companies were asked to keep them top secret. However, it was believed that the different price increases requested under varying justifications would amount to a tax-paid cost of about $2.57/Bbl for 40° API crude oil. The breakdown of the increase included a substantial increase in posted prices plus a special freight premium to reflect the specific advantage of the geographical location resulting from the closure of the Suez Canal, and, therefore, this premium would be implemented with a retroactive effect back to June 1967. Furthermore, the basic tax rate would be raised from 50 to 55 percent and oil companies would be given, once again, the opportunity of settling their 1965-70 retroactive obligations resulting from the September-October arrangements by one of the three alternative methods: (1) in cash payments with 10 percent discount, (2) in installments over five years including interest at current market rates, or (3) by an increase in the tax rate over and above the 55 percent minimum set by OPEC at the Caracas Conference. Other demands included accelerated payments of taxes and royalties on a monthly basis; the obligation of reinvesting 25¢/Bbl of crude exported in exploration, secondary recovery, or any other activity outside the oil industry in Libya; and supplying Libya with its domestic oil and gas requirements at cost.

The tough Libyan stand was a cleverly designed tactical move, repeating the successful strategy put into action a few months earlier, which had driven the oil companies, mainly the vulnerable independents, to a company-by-company capitulation. The Libyan Government was further encouraged by the direct sale of substantial quantities of its royalty oil at prices exceeding the posted price of $2.55/Bbl by as much as 35¢/Bbl. These were spot sales to some European brokers at a time when the actual value of the short-haul premium was substantially in excess of the sale overpricing of the crude resulting from the acute crisis in the tanker market.[23]

The oil companies were summoned to individual, separate talks; Occidental and Bunker Hunt were the first to be confronted with an over-all "take it or leave it" deal. The psychological pressure was further enhanced by more or less open threats of takeover presented in terms of "appropriate unilateral action" and by the emotion-charged nationalistic tone given to price talks by Jallud, who according to the Petroleum Intelligence Weekly "told oil executives in Libya that they must be made to feel the 'effect of Zionist aggression' in order to press Western governments to change their Middle East policies [and make them] more favorable to the Arab cause; 'when companies suffer, Jallud declared, 'they may be impelled to do something to alter United States policy.' "[24]

This renewed vigor in Libyan demands certainly played a determining role in pushing the independents to unite with the majors, with the encouragement of the White House, in signing the collective message to OPEC of January 16. But the Libyans understood the tactics of the "January 16 Group," and it was no later than the following day that the companies initiative was vigorously attacked and formally rejected in the joint communiqué issued in Algiers on January 17, following talks between Jallud and Belaid Abdesselam, the Algerian minister of Industry and Energy.

The Libyans stood very firm in their determination to negotiate separately with individual companies. George Piercy learned this at his own expense, when he appeared in Tripoli after being delegated by the "January 16 Group" of oil companies. He was abruptly told that he could talk on behalf of Standard Oil of New Jersey alone and that his presence as representative of other oil companies was not warranted. During the eventful days leading up to the Teheran settlement on February 14, the Libyans followed the development of the situation closely and very cautiously, with an increasing feeling of uneasiness resulting from persistent efforts by oil companies to squeeze them into a collective bargain with the Gulf producers. Moreover, they were further frustrated by the conditional support extended to them by other OPEC member countries, which limited their solidarity only in the case where "reasonable" additional Libyan demands that reflected a justified short-haul premium would not be agreed to by the oil companies.[25]

Under these conditions, the Teheran Agreement was received with anger and an increasing feeling of frustration in spite of its satisfactory achievements in the area of price improvement. The specific reference to Libyan oil and the limitation of the short-haul premium to 21.5¢/Bbl were regarded as an unacceptable provocation and an uncalled-for interference with Libyan affairs. The interests of Libya could not be debated or dictated by anyone but the Libyans themselves, and the Gulf states were almost suspected of complicity with the oil companies in their efforts to put Libya "in quarantine" by limiting, a priori, the reasonably justified short-haul freight premium to only 21.5¢/Bbl, which actually condemned Libya to accept this value dictated by the oil companies or to renounce active support from other OPEC member countries.

However, in order to safeguard the apparent unity and cohesion of OPEC and to reinforce the political strength related to the collective regionalization approach as decided in the Caracas Conference and achieved in the Teheran victory, representatives of OPEC member countries of the Mediterranean group—namely Algeria, Iraq, Libya, and Saudi Arabia—met together in Tripoli on February 23 and decided to delegate Libya to negotiate prices on their behalf. The joint communiqué stated that "Libya should undertake negotiations separately with each oil company operating in Libya." A two-week schedule was set for these talks, after which appropriate measures, including a stoppage of the flow of oil, could be undertaken in case the oil companies did not comply with minimum demands formulated by the four countries.

In this way, Libya got a carte blanche to go its own way without undermining the collective front of OPEC. The tough and maximalist Libyan stand was further reinforced and received more credibility with the swift Algerian takeover, on the following day, of 51 percent (majority operating control) of all French oil-producing interests. Separate negotiations started shortly thereafter, and the oil companies were faced with much higher demands than the initial package handed to them earlier in January. Jallud declared that the Teheran price settlement was far from meeting minimum Libyan demands and in fact, the price increase now put to the oil companies would raise the posted price of 40° crude oil from the prevailing \$2.55 up to \$3.75/Bbl. Other demands concerning the general increase in income tax by 5 percent and the obligation of reinvesting 25¢/Bbl exported, were emphatically reiterated. The oil companies were determined not to give the Libyans more than a straight extension of the Teheran agreement.

An atmosphere of apprehension and uncertainty prevailed throughout the eventful and heavily charged month of March 1971. To the surprise of all, Jallud, a 28-year-old officer, proved to be a tough and highly skillful negotiator who largely dominated the scene in the face of the legions of experts and executives from all companies. He was successful in keeping the situation permanently under control and in choosing the time and the appropriate style to call on one or the other of his designated negotiators, or victims. The first deadline fixed as March 10 was postponed to March 13, when an offer by the companies was rejected. A second conference of Oil ministers from Algeria, Iraq, Libya, and Saudi Arabia was held in Tripoli on March 15, and there a decision was made to continue supporting Libya in its solitary confrontation with the oil companies. A decisive and most confused week followed, during which almost nonstop negotiation sessions were held with representatives of the various companies. A draft agreement was finally reached around March 20, but its formal approval by the Libyan Revolutionary Command Council was withheld for about 10 days, during which alarming rumors were circulated and nobody could guess what would be the final outcome. The suspense continued right up to the last minute, and the big question was whether top Libyan officials were withholding their definitive endorsement of the deal raeched by Jallud and Oil Minister Mabruk because they considered its terms unsatisfactory and advocated more drastic action or because of minor difficulties over specific items such as the "Suez retroactivity."

It was with immense relief that the industry welcomed the announcement of formal approval of the agreement on April 2, although the oil companies had to add a further 2¢/Bbl increase to the initial deal. Looking back to the few weeks that had just passed, top oil executives visualized the Tripoli talks as probably the longest, hardest, and most acrimonious in the history of international oil. They realized that they had been confined to a tight defensive position all along the way, a situation they had never faced before. The Libyan Government, completely confident in its bargaining position, had approached the negotiations in an inflexible way, adopting "carrot and stick" tactics and waving, skillfully, the specter of embargo and/or takeover.

The initial gap separating the two positions was substantial indeed, but the two sides moved progressively towards a compromise after the Libyans realized their increasingly isolated position resulting from the limitation of the support they could expect from other OPEC member countries and after they became conscious that the acute freight crisis would ease away after the resumption of oil flow to the Mediterranean through the Tapline. An agreement to resume Tapline service had been reached on January 28, 1971 with the Syrian Government; the repair of the damaged line took only 27-1/4 hours, and the flow of Arabian oil to Sidon resumed after an interruption of more than 270 days. For their part, the oil companies were keen not to jeopardize the Teheran agreement by exposing themselves to a dramatic rupture in Libya so that they were inclined to allow the Libyans to obtain the victory the latter were determined to achieve at any cost, and thus they agreed to increase Libyan prices over and above the Teheran conditions.

The agreement was finally approved and signed on April 2, with an effective date fixed at March 20, by each individual company separately. Its main terms and conditions are outlined below, and its scope and significance will be investigated and analyzed later in this chapter:

1. <u>Operative Date</u>: March 20, 1971.
2. <u>Term and Quit-Claims</u>: Five-year term with quit-all clause on back government claims and stability guarantee for prices and government take for the duration of the agreement. In particular, this covers claims with respect to retroactivity and reinvestment obligations.
3. <u>Taxation</u>: Total taxes on companies' profits will stabilize at 55 percent (except for Occidental, which will add a further 5 percent liability against its specific commitment for the Kufrah agricultural project). Retroactivity for the period 1965-70, agreed on in September, will be paid in supplemental cash payments equivalent to what would

have been due under the previous surtax arrangement. The payment of royalties, taxes, and surtaxes will be made on a monthly basis. Total elimination of allowances and marketing expenses.

    4. Reinvestment: Each concessionary group will keep at least one exploratory rig in operation for the duration of the agreement. In case no favorable exploration prospects are available, then the group will undertake an equivalent financial obligation in secondary recovery or gas utilization.

    5. Supply of Oil: Oil for domestic refining and consumption will be supplied at cost plus a fee by each company pro rata with other producers in Libya.

    6. Posted Prices: Posted prices will be composed of the following:

1. Base Posting: The base posting for $40.0°$ API crude oil will be \$3.07/Bbl from the operative date through June 30, 1971.
2. Gravity Differential: The base posting will be subject to gravity differential adjustment as follows: It will be increased by 0.2¢/Bbl for each full 0.1 of API gravity above $40.0°$ and will be decreased by 0.15¢/Bbl for each full 0.1 of API gravity below $40.09°$.
3. Sulfur Differential: The base posting includes a 10¢/Bbl low-sulfur premium for crude oils with maximum sulfur content of 0.5 weight percent. This sulfur premium will be increased by 2¢/Bbl on the first of January of each of the years 1972 through 1975. The case of crudes with higher sulfur content (which do not exist currently in Libya) will be debated in accordance with the Petroleum law.
4. Escalation: The base posting will be increased, as from the operative date, by 2.5 percent plus a flat 5¢/Bbl. the same will apply again on the first of January of each of the years 1973 through 1975 (the 2.5 percent increase will apply to the base price prevailing on December 31 of the preceding year. It will be rounded to the nearest tenth of a cent).
5. Suez Canal Allowance: 12¢/Bbl will be added to the base posting as long as the Canal remains closed to navigation. It will be reduced to 4¢/Bbl effective on the first date that the Suez Canal is open for the passage of commercial ships to a draft of 37 feet and will be eliminated entirely when the draft reaches 38 feet.
6. Temporary Freight Premium: 13¢/Bbl will be added to the base posting for the period from the operative date through June 30, 1971. Subsequently, the value of

the premium will be determined quarterly on the basis of 0.058¢/Bbl for each 0.1 percentage point of Worldscale by which the assessed LR (long-range) 2 AFRA exceeds Worldscale 72.

As a consequence of all the above, the new posted price for 40° API Libyan crude oil worked out to $3,447/Bbl for the period March 20 through June 30, 1971. It should be noted that escalation was implemented from the operative date, as against June 1, 1971 in the Teheran Agreement. This presumably offsets the fact that the Teheran deal was effective earlier.

The Libyan performance throughout these negotiations was really outstanding. The base posted price (40° API) of $2.23 per barrel, which was applicable up to September 1, 1970, was increased by $0.320 up to $2.55 in the first Tripoli Agreement, and by $0.647 up to $3.197 per barrel on March 20, 1971. The total posted price is obtained by adding the Suez Canal allowance and the freight premium as long as applicable.

The magnitude of the increase may be visualized by stating that the per barrel income would rise by $0.637/Bbl from $1.378 to $2.015/Bbl and that estimated cumulative additional revenue would not be very far from $4 billion.[26]

FURTHER WORLDWIDE PRICE SETTLEMENTS

The price settlements reached in Teheran and Tripoli set new standards for oil economics throughout the world, with which crude prices not directly involved in these agreements would have to align. The necessary adjustments were made within a few weeks, particularly in the Gulf area, Nigeria, and the Eastern Mediterranean. Venezuela went its own way, increasing tax reference prices and modifying the over-all pricing structure. The Algerian oil crisis culminated in a dramatic break with the French and resulted in partial nationalization, in contrast to the compromise agreements reached in other producing countries.

The Gulf Area

The postings of crude oils not included in the Teheran Agreement of February 14, 1971 were adjusted upwards accordingly, with respect to prices, taxation, escalation, and so on, effective on February 14, as indicated in Table 16.1. The following remarks and comments should be made:

TABLE 16.1

Price Adjustments Under Teheran Conditions
for Crude Oils Not Included in Teheran Agreement

Crude Oil	Terminal	Gravity (°API)	Pre-Nov. 14, 1970	Posted Price Nov. 14, 1970	Feb. 15, 1971
Iran Offshore					
Rostam	Lavan	38.00-38.09	—	—	2.232[a]
Darius	Kharg Is.	34.00-34.09	1.63	1.63	2.16
Sassan		34.00-34.09	1.7	1.7	2.172
Bahrgansar	at terminal	32.00-32.09	—	—	2.135
	at buoy	32.00-32.09	—	—	2.12
Nowruz	at buoy	20.00-20.09	—	—	1.84[a]
Cyrus	Cyrus	19.00-19.00	1.34	1.34	1.83
Neutral Zone					
Hout	Ras Khafji	35.00-35.09	1.81	1.81	2.185
Khafji	Ras Khafji	28.00-28.09	1.46	1.55	1.97
Burgan	Mina Saud	23.5 -24.4	1.48	1.48	1.48[b]
Batawi	Mina Saud	23.5 -24.4	1.41	1.41	1.41[b]
Eocene	Mina Abdulla	16.5 -17.4	1.28	1.28	1.28[b]
Others					
Arabian H.	Ras Tanura	27.00-27.09	1.47	1.56	1.96
Dubai	Fateh	32.00-32.09	1.69	1.69	2.13
Oman	Fahal	33.00-33.09	1.82	1.82	2.205

[a]Effective April 1, 1971.
[b]Unchanged.

Source: Compiled by the author.

- The posted prices of Iranian offshore production were adjusted, or newly published, in accordance with the Teheran Agreement, although these crude oils are produced under specific joint-venture agreements, which stipulate that tax payments to Iran are based on realized prices and not on posted prices.
- Japanese-owned Arabian Oil Khafji crude oil was aligned with the Teheran Agreement, although the tax structure provided for a higher rate as stipulated in the original concession agreement. In fact, the tax liability of the Arabian Oil Company was raised by 5 percent as of November 14, 1970, and its financial obligations will now consist of either (1) 62 percent of profits (formerly 57 percent) based on posted prices without royalty expensing, or (2) 20 percent expensed royalty plus 45 percent tax (formerly 40 percent), whichever is the greater.
- The price adjustment of heavy crude oils with a gravity below $30°$ was not settled in the Teheran Agreement. Only the price of $27°$ Arabian heavy crude oil was adjusted thereafter, by Aramco. From a former $1.47 FOB Ras Tanura, the posting of that crude was increased to $1.56 as of November 14, 1971, to $1.96 as of February 15, 1971, and to $2.064 as of June 1, 1972. Prices of heavy crude oils produced by independent Getty and Aminoil in the neutral zone remained unchanged.

Moreover, the Gulf postings were adjusted upwards, effective June 1, 1971, by the application of the escalation clause calling for a 2.5 percent increase plus a flat 5¢/Bbl. The new prices remained unchanged until January 1, 1973.

Nigerian Crude Oils

The extension of the new conditions set in the Teheran and Tripoli Agreements was no more than a formal procedure, not only because of the "most favored African nation" clause in Nigerian concessions but also because these settlements were intended to set the new over-all framework for international oil economics. After a few weeks of negotiations, a new five-year agreement was reached in early May, 1971, effective the same March 20 date as the Tripoli Agreements. It provided for the same 10¢/Bbl low-sulfur premium escalating 2¢/Bbl annually through 1975, the same temporary short-haul premium (adjusted for Nigeria's geographical position), the same escalation provisions, and the same rise in the base tax rate from 50 to 55 percent along with a general price increase of 35¢/Bbl.

The new base posting for $34°$ API Nigerian crude oil was set at $3/Bbl, to which 21¢/Bbl were added with respect to temporary Suez and freight premiums, which set the total posted price at $3.21/Bbl. The base posting of $3 comprises a 35¢ flat increase a 10¢ sulfur premium, a 3¢ gravity adjustment, a 4¢ harbor dues, and the 2.5 percent plus 5¢ escalation, the same as in the Tripoli Agreement. The government income on March 20, 1971 was estimated at $1.775 per barrel as compared to $1,118 on September 1, 1970 and to $0.985 before that date. On January 1, 1975 the base posting of $3 will be increased to $3.47 as a consequence of the escalation clause.[27]

Eastern Mediterranean Crude Oils

The Tripoli negotiations were conducted by Libya, in principle, on behalf of Mediterranean-exporting OPEC member countries. Therefore, the agreement reached for Libyan oil could be expected to extend, without difficulty, to Eastern Mediterranean oil with adjustments in the gravity and freight elements. However, the issue proved to be more difficult than expected, and for a time the entire agreement might have been nullified. The root of the problem was the Iraqi refusal to recognize any specific quality premium for Libyan oil, which premium was not to be extended totally to Iraqi oil. This dealt, of course, with the 10¢/Bbl low-sulfur premium (plus 2¢/Bbl annually through 1975), which was recognized for Libyan crudes with less than 0.5 weight percent sulfur content and which, consequently, should not apply to Iraqi higher-sulfur crude (1.9 weight percent sulfur for 36° Kirkuk crude oil). The Iraqi refusal was contradictory to a previous understanding between the Oil ministers of Mediterranean exporting countries in their Tripoli meetings, during which the low-sulfur premium was specifically recognized for Libyan oil.

In fact, Iraq was not formally challenging Libya's right to a low-sulfur premium. Its claim was that if a quality premium or penalty is once admitted for any one crude oil characteristic, such as sulfur content for example, other than the traditional gravity yardstick, then all other quality characteristics should be considered equally. In any case, there should be no quality differential between Libyan and Iraqi crude oils, and if the former should get a low-sulfur premium, the latter should get exactly the same premium, under any qualification (the specific advantages of Kirkuk crude oil, as stated by the Iraqis, include high-lube oil yield, low pour point, low H_2S, and others). But, should the oil companies accept the Iraqi viewpoint, then the Libyans would be determined to get the same new advantage. This would be the end of the "no leapfrogging" stability that was the real reward for the oil companies in the costly price settlements. And thus the Libyans formally declined to issue any guarantee against further leapfrogging should Eastern Mediterranean oil get better financial treatment than the straight extension of the Tripoli Agreement. This stance was the result of a meeting between the Oil ministers of Iraq, Libya, and Saudi Arabia in Tripoli on April 9-10.

Meanwhile, Saudi Arabia had earlier declared the terms of the Tripoli Agreement satisfactory and stood to accept their extension to its 34° API crude FOB Sidon with adequate adjustment for gravity and freight only. However, in support of Iraq, it declined to sign any price agreement with Aramco until the price dispute in Iraq was settled.

In fact, the hard-line stand and flexible attitude of the Iraqis appeared later as a tactical move intended to strengthen their bargaining position with the IPC group. For the price dispute was not the only issue on the agenda, and it was hoped that the accommodating spirit of oil companies evidenced in the Teheran and Tripoli talks would provide an ideal climate for a satisfactory package settlement to the formidable problems operant between Iraq and IPC for more than a decade (such as Law No. 80, North Rumaila field, back payments, and royal expensing).

And, indeed, the final settlement reached in Baghdad on June 7, 1971 was a package deal in which the Iraqis dropped their claim for a quality compensation in the new posted price to balance the low-sulfur premium accorded to Libyan oil but gained a whole string of lucrative benefits, including a net lump-sum payment of about £13.9 million ($33.6 million) in settlement of oil cost-accounting issues (IPC used a "buffered fixed cost" formula in computing its costs, which henceforth would be calculated on an actual cost basis, while drilling expenses would be amortized at a newly agreed annual rate); an interest-free loan of £10 million on July 1, 1971, with repayment to begin within four years (while repayment of an earlier £20 million loan extended the previous October and due to begin on July 1, 1971 was deferred for another four years); and commitments to increase production and install a deepwater loading facility at Khor el Amaya terminal.

Two major conflicts, however, remained unresolved, namely the IPC expropriation of 99.5 percent of its concession area (including North Rumaila field) and royalty-expensing arrears held by the companies.

As stated above, the price settlement was an extension of the Tripoli Agreement. The base posting of 36° Kirkuk crude oil was fixed at \$2.85/Bbl FOB Banias/Tripoli, including a 35¢/Bbl general increase and 2¢/Bbl gravity and 7¢/Bbl fixed freight adjustments added to the pre-March 20 price of \$2.41. Escalation and temporary freight elements would raise the total posted price to \$3.211/Bbl. The temporary freight premium was calculated at 0.053¢/Bbl (as against 0.058¢/Bbl in Libya) for each 0.1 percent of Worldscale, by which LR 2 AFRA exceeded Worldscale 72, rounded to the nearest tenth of a cent per barrel. The Suez Canal premium was the same as in the Tripoli Agreement. The new Iraqi posting, like that for Libyan oil, was effective March 20, 1971.

The other benefits gained by Iraq in respect to East Mediterranean crude exports in line with the Teheran and Tripoli settlements and OPEC resolutions include

- Increase of the tax rate from 50 percent to 55 percent, effective November 14, 1970.
- Annual escalation of the permanent posting by 2.5 percent plus 5¢/barrel. The first escalation was accelerated to form part of the March 20 price. Subsequent increments will be due on January 1 of 1973, 1974, and 1975.
- Increase of the future fixed payment in lieu of royalty expensing to reflect the financial equivalent of full royalty expensing and the elimination of the marketing allowance (previously 1 percent of the posted price in Iraq). Meanwhile, Iraq continued to reserve its rights as regards past royalty expensing and the difference between 1 percent and 0.5 cents per barrel marketing allowance for 1963-71.
- Implementation of the new gravity differential system—0.15 cents a barrel per one-tenth of a degree API above or below the full degree span of posted price.
- Payment of taxes or royalties on a monthly basis, instead of quarterly as before.

One of the most significant aspects of this package agreement, which was designed to settle most standing problems and to provide favorable grounds for understanding and confidence for future relations between Iraq and IPC, was the implicit recognition by the company that the posted price of $34°$ Bai Hassan/Jambur blend, FOB Banias, had been unduly cut down by 10¢/Bbl since its first application in 1960. The new price was adjusted upwards to align with the $36°$ Kirkuk price after appropriate gravity adjustment by 0.15¢/Bbl/0.1 degree. Past underpricing was retroactively compensated for by the payment of about £4.9 million, this sum being a part of the £13.9 million quit-all payment.

The evolution of Iraqi postings and government oil revenue through the five-year duration of the agreement is shown in Table 16.2. It should be noted that the government income from oil exports from northern fields is not calculated on the basis of posted prices FOB Eastern Mediterranean terminals but, rather, on the Iraqi-Syrian border value estimated at $2.865/Bbl from March 20, 1971.[28]

Shortly afterwards, a similar agreement was signed between Saudi Arabia and Aramco, extending the Tripoli settlement to $34°$ API Arabian crude FOB Sidon. The base posting was set a $2.82/Bbl effective March 20, and the total posted price including escalation, temporary Suez allowance, and freight premium was set at $3.181/Bbl. Similarly to Iraqi postings, the temporary freight premium would be

TABLE 16.2

Evolution of Iraqi Posted Prices and Oil Income
(dollars/barrel)

36° Crude Banias/Tripoli	Posted Price	Govt. Take
Pre-September 1, 1970	2.210	0.940
September 1-November 13, 1970	2.410	1.040
November 14-December 31, 1970	2.410	1.139
January 1-March 19, 1971	2.410	1.199
March 20, 1971-December 31, 1972	3.211	1.671
January 1-December 31, 1973		
Permanent base	3.095	1.601
Variable premiums retained	3.335	1.746
January 1-December 31, 1974		
Permanent base	3.222	1.678
Variable premiums retained	3.462	1.823
January 1-December 31, 1975		
Permanent base	3.353	1.757
Variable premiums retained	3.593	1.902
35° Crude Khor al Amaya		
Pre-November 14, 1970	1.720	0.800
November 14-December 31, 1970	1.720	0.880
January 1-February 14, 1971	1.720	0.950
February 15-May 31, 1971	2.155	1.240
June 1, 1971-December 31, 1972	2.259	1.303
January 1-December 31, 1973	2.365	1.368
January 1-December 31, 1974	2.474	1.434
January 1-December 31, 1975	2.586	1.502

36° Banias/Tripoli: Pre-September 1, 1970—50 percent tax on Iraqi-Syrian border value estimate, $1.864 plus 7¢/Bbl payment as partial royalty expensing; September 1-November 13, 1970—50 percent tax on border value $2.064 plus 7¢/Bbl payment as royalty expensing; November 14-December 31, 1970—55 percent tax on border value $2.094 plus 7¢/Bbl payment as partial royalty expensing; January 1-March 13, 1971—55 percent tax on border value $2.064 plus 13¢/Bbl payment in lieu of royalty expensing; March 20, 1971 onwards—55 percent tax and equivalent 12.5 percent fully expensed royalty on border value adjusted as appropriate (border value on posted price $3.211 is estimated at $2.865).

35° Basrah: Pre-November 14, 1970—50 percent tax, 12.5 percent royalty credited not expensed; November 14-December 31, 1970—55 percent tax, 12.5 percent royalty credited not expensed; January 1-February 14, 1971—55 percent tax plus 7¢/Bbl payment as partial royalty expensing; February 15, 1971 onwards—55 percent tax and equivalent 12.5 percent royalty fully expensed.

Source: Middle East Economic Survey, June 11, 1971.

calculated at 0.053¢/Bbl for each 0.1 percentage point of Worldscale by which LR 2 exceeded Worldscale 72.

Venezuelan Oil

Venezuela was not a participant in the major crisis that shook the international oil industry and resulted in the profound changes in Eastern Hemisphere oil economics outlined throughout this chapter. World attention was focused on Teheran and Tripoli, so that Venezuela could enjoy a quiet position in the shadows while watching the evolution of the situation and preparing for its own action. Its tactics were completely different and did not bother with negotiations or companies' acceptance. Government measures were enacted unilaterally by legislation, and the oil companies were faced with a fait accompli and had no choice but to comply.

This "procedure" was successfully followed in December 1970 to raise the tax rate from 52 percent to 60 percent, and international attention was diverted for only a few moments from the Middle East to take notice of this change. The same procedure was effectively repeated once again on March 8, 1971, when a new pricing structure for Venezuelan crudes and products was published in the Official Gazette, introducing substantial increases in prevailing prices. Although the move was unilateral, the event was mentioned only briefly in specialized journals and passed rather unnoticed while turmoil whirled above the Libyan negotiations.

The Venezuelan decree of March 8, 1971 set "the minimum values FOB Venezuelan loading port . . . for the following types of hydrocarbons and their by-products exported until December 31, 1971," the effective date being March 18. The main features of the decree can be summarized as follows:

1. Crude Oils: Export values, or tax reference prices of Venezuelan crude oil were determined by the relation

$$\text{Tax reference price, \$/Bbl} = 0.0364\, G + 1.070$$

The new price structure changed the value of the basic gravity differential from 3.64 to 1.5¢/Bbl/degree for crude oils of 20° API gravity and above through 53°, and to 3¢/Bbl/degree for crude oils of 19° gravity and below to 7°. Moreover, gravity adjustment would be made in terms of each full tenth of a degree API. The relatively lower value for heavy crude oils was compensated for by a specific premium of 1.5¢/Bbl/degree so that its combination with the 3¢/Bbl/degree basic gravity differential would result in a total 1.5¢/Bbl/degree, as for lighter crude oils. This distinction was actually made to facilitate the marketing of heavy crude oils of 20° and below, which account for roughly 16 percent of Venezuela's total crude production. The competitive position of these crude oils, which have very specific markets (mainly for the manufacture of bitumen), may not support the full increase of prices so that, according to the decree, "the Ministers of Finance and of Mines and Hydrocarbons may decide to grant partial or total exoneration . . . of the premium . . . when deemed advantageous to the national interest."

The tax reference prices of Venezuelan oil are valid only until December 31, 1971, and the government was understood to be preparing new price schedules for subsequent years reflecting escalation related to inflation and other factors. The new Venezuelan pricing structure can be represented by the relations:

$/Bbl, 0.0150 G + 2.200   G = 20.0° API and above
$/Bbl, 0.0300 G + 1.900 + 0.0150 (20.0 - G)A   G = 19.9° API and below

Crude gravity (G) being determined with a precision of one-tenth of a degree API, (A) is a factor in the premium (0.0150 (20.0 - G)A) affecting heavy crude oils, the value of which varies between 0 and 1 so as to reflect the degree of its optional elimination by the government.

2. <u>Natural Condensate and Reconstituted Crude Oils</u> were also priced: $2.30/Bbl for natural condensate and weighted average of constituents' prices making up the blend of reconstituted crude oils.

3. <u>Refined Products</u>: Tax reference prices were set for refined products, which account for about half of Venezuelan exports. A comprehensive and large range of refined products was enumerated, and detailed sets of prices were indicated. Comparison with previous tax reference prices is very difficult because the latter were not published, but it is generally believed that the average increase for refined products as a whole was about 83¢/Bbl, as against an average increase of about 59.5¢/Bbl for crude oils.[29]

The most significant feature of the new products' pricing pattern was the formal introduction of a low-sulfur premium. The tax price of diesel oil/gas oil would be increased by 5¢/Bbl whenever its sulfur content is less than 0.3 percent. More significant was the sulfur premium added to the value of fuel oils with a sulfur content below 2 percent, as shown in Table 16.3. For residual fuel oils, which account for over 30 percent of Venezuelan total oil exports and of which some 80 percent go to the U.S. market, the increase in tax reference prices ranged between 58 and 82¢/Bbl.

All these export values for crude oils and refined products would be increased by a freight premium calculated quarterly, on the first day of the first month of each quarter, by applying a complicated formula, as indicated in Table 16.3.

TABLE 16.3

Selected New Venezuelan Tax Reference Prices
(Effective March 18–December 31, 1971)

Light Products		Residual Fuel Oils (2 percent or more sulfur)		Low-Sulfur Fuel Oils	
Fuel	($/Bbl)	Fuel	($/Bbl)	percent of Sulfur	($/Bbl)
Propane	2.220	Heavy residual	2.310	1.9–1.7	2.818
Butane	2.260	Medium residual	2.410	1.6	2.835
Isobutane	3.205	Light special	2.610	1.5	2.945
Natural Gasoline	2.450	No. 4	2.818	1.4	3.035
Motor Gasoline				1.3	3.115
100 RON[a]	4.221			1.2	3.162
90 RON	3.254			1.1	3.205
80 RON	2.671			1.0	3.235
75 RON (and less)	2.450			0.9	3.275
				0.8	3.315
				0.7	3.355
				0.6	3.399
		Freight Premium		0.5	3.439
		0.33 [(AFRA LR II–72.5) 100 W RTR–(AFRA LR I–100) 100 W PRC][b]		0.4	3.479
				0.3	3.519

[a]Research Octane Number.
[b]AFRA LR II: AFRA value for Long Range II tankers, expressed in terms of percent of Worldscale, whenever equal or higher than 72.5; AFRA LR I: AFRA value for Long Range I tankers, expressed in terms of percent of Worldscale, whenever equal or higher than 100; 100 W RTR: Worldscale flat rate between Ras Tanura–Rotterdam via the Cape, in $/Bbl, for 34° API crude oil; 100 W PCR: Worldscale flat rate between Punta Cardon–Rotterdam, in $/Bbl, for 25° API crude oil.

Source: Compiled by the author.

It was estimated that the government would get additional revenues in 1971 of about $420 million from the higher tax prices and the 60 percent income tax rate, raising its per barrel "take" to $1.42. Including royalties, total oil income in 1971 was expected to be about $1.7 billion.

On December 23, 1971, a new price list to be applicable to Venezuela oil in 1972 was announced in Caracas. Its impact on the industry is best illustrated by the following quotation from Petroleum Intelligence Weekly:

> Once again last week, as it did last December, Venezuela dropped a twin bombshell on the oil companies. This time it is another stiff increase in tax prices on oil exports—on top of those levied for most of 1971—and, in addition, a tax penalty for variations in 1972 oil exports above or below "base" volumes in 1970.
>
> In addition, Venezuela revalued the bolivar and the so-called "oil dollar" rate by about 2.2%, and this too will cost the oil companies more (in terms of Dollars) for their tax payments, which are stated and paid in bolivars. This will be offset to some extent by lower costs for royalty payments, since those are set in dollars but paid in bolivars. The new bolivar rate is changed to 4.40 to $1, from 4.50; and the "oil dollar" to 4.30 from 4.40.[30]

It would be beyond the scope of this study to discuss the economics of Venezuelan oil in more detail, but its impact on pricing patterns is illustrated in Figure 16.1. It should be noted, however, that basic gravity differentials governing tax reference prices were set as follows: 0¢/Bbl/0.1 degree for crudes below 12° and above 40° API gravity; 0.1535¢/Bbl/0.1 degree for crudes between 12° and 29° gravity; and 0.4400¢/Bbl/0.1 degree for crudes between 29° and 40° gravity.

The financial benefits of the new deal were noted by Petroleum Intelligence Weekly as follows:

> The Oil Minister says the average increase in tax prices for crude oils is 26¢ a barrel plus 6¢ for the bolivar revaluation. For products, he put the average hike at 36¢. As a result, the government's per-barrel income in 1972 will jump to $1.61, from $1.41, and its total additional annual income will be an estimated $245 million, including $195 million from the new prices plus $50 million from revaluation.[31]

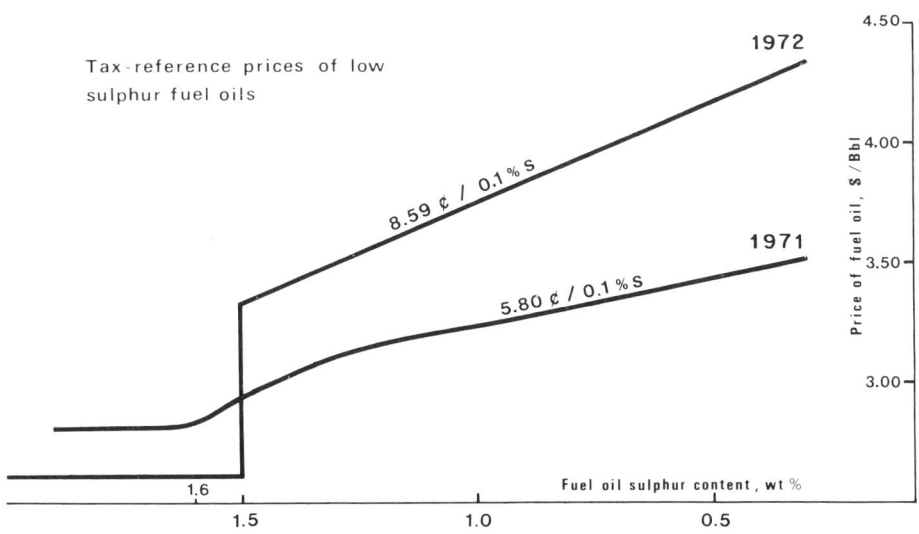
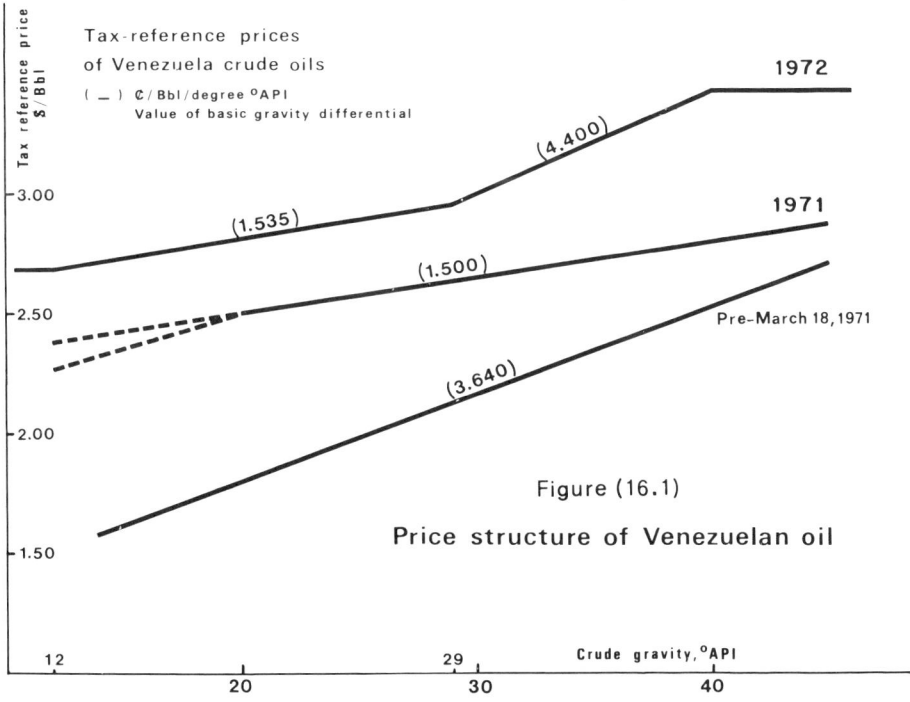

Figure (16.1)
Price structure of Venezuelan oil

THE ALGERIAN CASE

The specific background of Algerian oil was explained and analyzed in some detail in Chapter 8. It was pointed out that Algerian oil was not governed by strictly professional considerations, but that it was only a part, though a major part, in a complex over-all package between Algeria and France, which aimed to establish cultural, economic, and political relations between the two countries on a new basis geared to "cooperation" rather than commercial considerations. The oil sector was intended to be the symbol of the innovative policy designed by Charles de Gaulle and welcomed by the Algerians as a significant step in the evolution of French philosophy towards a new approach to relations with developing countries. The new Algerian oil industry was shaped in the 1965 cooperative agreement, which recognized Sonatrach as a full active partner.

The Algerians claimed to have been disappointed and disillusioned when the cooperative agreement went into effect in the second half of the 1960s. A nationalist, far-reaching strategy, aiming for total or at least majority control of the Algerian oil industry by Sonatrach began to emerge, and within a few years these objectives were attained for non-French assets in Algeria. French interests seemed to be effectively protected by the provisions of the 1965 agreement and insulated against any unilateral Algerian action because oil was only a part of the game and the French had disposed of serious dissuasive and retaliation possibilities that could have proved very harmful to the Algerian economy.

The cooperative agreement was to be renegotiated in 1969. Negotiations were actually begun, at the experts' level, but broke down in June 1970. The failure of these technocratic talks did not surprise anybody. Crude oil prices, by themselves, are mainly governed by strategic and political considerations, as demonstrated throughout this study, and the Algerian-French political and economic imbroglio only added new and complex dimensions to the debate. Therefore, technical and economic arguments could lead nowhere but to a total and profound disagreement when the philosophy and political motivations of the two parties were so far apart; the Algerian starting point was to reject the framework and grounds of what they described as the former "colonial pact," whereas the French were anxious to prolong the status quo.

The Algerians realized more and more clearly that they were in a deadlocked situation with the French, and though the temptation to blow up the whole system was growing, the over-all consequences of such a total break seemed to be dissuasive enough to keep the talks within professional limits. Their failure in June 1970 was a challenge to the dynamic and aggressive Algerian nationalist policy and provoked a bold move on July 27, when tax reference prices were unilaterally raised from $2.08 to $2.87/Bbl.

The French seem to have realized that oil discussion should be made within an overall political approach, and it was precisely on such grounds that talks resumed in Algiers in late August 1970 at the governmental level. The negotiations that extended over the following months were headed by M. Ortoli, the French minister of Industrial and Scientific Development and by Abdelaziz Bouteflika, the Algerian minister of Foreign Affairs. When considered now, long after their dramatic issue, these talks seem to have been inevitably destined to fail. The Algerians seem to have adopted their new policy and strategy from the beginning, and their claims and bargaining tactics were always one or two moves ahead, whereas the French proved to be always one or two battles behind.

The Libyan settlement in September-October and its early extension into the Middle East gave new impetus to the escalation. The French began to say that they would not give the Algerians more than the Libyan terms, which would put the Algerian oil at about $2.73-2.78/Bbl, although the French did not officially offer more than $2.65/Bbl. The Algerian minister of Industry and Energy, Belaid Abdesselam, stated on October 21 that the tax reference price of Algerian oil should be over $3/Bbl, and that this price ought to rise to $3.35/Bbl by 1975. Indications of the new Algerian policy could be indirectly found in the close cooperation and policy harmonization between Algeria and Libya, in the active Algerian role within and outside OPEC, and occasionally in some official statements.

Profound misunderstanding continued to overshadow government talks held in Algiers and Paris, and the negotiating climate kept on deteriorating in spite of apparent official optimism. The new unpredictable prospects provided by the OPEC Caracas Conference and the escalating Libyan demands of January 3, 1971 actively militated against any possible ageement. The French reached a stage where they were convinced that the talks were useless since the Algerians would never be satisfied and would keep on escalating their demands each time the French made special efforts to improve their offers. A final French move offered the Algerians a tax price of $2.65/Bbl for 1969 and 1970, and $2.75/Bbl thereafter, and participation in some French assets. This offer was flatly rejected and Algerian counter-demands included higher prices as well as 51 percent of all production interests. On such terms, French companies declared that it would be impossible for them to operate on a commercial basis, particularly since Algeria made two further demands: substantial French financial participation in a big new exploration effort, and an obligation to domicile the total receipts from their crude oil export sales in Algeria.

Shortly afterwards, the point of no return was reached and then passed by the French Government, which declared, on February 5, the suspension of talks until the results of the Teheran negotiations were known. The only way the French saw to shield themselves against the ever escalating Algerian demands, which then went beyond the price issue to include concession ownership, was to join the united front of the international oil community represented by the "January 16 Group" of oil companies. In any case, the French could not escape aligning with any prospective Teheran settlement, but they wanted to be sure that they would not be committed to more than that. The Algerian reaction described the French move as "dilatory," and Bouteflika officially requested, on February 8, the immediate reopening of serious negotiations.

On February 24, the day after the Oil ministers of Mediterranean exporting countries met in Tripoli to delegate Libya to negotiate on their behalf with oil companies, Algerian President Houari Boumédienne announced that Algeria had unilaterally decided to take over a majority holding (51 percent) in all French oil interests in Algeria and to nationalize fully all hydrocarbon transportation as well as all gas resources. As far as oil was concerned, these measures, following the previous takeover of all non-French interests, realized Algeria's long-held ambition to bring all oil production within the country under state control. There were about a dozen companies, but the major shareholders were the state-owned Elf-Erap and Compagnie Française des Pétroles (Algerie). In 1970, Algerian oil production amounted to some 47.5 million tons, of which 33 million tons were produced by the French: 16.5 million tons by Elf/Erap (representing about 80 percent of its total crude production), 11.6 million tons by CFP (representing about 20 percent of its total crude production), and 3.9 million tons by different French private interests. However, the degree of transfer of ownership varied from one company to another because Sonatrach was already a minority partner, at different rates, in most of them.

The French Government was not really surprised by the Algerian decision, although its sudden timing was rather unexpected. The official French position was formulated on March 9 in a memorandum that recognized Algeria's sovereign right to nationalization but stressed the necessity of fair compensation. The desire for continued cooperation, with or without oil, was reconfirmed and emphasized. Thus, the French seemed to capitulate to the fait accompli and did not implement the widely publicized retaliatory measures. The new official position was to consider oil affairs on a strictly commercial basis, and the role of the government would be to ensure a satisfactory overall settlement, with fair compensation, so that a new era of cooperation could be initiated. In particular, sufficient guarantees should be granted to French companies then holding minority shares in Algeria. But Algeria considered these demands incompatible with its sovereignty, the tension mounted, and there occurred a series of minor conflicts and confrontations with French companies operating in Algeria. The final blow came on April 13, when Algeria took unilateral decisions settling standing problems on its own conditions. These were

- A rise in the tax reference price for Algerian crude from the $2.08/Bbl level set in the 1965 Algiers Agreement to $3.60/Bbl and the introduction of OPEC fiscal terms (12.5 percent expensed royalty, 55 percent income tax rate), rather than the straight 55 percent income tax paid under the 1965 agreement.
- A new oil code abolishing the concession system in Algeria and transferring all rights to Sonatrach, which would henceforth be permitted to take foreign partners on a 51/49 basis.
- A unilateral assessment of compensation for the French companies wholly or partly nationalized on February 24 at $100 million, to be paid in crude oil.

The first two points were already accepted in principle by the French, but the Algerian assessment of compensation at $100 million was about one-seventh of the French evaluation of their interests in Algeria. And worse, the compensation was to be more than offset by Algerian claims for tax arrears. On such conditions, the French Government decided, on April 14, to end government negotiations and to let the companies concerned take care of their own affairs on a strictly legal and professional basis. The latter took a number of retaliatory and protective steps including withdrawal of technical personnel from Algeria, stoppage of oil liftings, and addressing formal warning to international companies and banks that might be involved in the purchase of "hot" oil or gas from the nationalized fields. The most important move was to prevent the finalization of a big gas deal with El Paso. Requests were made to the U.S. Government to withhold its approval through the Federal Power Commission and to the World Bank, from which Algeria was seeking financing for its giant $300 million project. Algeria counterreacted by active public relations campaigns, and Sonatrach publicized its case in lengthy articles and studies published, as an advertisement, in major world newspapers.

However, the battle was unequal, especially after the political withdrawal of the French Government, and the long-term effectiveness of company measures were very doubtful. GFP, which was less emotionally motivated than Elf/Erap, took the initiative of resuming the dialogue with Sonatrach on May 25, and an agreement was quickly reached on June 30. Its main provisions were as follows:

- Algeria's February 24 partial nationalization of French oil interests is confirmed. CFP is to establish a company in Algeria that will hold 49 percent of the capital of Alrep, the company set up by Sonatrach to take over CFP's interests, while Sonatrach will retain a 51 percent majority interest.
- Compensation for nationalized CFP assets is set at 300 million dinars ($60 million), payable in seven equal, interest-free annual installments beginning in May 1972.
- CFP will remit to Algeria $2.75 for each barrel of oil exported in 1971 and 1972. For subsequent years the sum to be remitted will be fixed annually by the Algerian Government and will, presumably, take into account any changes affecting the market value of Algerian crude.
- Back taxes owed by CFP for 1969 and 1970, which are estimated to be in the area of $25-$30 million, will be paid in six monthly installments between August 1 and December 31, 1971. Meanwhile, taxes on current production will be assessed on the fiscal basis announced by President Boumédiene on April 13: a reference price of $3.60/Bbl, 12.5 percent expensed royalty, and a 55 percent income tax.
- An agreed investment program has been drawn up for the period until 1975 that will involve expenditures by CFP of some $100 million, notably in raising annual production from the Hassi Messaoud field to 30 million tons in 1973.
- The agreement expires on December 31, 1980. Nonetheless, the provisions relating to taxation and remittance and investment obligations will be renegotiated in 1975, and if, by October 31 of that year, the two sides fail to agree on revised terms, the agreement will lapse on December 31, 1975.

Clearly, the agreement represented an acceptable compromise for both sides. In return for a break in the boycott of its oil exports, which is reported to have been unexpectedly successful, Algeria moderated its initial demand for a remittance of $2.95/Bbl and raised its compensation offer by an undetermined amount from the original $100 million for the nationalized assets of both CFP and the French state oil group Elf/Erap. CFP, on the other hand, regained access to its Algerian crude supplies, which amount after nationalization to some 6.7 million tons annually, in exchange of accepting and confirming Algeria's sovereignty in oil matters.

An agreement with Elf/Erap followed a few months later along the same lines, but the details of its provisions were more complex because of the intricate structure of Elf/Erap's assets and holdings in Algeria.

The parties agreed that the exploitation of Algerian oil would be under majority control of the Algerian Government through Sonatrach and would be governed by the Algerian Petroleum Legislation issued on April 12, 1971, which made fundamental changes in the previous Petroleum Code. Its main features may be outlined as follows:

1. New posted prices, with annual increases until 1975, along with an OPEC tax and royalty rate and a variable short-haul premium of 25 cents.
2. The transfer of all mining titles to Sonatrach and the restriction of foreign participation in oil ventures to a maximum of 49 percent. Gas and condensate to be the exclusive property of Sonatrach.
3. Foreign minority partners are entitled to their share of oil production at cost but must repatriate to Algeria the entire proceeds of export sales. These may then be remitted abroad after fulfillment of the company's financial obligations.
4. Producing companies must invest sums for a continuous replacement of reserves and must sell oil at prices consistent with prices on the international market.
5. In the event of a dispute on tax matters, detailed provisions are set out for arbitration by a Conciliation Board, with recourse to the Supreme Court of Algiers, if necessary.

The new price structure was set as follows:

1. Retroactive increase of the tax reference price for the period 1969-70 from \$2.08 to \$2.77/Bbl (FOB Bejaia), without royalty expensing.
2. Royalty expensing, as of January 1, 1971, at 12.5 percent of tax reference prices, plus income tax at 55 percent.
3. Introduction of a new scale of gravity differential as follows: increase by 0.2¢/Bbl for each full one-tenth of a degree above $44°$ API; decrease by 0.2¢/Bbl for each full one-tenth of a degree between 44 and $40°$; decrease by 0.15¢/Bbl for each full one-tenth of a degree below $40°$.
4. The tax reference price FOB Bejaia for the period January 1-March 19, 1971, fixed at \$2.70/Bbl.
5. As of March 20, 1971, the base price of oil of $44°$ API to be
    (a) until 31 December 1971, equivalent, for oil of $44°$, to \$3.35 U.S. per barrel FOB Bejaia and FOB Skikda; \$3.365 U.S. per barrel FOB Arzew; \$3.320 U.S. per barrel FOB La Skhira (Tunisia).
    (b) from January 1, 1972 the values fixed in paragraph a above to be increased by 2 cents.

(c) from January 1, 1973, 1974, and 1975 the values fixed in paragraph a above and increased as stated in b above, to be each year further increased (1) by a sum, calculated to the nearest one-tenth of a cent, equal to 2.5 percent of the base element in force on December 31 of the preceding year; and (2) by a sum equal to 7 cents.

6. A complementary freight element of 25¢/Bbl to be added to the base price for the period March 20-June 30, 1971. The method of calculating this element for subsequent periods to be determined by decision of the minister of Industry and Energy.

OIL TRANSIT COUNTRIES

Oil transit countries benefited from the general increase and improvement of oil economics in the Middle East, and particularly at the Eastern Mediterranean, where the terminals of the Tapline and IPC lines are located. Crude pricing is not connected with transit economics and terms, and the level of Eastern Mediterranean new postings was determined as indicated earlier in this chapter. Transit fees and payments are considered as expenses among others and are deducted from posted prices in order to calculate the taxable income in producing countries. Nevertheless, they are of importance in evaluating the tax-paid cost of Eastern Mediterranean oil and could thus influence, to some extent, the economics of lifting programs.

On the professional side, Syria played an active role in the rapid development of the major oil crisis. On May 3, 1970, the Tapline was slightly damaged in Syrian territory, and its repair required a quite simple operation. Nevertheless, the Syrian Government firmly opposed repair under different pretexts. The stoppage of the flow of some 25 million tons of Saudi Arabian oil, coupled with production cuts in Libya, caused freight rates to shoot up, as indicated earlier, and the continuity of supplies to Western Europe seemed, for a while, uncertain. The freight crisis played a major role in determining the outcome of the crisis as formulated in the Tripoli and Teheran Agreements.

It was not until January 28, 1971, that an agreement was reached that enabled the repair to be done rapidly, and the flow of oil resumed on January 30. The first full day of operation, January 31, saw 497,000 barrels received at the tank farm in Sidon.[32]

Shortly afterwards, similar arrangements were reached with the other transit countries, namely Jordan and Lebanon, taking effect retroactively as of January 30, 1971. The main terms of these agreements may be summarized as follows:

1. Payment of a lump sum of $9 million, to each government, of which $4 million will be paid immediately and the remainder in six-monthly installments thereafter.

2. Increase in the general pipeline transit tariff by 1.3¢ for 100 barrel/miles—from the previous rate of 1.8¢ set in 1962 to 3.1¢ for 100 barrel/miles. This increase would boost payments by 72.2 percent.

3. Annual payment of $800,000 to Lebanon and Syria, under specific conditions, in connection with Tapline's commitment to supply crude for local refining.

4. Increase in two types of fees at Tapline's loading terminal at Sidon-Lebanon, which are subject to a benefit-sharing agreement between Syria and Lebanon (concluded on October 28, 1949). The terms are as follows:

	Terminal Fee (¢/Bbl)	
	Old Rate	New Rate
	2.00	4.000
Syrian share	0.50	1.154
Lebanese share	1.50	2.846
Loading Fee (¢/ton)	3.50	5.250
Syrian share	1.75	2.625
Lebanese share	1.75	2.625

Taking into account the specific mileage of the Tapline through each country, transit fees indicated in 3 above would be as follows:

Syria	2.451¢/Bbl
Lebanon	0.801¢/Bbl
Jordan	3.568¢/Bbl
	3.379¢/Bbl (for crude delivered to the government-owned Zarka refinery)

On July 6, 1971, another agreement was concluded between Syria and the IPC group substantially increasing transit fees and dues. Its main terms may be summarized as follows:

1. Transit Rates

	1971 Agreement	1967 Agreement	Percent Increase
Transit fees	17.808¢/Bbl	10.77¢/Bbl (6s.92/3d./ton)*	65.3

	1971 Agreement	1967 Agreement	Percent Increase
Terminal fees	4¢/Bbl	3.69¢/Bbl (2s.4d./ton)*	8.4
Port dues	0.692¢/Bbl	0.46¢/Bbl (3.5d./ton)*	50.4
Total	22.5¢/Bbl	14.92¢/Bbl	50.8

2. In addition to the agreed new rates, IPC is to pay to Syria a temporary Suez Canal premium of 0.5 cents a barrel of oil transiting Syrian territory.

3. IPC has agreed to pay the transit royalties to Syria on a monthly basis instead of on a quarterly basis as before.

4. IPC has agreed to pay a lump sum of £14 million ($33.6 million) to Syria.

5. IPC has agreed to double the discount off posted prices on crude oil supplied to the Syrian state-owned Homs refinery from 22¢/Bbl to 44¢/Bbl on the first 600,000 tons a year. (An additional discount of 3.85¢/Bbl is applicable on supplies in excess of the first 600,000 tons a year.)

A similar arrangement was concluded between Lebanon and IPC a few months later.

ASSESSMENT AND CONSEQUENCES OF THE CRISIS

The major crisis that shook the international oil business to its foundations, and the Teheran and Tripoli Agreements it led to, produced basic and fundamental changes in the many aspects of the industry so that the industry will never again be as it was before 1970. The most significant change is to be found in the political area, where OPEC was formally recognized as representative of the community of oil-producing countries, which have demonstrated that, by exercising their sovereign rights over their natural resources and by uniting their efforts and wills, they can impose themselves as full, if not preponderant partners in the mechanisms of policy and decision-making of international oil. The time of oil companies enjoying the exclusive

---

*The sterling rates were raised by one-sixth to compensate for the devaluation of the pound sterling in November 1967. The rates as originally given in the 1967 agreement were 5s.10d./ton, 2s./ton, and 3d./ton, respectively.

privilege of unilaterally setting crude oil prices has passed into history. Furthermore, the victory of the producing countries was so decisive that it encouraged them to proceed a step further and challenge the philosophy and the very nature of conventional concession agreements by requesting direct state participation.

However, a profound analysis of the political aspects of the crisis and an evaluation of its significance and consequences from the viewpoint of long-term strategy will lead to more complex conclusions, as will be demonstrated in the following chapter. We will confine our investigation hereafter to a "technocratic" approach; the impact of the Teheran and Tripoli Agreements on crude oil pricing patterns in the Eastern Hemisphere will be investigated before we evaluate their fiscal and financial consequences.

A significant modification in crude oil pricing patterns in the Gulf area was the change in the value of the basic gravity differential from 2 to 1.5¢/Bbl/degree, applicable to crude oils of $40°$ gravity or below. This added 0.5¢/Bbl for each degree below $40°$, over and above the general increase of 35¢/Bbl. For example, the posted price of $34°$ light Arabian FOB Ras Tanura was increased by 38¢/Bbl (from $1.80 to $2.18/Bbl), of which 3¢/Bbl accounted for additional gravity adjustment. Furthermore, gravity escalation will be made henceforth in terms of one-tenth of a degree API gravity (prices will be rounded to the nearest tenth of a cent), instead of having the posted price constant over the full range of one degree gravity. This will add a further although minor increase to government oil revenues.

This move actually represented the ultimate consecration of the fiscal motivation and significance of posted prices and the formal end of explaining or justifying their pattern or structure on economic or professional grounds. The change in the basic gravity differential— corresponding to a change in the slope of linear price scales in the Gulf—was merely intended to increase the prices of crude oils heavier than $40°$ without affecting the value of lighter crudes. This was effectively achieved by retaining $40°$ as the reference gravity. In fact, this provision of the Teheran Agreement was only the application of one of the specific innovations of the September 1970 deal in Libya, which had been introduced in the price arrangement with Occidental Petroleum, as described earlier in this chapter.

This kind of artificial price manipulation is not new to the industry. It was utilized by major oil companies in the early 1950s, when they "arbitrarily" defined the pricing pattern of the Gulf area oil on the basis of a basic gravity differential of 2¢/degree with respect to a reference gravity of $36°$, as noted in chapter 15. Their move was intended to recover a part of the price reductions they were forced to accept under the pressure of the Economic Cooperation Agency. No economic significance should be sought for this artificial (mechanical) price manipulation, and the Gulf countries should be grateful to Libya for having set the precedent in September 1970, the benefits of which were minor to Libya itself but proved to be very significant to Middle Eastern oil.

Nor should economic significance be sought in the narrowing of the range of price indetermination down to one-tenth of a degree API. We have shown, in Chapter 14, that keeping prices constant over the range of one full degree gravity represented an indirect way of expressing normal fluctuations in the economics of the industry and the normal indetermination they induce in the over-all correlations and approximations upon which oil economics and pricing patterns are based. The move is no more than a further "mechanical" manipulation designed to increase prices and should not be construed to mean that the over-all economics of the industry, as reflected in the pricing structure, can be determined with an approximation 10 times more precise than was previously the case.

Furthermore, specific price adjustments were introduced to eliminate politically induced price disparities and penalties in the Gulf area. In particular, the 10¢/Bbl investment discontinuity penalty affecting the prices of heavy crude oils was ended, as were the 6¢/Bbl penalty affecting the posted price of Basrah crude oil FOB Khor el Amaya and the 10¢/Bbl penalty affecting the price of Bai Hassan/Jambur blend FOB Banias. The companies' motivations in these "voluntary" moves are political since the companies' main objective in the price agreements was to secure stability, and stability could be endangered by potential sources of conflicts resulting from nonprofessional price discriminations.

The new pricing pattern might be investigated and checked by calculating prices as per FOB Quoin Island in order to eliminate specific freight disparities in the Gulf. The results are illustrated in Figure 16.2, which confirms the structure of linear price scales with a slope of 1.5¢/Bbl/degree. It should be noted, however, that the problem of heavy crude oils of less than 30° API has not been fully settled. The prices of Arabian heavy and Khafji crude oils display a specific penalty of about 7¢/Bbl over and above their high-sulfur penalty, and the posted prices of crude oils from the Neutral Zone (Burgan, Ratawi, and Eocene) are still at their previous levels.

Posted prices in the Gulf, effective February 15, 1971, after specific adjustments, whenever required, for high-sulfur and/or other penalties, may be represented by Equations 16.1 and 16.2 applicable to prices as per FOB Quoin Island:

$$X = 0.015\, G + 1.670 + 0.043\, W \qquad G \leq 40 \qquad (16.1)$$
$$X = 0.020\, G + 1.470 + 0.043\, W \qquad G \geq 40 \qquad (16.2)$$

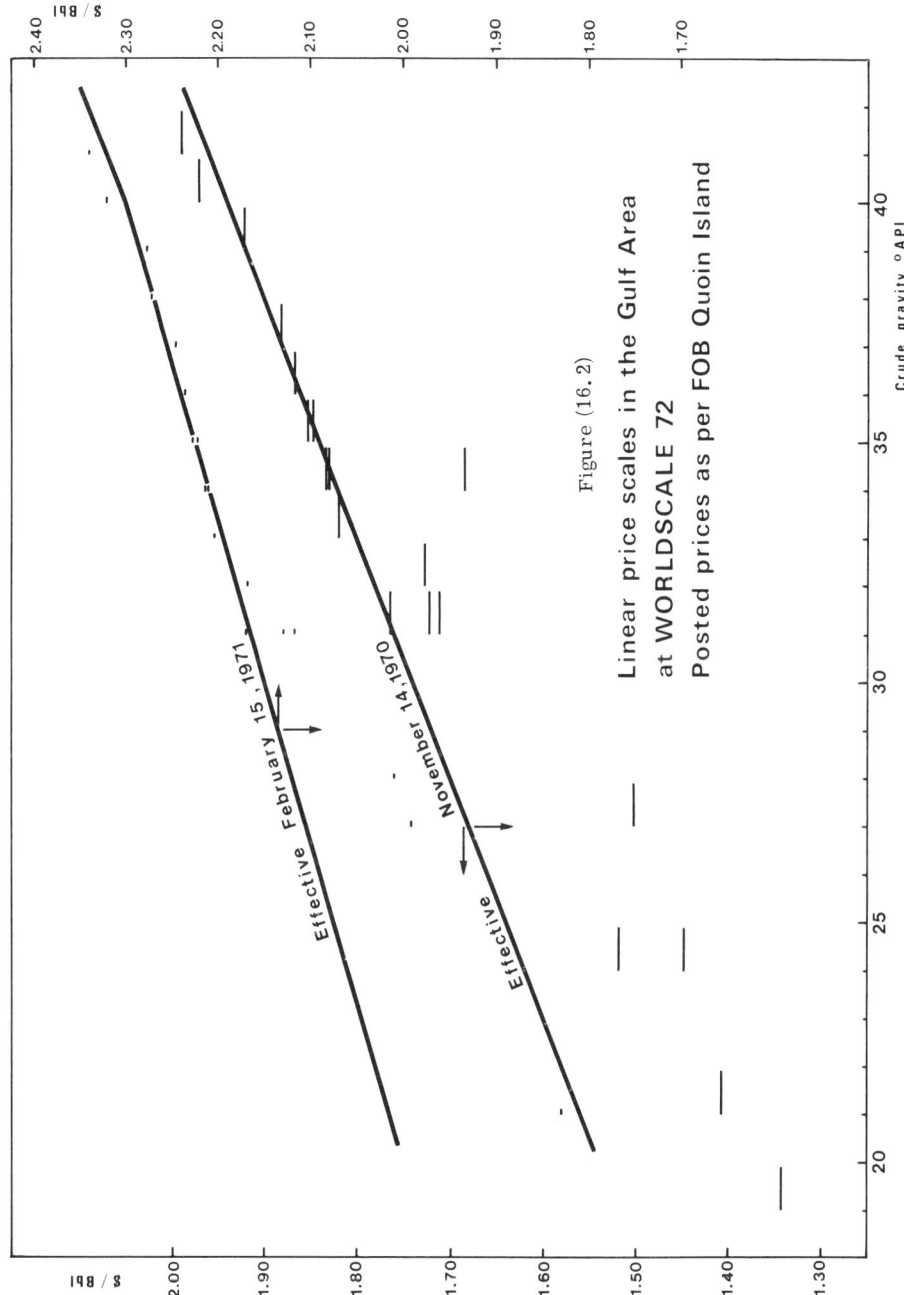

Figure (16.2)

Linear price scales in the Gulf Area at WORLDSCALE 72
Posted prices as per FOB Quoin Island

where:
X, $/Bbl, is crude price as per FOB Quoin Island;
G, °API, is crude gravity expressed with a precision of one-tenth of a degree; and
W is freight variation as a percentage of Worldscale.

As already noted in Chapter 6, the linearity of price scales is not significantly affected by freight variations. Moreover, the pricing pattern may be represented as an almost continuous function of gravity (G) measured in such a small unit as one-tenth of a degree.

Equations 16.1 and 16.2 will prove helpful in illustrating the impact of escalation arrangements provided for in the Teheran Agreement. Prices FOB export terminals would be increased by 2.5 percent plus 5¢/Bbl on June 1, 1971, and on January 1 of the years 1973, 1974, and 1975. The general flat increase of 5¢/Bbl would not alter the pricing pattern but would represent a collective move reflected in the vertical sliding of linear price scales. By contrast, the percentage increase would change the value of the basic gravity differential, which change would be reflected in a change in the slope of linear price scales as well as in their vertical sliding. Consequently, pricing patterns as per FOB Quoin Island, over the five-year period of the agreement, would be as follows:

Period	Pattern	
2/15-5/31/71	$X_1 = 0.0150\,G + 1.670 + 0.043\,W$	$X_1 = 0.0200\,G + 1.470 + 0.043\,W$
6/1/71-12/31/72	$X_2 = 0.0154\,G + 1.762 + 0.043\,W$	$X_2 = 0.0205\,G + 1.557 + 0.043\,W$
1973	$X_3 = 0.0158\,G + 1.856 + 0.043\,W$	$X_3 = 0.0210\,G + 1.646 + 0.043\,W$
1974	$X_4 = 0.0162\,G + 1.952 + 0.043\,W$	$X_4 = 0.0215\,G + 1.737 + 0.043\,W$
1975	$X_5 = 0.0166\,G + 2.051 + 0.043\,W$	$X_5 = 0.0221\,G + 1.830 + 0.043\,W$

These price patterns underwent a further significant modification on January 20, 1972, as a consequence of the Geneva Agreement, which increased base postings by 8.49 percent to account for the devaluation of the dollar on December 17, 1971. The conjugated impact of the Teheran and Geneva Agreement will thus set pricing patterns as per FOB Quoin Island as follows:

Year	Pattern	
1972	$X_2 = 0.0167\,G + 1.911 + 0.0466\,W$	$X_2 = 0.0224\,G + 1.689 + 0.466\,W$

Year	Pattern	
1973	$X_3 = 0.0171\ G + 2.013$ $+ 0.0466\ W$	$X_3 = 0.0233\ G + 1.884$ $+ 0.0466\ W$
1974	$X_4 = 0.0176\ G + 2.118$ $+ 0.0466\ W$	$X_4 = 0.0233\ G + 1.884$ $+ 0.0466\ W$
1975	$X_5 = 0.0180\ G + 2.225$ $+ 0.0466\ W$	$X_5 = 0.0240\ G + 1.985$ $+ 0.0466\ W$

In the meantime, crude prices will continue to be adjusted, individually, by 0.15 ($G \leq 40°$) and 0.20 ($G \geq 40°$)¢/Bbl per each one-tenth of a degree API gravity. In other words, the gravity fluctuation differential will stay constant throughout the five-year period, whereas the basic gravity differential will change every year so as to reach 0.18 ($G \leq 40°$) and 0.24($G \geq 40°$)¢/Bbl/0.1 degree. This amounts to another illustration of the difference between the two concepts of gravity differential.

The financial consequences of the Teheran and Tripoli Agreements are quite impressive. In May 1971, total government revenue of the OPEC member countries throughout the five-year period 1970-75 was estimated at about $90 billion,[33] of which some $31.5 billion represents the incremental revenue in connection with the increase of the government take. The actual figure of government revenue will be higher as a consequence of the Geneva Agreement and of the acceleration of production expansion, especially in the Gulf area. Now, if we recall the breakdown of the consumer barrel as elaborated by OPEC and given in Table 4.1, the relative part of payments to producing countries in the ultimate total bill charged to consumers could be set at about 10 percent (against 7.9 percent before 1970), so that the total cost to be paid by the ultimate consumers for products derived from oil exported by OPEC member countries would be around $900 billion. According to the OPEC breakdown, about 45 percent of this sum, say $405 billion, would represent government revenue in the consuming countries themselves. This figure should be mentioned to anybody who may claim that the "fantastic" payments made to producing countries represent an unreasonable and unbearable burden to "poor" consumers and a factor contributing to inflation.

Crisis periods have always been beneficial to the oil companies, and, as in 1957 and 1967, the 1970-71 upheavals proved to be very "juicy" for the international majors, as shown in Table 16.4. The financial statements of the five U.S. international major oil companies show substantial improvement in net earnings for the first and second quarters of 1971, compared with the same periods of 1970. If we consider figures for the first half of the year, it appears that the profits of the five U.S. majors increased by an average of 17 percent as a group, compared with an over-all decline of 4.2 percent in 1970.

TABLE 16.4

Comparative Net Earnings of U.S. Oil Companies

Major U.S. Internationals	First Quarter Net Earnings (millions of dollars)		Percent Change	Net Earnings, 1971 (millions of dollars)			Half-Yr. Change (percent)
	1971	1970		Second Quarter	First Half		
Standard (N.J.)	374.0	319.0	+ 17.2	352.0	726.0		+ 25.0
Texaco	236.8	205.4	+ 15.3	203.6	440.4		+ 15.4
Gulf Oil	146.2	139.2	+ 5.0	149.8	296.0		+ 6.1
Mobil	132.5	117.8	+ 12.5	124.7	257.2		+ 13.1
Standard (Cal.)	119.9	103.6	+ 15.7	127.0	246.9		+ 17.0
Subtotal	1,009.4	885.0	+ 14.1	957.1	1,966.5		+ 17.0
Mainly domestic							
Standard (Indiana)	92.5	85.6	+ 8.1	82.9	178.1		+ 11.7
Atlantic Richfield	53.7	53.8	- 0.2	50.1	103.9		+ 0.7
Shell (U.S.)	45.1	55.0	- 18.0	47.2	92.3		- 19.2
Continental	37.6	33.0	+ 13.9	37.4	75.0		+ 7.7
Sun Oil	37.0	30.1	+ 22.9	37.8	74.9		+ 16.6
Phillips	36.1	32.0	+ 12.8	27.6	63.7		+ 10.4
Cities Service	34.3	38.5	- 10.9	28.4	62.7		- 7.2
Getty Oil	31.2	24.0	+ 30.0	31.7	62.7		+ 38.2
Union Oil	29.7	23.2	+ 28.0	26.9	56.6		+ 6.8
Marathon	21.1	19.1	+ 10.5	22.0	43.1		+ 9.4
Subtotal	418.3	394.3	+ 6.1	392.0	813.2		+ 9.4
Total	1,427.7	1,279.3	+ 11.6	1,349.1	2,779.7		+ 14.5

Source: Petroleum Intelligence Weekly, May 3 and August 2, 1971.

The most dramatic turnaround was a 25 percent jump for Standard Oil of New Jersey, as against a 4.6 percent decline in the first half of 1970. Second follows Standard of California, with a 17 percent increase, as against a 7.8 percent decline the year before. The poorest performance was registered by Gulf, which increased first-half 1971 profits by only 6.1 percent, as against a record drop of 13.9 percent in 1970.

These promising performances of the U.S. international majors are mainly due to their worldwide business outside the United States, especially in the Eastern Hemisphere. This is in marked contrast to the first-half 1971 performances of other big U.S. oil companies whose activities are mainly domestic. The average increase in net earnings from domestic operations of these latter companies for the first half of 1971 over 1970, was only 9.4 percent, as against 17 percent for the international majors, thus bringing the over-all U.S. average increase down to 14.5 percent.

However, the best performance by far was registered by British Petroleum: Its net profits for the second quarter of 1971 were almost triple its net for the same quarter in 1970. Nonetheless, second-quarter profits of $90 million were below $117 million for the first quarter. Total net profits for the first half-year ($207 million) represented a 127.5 percent increase over the same period of 1970.

NOTES

1. A contemporary statement quoted in Libyan newspapers.
2. Petroleum Intelligence Weekly (PIW), April 13, 1970.
3. PIW, May 25, 1970.
4. PIW, June 15, 1970.
5. PIW, July 27, 1970.
6. Oil and Gas Journal (OGJ), September 14, 1970, p. 64.
7. Middle East Economic Survey (MEES), October 16, 1970.
8. MEES, October 2, 1970.
9. PIW, November 30, 1970.
10. Ibid.
11. PIW, December 14, 1970.
12. PIW, November 23, 1970.
13. Fortune Magazine, August 1971, p. 113.
14. MEES, May 9, 1969.
15. MEES, May 23, 1969
16. Ibid.
17. PIW, September 7, 1970.
18. MEES, December 18, 1970, Supplement.
19. Specifically, the highlight of the conference was Resolution XXI-120, which read in summary:

"Considering the general improvement in the economic and market outlook of the international oil industry, as well as its competitiveness with other sources of energy:

[the Conference] resolves that all member countries adopt the following objectives:

(1) To establish 55 percent as the minimum rate of taxation on the net income of oil companies operating in the Member Countries.

(2) To eliminate existing disparities in Posted or Tax Reference Prices of crude oils in the Member Countries on the basis of the highest Posted Price applicable in the Member Countries, taking into consideration differences in gravity and geographic location and any appropriate escalation in future years.

(3) To establish a uniform general increase in the Posted or Tax Reference Prices in all Member Countries to reflect the general improvement in the conditions of the international oil industry.

(4) To adopt a new system for the adjustment of gravity differential of Posted or Tax Reference Prices on the basis of $0.15¢/Bbl/0.1°$ API for crude oil of $40°$ API and below, and $0.2¢/Bbl/0.1°$ API for crude oil of $40.1°$ API and above.

(5) To eliminate completely the allowances granted to oil companies as from the first of January, 1971."

20. MEES, January 15, 1971.
21. MEES, January 29, 1971.
22. MEES, February 19, 1971.
23. PIW, January 25, 1971 gives some details about these deals.
24. PIW, January 18, 1971.
25. Resolution XXII-131, OPEC Extraordinary Conference, Teheran, February 4, 1971.
26. PIW, April 5, 1971, Supplement.
27. PIW, May 3, 1971.
28. MEES, June 11, 1971.
29. PIW, March 15, 1971.
30. PIW, December 27, 1971.
31. Ibid.
32. OGJ, May 10, 1971, p. 89.
33. PIW, May 24, 1971.

CHAPTER 17

FUTURE OUTLOOK

THE AMBIGUITIES OF A MOST PECULIAR CRISIS

In the last nine months there have been radical changes in the world of oil and 1970 may appear historically as a watershed in the history of the industry. This is not because the basic elements of the scene have changed dramatically overnight; such things do not happen in the world of energy economics. It has come about because certain happenings, some fortuitous, some man-made, have enabled a totally new practical interpretation to be put on these basic elements in such a fashion that we need to rethink the future in a different fashion from the past.[1]

Several attempts were made to "explain and justify" the crisis and its consequences, and arguments were mostly based on professional and economic grounds. For some observers, the most important change that took place in 1970 and that prepared the way for the upheavals, was "the shift from a buyer's to a seller's market and intimately tied into that [was] the parallel shift of power from the international oil companies to the governments of oil-producing countries."[2] M. A. Adelman noted that "the power over oil has shifted dramatically to the OPEC countries, ending a 15 year buyer's market. . . . But nobody foresaw the extent to which [the foreign producing nations] would act together and threaten to withhold supply if their terms were not met."[3] On the other hand, it was alleged that economists in consuming countries and oil companies were mistaken in their forecast of oil needs since "there has been an unprecedentedly rapid rise in the demand for oil in Western Europe reflecting both a higher than expected economic expansion and a failure both of coal production and nuclear energy to meet their targets."[4]

The direct and immediate cause of the crisis was the acute tanker shortage that followed the sabotage of the Tapline and the cuts in Libyan production resulting in a sudden gap of about 50 million tons in Mediterranean supplies. Freight rates shot up to unprecedented highs, and even the continuity and security of European supplies seemed to be threatened. However, the determining factor is to be found in the political arena. The revolutionary regime in Libya showed, for the first time, "the determination and preparedness of a government enjoying geographical advantage and financial strength to use its political power in order to enforce a change of existing agreements by coercion."[5] Moreover, OPEC proved to be politically mature and professionally prepared to unite as a "producers' cartel" in the wake of the Libyan victory. Unanimous resolutions were passed once again, in the Caracas Conference in December 1970, but this time, they were followed by action and inflexible determination. In the face of this united front of producers, the oil companies seemed very eager to achieve one objective: to capitulate in an orderly and rapid fashion. The united front of the "January 16 Group" of oil companies went to Teheran to sign an agreement, not to engage in a battle, and within few weeks the crisis had ended, in spite of some suspenseful moments in the Tripoli episode.

But one cannot refrain from feeling quite uneasy, in a rather confused way, about explanations suggested by various observers. This intellectual reluctance results from the fact that the arguments noted above mainly concentrate on "tactical" aspects—which, alone, cannot be expected to have induced change—and that they ignore the fundamental strategic aspects of the crisis. Moreover, they fail to point out the direct involvement of U.S. strategic interests or objectives; the only American interference even indirectly referred to consists of discreet information about the travels of John Irwin II, U.S. under-secretary of State, throughout the Middle East prior to the Teheran settlement. The pricing philosophy that has been developed in this study would suggest that the new price arrangements could not have taken place, a priori, if they were not commanded by a new reassessment of American domestic oil strategies. We could even add that these arrangements would have been imposed on world oil economics in one way or another. And, in fact, it is obvious that the U.S. Government brought all its weight into play so as to make the Teheran price settlement materialize without delay. The active role of the Nixon Administration is outlined in Chapter 16: consultative meetings, the suspension of antitrust proceduers, Irwin's travels, and so forth.

Now, bearing all the above in mind, we cannot refrain from regarding the tactical arguments with suspicion and scrutinizing them cautiously in order to check the validity and accuracy of the explanations they are supposed to provide. And, indeed, such a critical approach would result more in raising ambiguities than in providing over-all satisfactory explanations. Several questions emerge unexpectedly and remain unanswered as long as the strategic dimensions of the pricing issue are not fully taken into consideration.

Among the most significant ambiguities of the crisis, we might consider the following:

1. The acute freight crisis that developed in the spring of 1970 is alleged to be the direct cause of the changes and upheavals of the industry. Because of the combination of the closure of the Tapline on May 3, 1970, the cutbacks of Libyan production, and the fuel oil shortage on the U.S. East Coast, the tanker market became very tight all of a sudden and spot rates shot up within a few weeks to set unprecedented records approaching Worldscale 300, followed by high AFRA rates for all categories of tankers. Specialists have no difficulty in describing the tight supply conditions that can fully justify the freight crisis. But they fail to explain the following subjective and underlying considerations: There were certainly objective reasons for a tanker shortage in the spring of 1970, but the consequences that resulted in the freight market were disproportionately high and exaggerated in comparison with what would have been expected under the then current professional conditions. First, the crisis developed in springtime, when oil movements generally start to drop to seasonable lows. Secondly, the magnitude of the shortage that had developed progressively was far less important than the abrupt drop in European supplies resulting from the sudden closure of the Suez Canal and the stoppage of oil flow from Eastern Mediterranean pipelines during the major Middle East crises of 1956 and 1967. And yet, the industry demonstrated, on both occasions, its formidable potential and fantastic ability to overcome an acute shortage in tanker tonnage and to adjust operations so that the freight market would be affected within reasonable limits far below the skyrocketing rates that developed in the spring of 1970. It is our subjective feeling, a feeling that cannot actually be supported by any objective argument, that the industry would have been capable of keeping the freight crisis within tight limits if the major oil companies had been willing to do so on purely professional grounds, especially since their formidable potential had been fully mobilized since the 1967 alert. In reality, it seems as if such an acute crisis was actually a desired part of an emerging strategy, so that the negative aspects were allowed to degenerate even though they could have been rather easily throttled. In support of this view, we might point out that the crisis disappeared as abruptly and surprisingly as it had been generated less than a year earlier, after having played its specific role in the game.

2. The prospects of inadequate oil supplies in Western Europe, which were feared throughout the crisis, were mainly due to the tanker

shortage, as outlined above, and not to shortage at the source. On the contrary, over production capacity was never as great in the Gulf, where there was active competition among producing countries to increase liftings by operating companies and through direct sales by national oil companies. The Iranians were very active in this respect, and during the Teheran confrontation, the key element to official Iranian oil policy was stated by A. A. Hoveida, Iranian prime minister, in the following terms:

> The government has also made it absolutely clear to member companies of the Oil Consortium that Iran is determined to maintain her historical role as the biggest exporter of oil in the Middle East through exports of oil from the Agreement Area. Besides Iran's historical role, this position is regarded by the Government of Iran as an undeniable and a conclusive fact in view of the country's population, its size, the needs of its development plans, as well as the regional commitment of Iran, which serves as a factor of peace and stability in this part of the world. This principle forms the basis of the country's oil policy, and the Oil Consortium has recognized it and assured that it will be reflected in Iran's oil revenues.[6]

In light of the above, it appears that the price-negotiation episode was to some extent an unpremeditated "accident" in the Iranian oil strategy. The prospects for joint production programs in the Gulf are as gloomy as ever, and one may legitimately wonder to what extent producing countries would have effectively put into action their threat of holding exports, had oil companies resisted a satisfactory price arrangement. In fact, the expansion of the Gulf production has never been as great as was the case just after the crisis, when, for example, Aramco unveiled its plans to raise Saudi Arabian production to some 10 million barrels a day by 1977.[7] Aramco production in January 1972 reached an average of 5,336,500 barrels per day, which represents a fantastic increase of 33.5 percent over the average production of January 1971.

3. The attitude of oil companies vis-à-vis claims for higher prices changed strangely during the crisis period. In the early Libyan negotiations in January-February 1970, a top executive of a leading major oil company operating in Libya stated that since his company was forced to accept a price increase, all it could afford would be about a 5¢/Bbl increase, beyond which the company would incur losses in its Libyan operations. A few months later, the same major company spontaneously announced unilateral price increases of much greater magnitude, not only in Libya but also at the Eastern Mediterranean, where it was not subject to any specific claims.

On the other hand, the front of oil companies represented by the "January 16 Group" did not show any significant resistance to OPEC claims, and the Teheran negotiations almost seemed to be "club discussions" for drawing up the details of a formal agreement rather than to challenge its basic components. The oil companies were there to sign, not to fight.

4. If we keep in mind the professional significance of the items indicated above, the statement that international oil economics has shifted from a buyer's to a seller's market seems rather meaningless, since no fundamental change has affected market conditions. In fact, such a statement is applicable only to the U.S. domestic market, where a large part of the oil is produced by independents and sold at the wellhead. In the Eastern Hemisphere, and especially in the Gulf, the oil market is neither a buyer's nor a seller's market for the simple reason that such a market does not actually exist since the largest part of oil is produced and moved within the integrated framework of the majors' operations. The largest bulk of oil is transferred and not bought or sold; the role of independent oil has been and still is only marginal.

5. The only change that might be considered indisputably significant is the modification of the balance of power to the benefit of the "producers' cartel," which emerged from the OPEC Caracas Conference. It is unquestionable that the bargaining power of OPEC was greatly enhanced and strengthened by the unanimous decisions of the Caracas Conference and the firm determination displayed to achieve the objective of price increases at any cost, including a total embargo on crude exports. However, it is highly questionable that the metamorphosis enabled OPEC to force the front of oil companies and European consuming countries into a prompt capitulation, since they had long resisted, and even ignored, previous claims for price increases by OPEC member countries. As noted above, there was not the slightest indication of any desire to resist, and the very short time that elapsed between the Caracas Conference and the Teheran Agreement could physically allow only for the implementation of mutually accepted sequences. It is rather naive to believe that the oil companies and the Western interests involved in the price confrontation actually disposed of the dissuasive and retaliatory means that would have enabled them, at least, to enter the battle with confidence.

6. The final surprise, which came from the unexpectedly high level of the price increases, should be appraised and analyzed on psychological grounds. One of the main objectives of the creation of OPEC was the restoration of crude prices to their pre-1960 level, involving a price increase of about 15-20¢/Bbl. Not only was this objective not attained, but it had been dropped from the OPEC vocabulary for many years. Were we to believe that the political changes

were important enough to put OPEC in a position to force the oil companies to accept a price increase, the magnitude of the readjustment would have been in the range of 15-20¢/Bbl and such an achievement would have been a big, big victory. The actual magnitude of the flat price increase and the future price escalation appears to be disproportionate to the psychological mood and thinking of the producing countries. This magnitude is not consistent with the conditions and framework of the crisis and seems to be almost superimposed on it.[8]

THE U.S. ENERGY CRISIS AND WORLD OIL ECONOMICS

On March 10, 1959, President Dwight D. Eisenhower, under the provisions of the Trade Agreements Extension Act of 1958, issued Presidential Proclamation 3279, establishing a mandatory oil import quota. The president stated,

> the new program is designed to insure a stable, healthy industry in the United States capable of exploring for and developing new hemisphere reserves to replace those being depleted. The basis for the new program . . . is the certified requirements of our national security which make it necessary that we preserve to the greatest extent possible a vigorous, healthy petroleum industry in the United States.[9]

During the decade since the oil import quota was first instituted on a mandatory basis, disagreements between the supporters and opponents of the program grew increasingly sharp, spurred by the growing disparity between world and U.S. domestic prices. It was generally agreed that ensuring an adequate supply of petroleum for essential uses in the event of a national emergency is a legitimate national objective. It was also generally recognized that absolute security of petroleum supplies for all civilian as well as military markets probably cannot be obtained at a reasonable cost.

Those supporting the quota system argue that the nation cannot afford to become heavily dependent on external sources of oil because these sources are subject to both the risks of wartime and peacetime disruptions. Therefore, it was said that the quota program is needed to ensure adequate domestic supplies and that, while the quota does impose some costs on U.S. consumers in the form of higher prices for petroleum products, alternative means of ensuring adequate supplies in a national emergency would be more costly and/or less effective. Moreover, supporters of the program argue that the apparent savings to consumers from liberalizing the quota are overstated because the price of foreign oil would rise once the U.S. market became dependent on foreign supplies.

The opponents of the quota policy argue that the costs of the quota are exceedingly high. Some feel that the security needs for oil and the risk that supplies will be interrupted have been overstated and question the effectiveness of the current program in meeting its stated security objectives. Many opponents question whether alternative means cannot be found for providing the same security of supplies at lower cost.

The literature published in connection with these controversial issues is fantastically large, especially since they became intimately involved in complex U.S. domestic politics. However, in the whirl of conflicting interests and opinions, one objective fact has emerged unchallenged: The import quota system was designed as a strategic move to fill the gap in U.S oil supplies while protecting the U.S. domestic oil industry, but it was unable to prevent the gap from increasing. In 1958, U.S. imports of crude oil and products were 619 million barrels, or about 18.7 percent of domestic consumption. In 1968, total imports were 1,038 billion barrels, representing about 21.7 percent of total consumption. The United States was on the road to a major energy crisis, a fact that became obvious in 1969-70 but that had been materializing in different forms throughout the 1960s.

And it is precisely an over-all energy crisis that the most advanced industrial nation is heading for and not only an occasional oil shortage. The decline in gas reserves is as acute as in oil reserves, and the development of nuclear energy has been quite disappointing. The most exposed area is the East Coast of the United States, where a potential fuel shortage was recognized in the early 1960s. Following a finding by the Office of Emergency Planning that imports of residual fuel in District I (the East Coast) did not endanger national security, the secretary of the Interior raised the quota of fuel oil imports in District I by 10 percent for the year beginning April 1, 1963. This ceiling continued to be lifted until 1966, when the controls on fuel oil imports to District I were abandoned. And yet an acute crisis burst out in the winter of 1970, when spot shortages of fuels—all kinds of fuels—were registered on the East Coast. Increasing concern over sulfur pollution and the effective enforcement of tough regulations caused many of the utilities' producers to switch from high-sulfur coal to natural gas. Because of the failure of nuclear power generation to come on stream as expected and the steady increase in energy demand by about 5 to 6 percent annually, and much more in some particular sectors with demand for power increasing 7.5 to 9 or 10 percent annually, the pressure on the residual fuel market reached critical levels. Natural gas suppliers were at the very limit of their producing capacity, and it became public knowledge that they would be unable to meet potential demand. Emergency peak-shaving LNG supplies were secured from Algeria, and supplies to some customers were curtailed even in the summer in the Southwest, in the heart of the producing region.

The magnitude of the U.S. energy shortage was emphatically outlined in October 1970 by Dr. W. L. Laird, director of the Office of Oil and Gas in the Department of the Interior, who declared that the spare oil cushion would be gone by the end of 1971, if not sooner. Furthermore, he added, "from now until 1985, the U.S. will need 100 billion barrels of oil, and by the end of that period, the nation will be using 22 million barrels per day. Gas demand will exceed 400 trillion cubic feet—if that much can be found. This is an average annual need of 27 trillion over the period, with a 1985 demand projected for 32 trillion cubic feet."[10]

In the early days of the 1970s, the United States stood at a crossroads in its energy history and everybody had become aware of this fact of life in the highly passionate atmosphere of crisis and politics. As usual in the dynamic and aggressive American scene, contributions to the debate were numerous and challenging, and different groups set to work to evaluate the real magnitude of the crisis and to suggest the best ways and means of facing it. Expectedly, the forecasts are different, but the differences are limited when compared with the magnitude of the needs and shortages in the 1970s and 1980s.[11]

Trouble with the import quota system began well before the possibility of a fundamental energy crisis in the United States could develop in the imagination of the most foresighted observers of the industry. According to an article in <u>Petroleum Press Service,</u>

> Over the decade that mandatory controls have been in existence, the whole nature of the import program has changed. What began simply as a means of protecting domestic producers against imports, in the name of security of supply, is now being used as a means to bolster the economy of U.S. territories in the Caribbean, to help the U.S. chemical industry against foreign competition, to make up shortages of products where they can be shown to exist, and even to help combat atmosphere pollution.[12]

Apart from the specific allocations for fuel oil imports into District I mentioned earlier, the program worked rather smoothly up to 1965. In that year, however, the Interior Department approved a special quota for Phillips Petroleum to import crude and unfinished oils into Puerto Rico for a large petrochemical plant Phillips was building, a scheme backed by the State Department since it would reduce unemployment on the island. At the same time, the company was permitted to ship nearly 25,000 Bbl/day of motor gasoline from there to the U.S. mainland.

The 1965 decision opened the way for other "special cases," and many other claims were put to the government by oil and chemical companies, which were also permitted to ship products to the mainland, Very rapidly, the chemical industry was pressing for unrestricted imports for petrochemical plants in "foreign trade zones." Another important break in the import quota program was contemplated by the Interior Department under its plan to provide "bonus" quotas to encourage refiners in Districts I-IV to make low-sulfur residual fuel oil. Only three quotas were granted under this plan before it was suspended in May 1969.[13]

The oil industry, in general, was fiercely opposed to any modification of the existing system, not only because the system provided effective protection for domestic producers but also because an import allocation quota is worth about $1.25 for each barrel it covers. This "import ticket" results from the fact that foreign oil could usually be delivered, in the mid-1960s, to a U.S. coastal refinery at a cost of $1.25 or more per barrel lower than the price of U.S. source crude oil delivered to the same refinery. Any producing company receiving an import allocation under the plan would be free either to import the foreign oil itself or to sell its import allocation to another company.[14]

By insulating the U.S. domestic market from international oil economics, the import quota program not only made it possible to keep prices high and stable but also permitted their increase when such a move was warranted. A first round of limited price increases took place in mid-1968 involving a general average increase in postings of about 5¢/Bbl and the elimination of the gravity penalties that had affected the prices of high-gravity crude oils. The move was initiated by Shell Oil, which eliminated the 2¢ downward scale for each degree climb in gravity above 44.9° API.[15]

The second round of price increases, much larger in scope, took place in March 1969. This time, Texaco initiated price hikes that rapidly spread throughout the country. This was the first nationwide round of price increases in 12 years.[16] The size of the increase ranged from 10 to 20¢/Bbl and averaged 16.5¢/Bbl in Districts I-IV. The hike was a flat 20¢/Bbl in California. These increases were justified by the industry on the grounds of increasing costs and salaries without even taking inflation into consideration. A further nationwide price increase of 25¢/Bbl was achieved on November 11, 1970, at a time when the price confrontation was nearing its culmination in the Eastern Hemisphere.[17]

The U.S. domestic oil industry started to run into serious trouble with the advent of the Republican administration of President Richard

Nixon in 1969, with Walter Hickel as Interior Secretary. At the time, the Oil and Gas Journal wrote,

> Unlike in the previous administration, oil will not be a forbidden word in the White House. Whereas President Johnson went to extreme lengths to dissociate himself from oil matters, President Nixon is taking the opposite tack. The first step was to recapture oil import policy from the Secretary of the Interior, to whom Lyndon Johnson delegated the authority in 1963. . . . But the most important issue under White House control is oil-import policy. The full review ordered by the President on February 20 is in competent hands. . . . The first decision that will be made as a result of his imports review will be whether there is still a national-security basis for limiting oil imports, which were controlled for this reason in March 1959.[18]

However, an immediate danger, and a far more serious one, was threatening the industry on another front. A vigorous campaign was launched against the exorbitant privileges enjoyed by the oil industry in terms of taxation, which enabled oil millionaires to escape tax payments or to bring the payments down to insignificant levels. An over-all tax-reform bill was put to the Congress, and the Ways and Means Committee started hearings on 17 different reform topics as early as February 18, 1969. One of these items included percentage depletion, production payments, intangible drilling costs, and exploration and development expenditures.[19] The most important item was actually the attack against the depletion allowance, regarded suspiciously as a tax loophole. The objective of the administration, as approved by the Congress, was to cut the depletion allowance from 27.5 down to 20 percent. The Senate raised the percentage to 23 percent, and the final verdict was set at 22 percent in a compromise by the House-Senate Conference on December 19, 1969. Details about other items in the tax reform will be found in the literature.[20]

In fact, governmental machinery went into full gear with the creation on February 20, 1969, by Nixon of a special cabinet-level task force under the chairmanship of then Secretary of Labor George Shultz to conduct a review of oil import controls. After extensive studies and investigations, the task force came to the conclusion that maintaining the import-quota program was not essential to national security and that the U.S. consumer was paying too high a price for the illusion of security. This line was taken in a report by R. McLaren, head of the Justice Department's Anti-Trust Division: "The present import control system imposes serious costs on the economy. These costs would not appear to be necessary to the attainment of any reasonable national security goal."[21] The cost to the economy was put as low as $3 billion a year and as high as $4 billion. Several studies undertaken by distinguished economists with no direct connection to the oil industry recommended complete abolition of the import control system.

The cabinet task force submitted its report to Nixon in the early days of 1970. This report showed a basic split in opinion between its members. The main recommendations were made by a majority of five out of seven, but they were vigorously denounced by the minority group.[22] In the opinion of the majority, "the existing quota system is no longer acceptable [and] a phased-in liberalization of import controls would not so injure the domestic industry as to weaken the national economy to the extent of impairing our national security." While suggesting the extension of quotas through 1970, the report recommended their replacement by differential tariffs, no later than January 1, 1971, as a means of restricting imports. The proposed initial levels of the tariffs would be $1.45/Bbl for the Eastern Hemisphere, imports from which would not be allowed to exceed 10 percent of U.S. demand, to avoid undue reliance on insecure supplies; $1.25/Bbl for a specified volume from Latin America; and 10.5¢/Bbl—the existing tariff—for a specified volume of imports from Canada and Mexico. This tariff structure would immediately result in a reduction of well-head prices for 30° gravity crude in southern Louisiana by about 30¢/Bbl down to $3/Bbl. Existing quotas would be progressively phased out in three years, with the tariff-free quota volume being reduced steadily. Tariff levels would be reduced in parallel fashion to induce an ultimate drop of domestic prices down to $2.50/Bbl for 30° gravity oil in southern Louisiana. This target price would be reached by setting a tariff of 87¢/Bbl for Middle East oil at the end of the phase-in period.

The majority views and recommendations were vigorously opposed and challenged by the industry, thus supporting the minority opinions. One of the best appraisals of the task force report was made by James E. Akins, director of the Office of Fuels and Energy, U.S. Department of State:

> The Task Force projections were spectacularly wrong. Total imports this very year, 1973, will be well over six million barrels per day—substantially above the level the Task Force predicted for 1980. Imports from the Eastern Hemisphere constituted 15 percent of comsumption in 1972, and are expected to rise in 1973 to 20 percent of a total consumption that will already be around 17 million barrels.

The errors of the Task Force were not those of isolated academics, as its critics were (and still are) wont to charge. The staff based its projections on information provided by the major oil companies, by the National Petroleum Council and by the Department of the Interior. There were two main reasons for their errors. Perhaps both should have been avoided, but as always, hindsight is clear and uncluttered. The first error was an uncritical acceptance of oil company and well owners' estimates of the capacity of their own domestic producing wells. These were almost always exaggerated. The second was the ignoring, or at least the deemphasis, of the decline in natural gas supplies and its effect on oil demand. We were, by 1970, already consuming far more gas than we were finding, and demand for gas continued to grow unabated, while domestic gas production leveled off (1973 production will actually be below that of 1972). The unsatisfied demand for gas was, of course, a real demand for energy. It could be covered only by oil—in fact, only by imported oil.[23]

In our opinion, however, the most flagrant shortcomings and misassessments of the majority views were to be found on the critical issue of national security. The strategic importance of oil is too vital to make future U.S. supplies dependent on the intellectual speculations of the technocratic approach displayed by the majority group of the task force, the opinion of which seems to have been influenced by short-term tactical considerations and predominately academic, economic, and social approaches to the delicate issue of adequacy and security of supplies. Backed up unanimously by the oil industry, which would stand to suffer substantial losses following a price reduction, the emphasis on national security advocated in the minority group was the determining factor in Nixon's decision on February 20, 1970, to reject the majority recommendations and order a new reassessment of American strategies and policies to be carried out by a new Oil Policy Committee. It is significant that the new committee was headed by General G. A. Lincoln, director of the Office of Emergency Preparedness, who had been sympathetic to the minority group, although he joined in the majority recommendations. It is obvious, however, by the very nature of his responsibilities in the administration, that he was well placed to appreciate the preponderance of strategic objectives over short-term economics.

The timing of the president's decision as well as the early investigations of the new Oil Policy Committee are very significant to an understanding of the evolution of the 1970 Eastern Hemisphere crisis. It is the author's opinion that the two issues are intimately interlinked, and this is why the complex of U.S. domestic oil politics has been presented and analyzed in such detail above. The following analysis may be considered as mere intellectual speculation since it cannot be supported by any objective evidence. Nevertheless, it can provide a comprehensive over-all understanding of the real significance of the Eastern Hemisphere oil crisis and may explain and dissipate most of the ambiguities and questions already oultined in this chapter. Furthermore, it enables us to understand the pricing mechanism that controlled the Teheran and Tripoli Agreements, to appraise the significance of the price stability they ensured, and to predict future evolution.

Bearing all the above in mind, we may imagine that the reassessment of U.S. oil strategies and policy, as carried out by the strategy-minded Oil Policy Committee, would have undergone something like the following scheme of brainstorming:

1. The U.S. energy crisis is a real and profound one, the magnitude of which seems to have been largely underestimated in the assumptions and the calculations of the cabinet task force.

2. Dependence on imports will be very great indeed, although to varying extents in relative percentage terms according to different estimates. However, even a small difference of few percentage points would represent such huge volumes in absolute terms that the nation cannot afford to take the risk of facing such a gap. Oil strategy should consider emergency conditions according to pessimistic assumptions rather than average or optimistic conditions.

3. The only producing areas that can support the fantastic future expansion of demand are the Middle East and North Africa, where reserves of Arab oil (excluding Iran) have been estimated by the State Department at 350 billion barrels, say 69 percent of the world total.[24] However, the department admits that these figures are greatly undervalued (for example, Saudi Arabian reserves are put at about 150 billion barrels, whereas most authorities are convinced the figure should be at least twice that).

4. Stability and security could be doubtful when we consider Arab oil, whose future is highly charged with the unpredictable and emotional Arab-Israeli conflict, with the ever increasing anti-Western feelings sweeping over Arab nations, and with political uncertainty connected with the British withdrawal from the Gulf at the end of 1971. Alternative sources of secure supply, such as Alaska and the North Sea, will be only marginal and would account for only a part of future increase in demand. The six-day war of 1967 and the alarming 1970 crisis provide some examples of the type of future hazards that could seriously endanger U.S. supplies.

5. Therefore, energy autonomy is, and should remain, the most important strategic objective of the United States. In the coming years, when imports will be increasing and inevitable, dependence on oil supplies from insecure Eastern Hemisphere sources should be kept to a minimum by maintaining adequate domestic reserves. Furthermore, all efforts should be devoted to making Middle Eastern oil as secure as possible by reinforcing the political and military position of friendly oil-rich countries in the region.

6. The relative importance of economics has been overestimated by the task force in its persistent drive to cut domestic prices down. The battle waged by the administration against inflation should not be restricted to the conventional technocratic remedy of a price reduction incompatible with national security but should envisage unconventional approaches to the U.S. economic difficulties and to its imbalance of trade and payments. Strategy should not be sacrificed for the sake of economics, but, on the contrary, economics should be modeled to achieve strategic objectives in optimum conditions.

7. At any rate, the import quota system is no longer well-adapted to the situation and should be replaced by a new, more adequate mechanism of control. It was originally designed to limit imports at a time when they were marginal. In the coming years, it could become an unacceptable bottleneck to necessary imports in substantial and increasing quantities. The new system should make it possible to achieve the following contradictory objectives:

• To ease if not eliminate import limitations so as to avoid any unwarranted shortage, and to have reliability of supply and enough flexibility to face emergency conditions;

• To maintain the level of U.S. domestic prices and if possible to pave the way for their increase;

• To create the necessary environment and conditions to limit dependence on insecure sources, and to strive to recover energy self-sufficiency to the greatest extent possible, compatible with the struggle against inflation and with a healthy balance in the national economy.

The situation would appear hopelessly inextricable when one considers the complexity and contradictions of U.S. problems within the existing framework of international petroleum economics. The task of the Oil Policy Committee was made particularly delicate and urgent by the existence of the unacceptable solution proposed by the cabinet task force and by the mounting crisis in the Eastern Hemisphere initiated by the Libyan revolution. The search for a new strategy started virtually at the same time the first round of Libyan negotiations, in March 1970, came to a deadlock after the refusal of the oil companies to comply with Libyan demands. The situation rapidly deteriorated thereafter with the enforcement of retaliatory measures envisaged by the Libyans, by the sabotage of the Tapline, and by the onset of an acute fuel oil and gas shortage on the East Coast of the United States. The aggravation of the situation forced the White House and U.S. strategists to give oil affairs first priority. The National Petroleum Council was told that the government was anxiously waiting and watching as crises developed in the Middle East. Late in June 1970, the Petroleum Security Subcommittee of the U.S. Congress met with representatives of the Office of Oil and Gas in the Interior Department and other federal agencies to consider requirements and supply for national security. A classified report of the subcommittee's findings was sent to the Defense Department.[25]

In the face of the extreme urgency of the explosive situation, a new miracle strategy had to emerge rapidly from the active brainstorming of emergency planners, which would conciliate all the contradictory objectives outlined above. But miracles do not happen. The fundamental difficulty and contradiction were represented by the large difference between U.S. and international crude prices, which was aggravated by the mounting inflation in the United States and by the dynamic opposition in the trends of both sets of prices—that is, U.S. domestic prices should be kept high and be allowed to move upwards only if necessary, whereas Eastern Hemisphere prices were very low and were under sustained pressure to go further down. But the miracle remedy to all problems would become a reality if international prices were allowed or even forced to catch up with American prices. Strategists could envisage such a possibility because the strategic nature of the over-all framework of international oil pricing could be altered if the prevailing strategy were to be reassessed, whereas the economic-minded technocrats of the cabinet task force were prisoners of the existing pricing structure within which crude prices would be immovable or, at least, would change according to economic laws and concepts.

And indeed, once you dare consider changing the nature of things, provided that you possess necessary ways and means to induce and/or to impose such a change in spite of its possible contradictions with actual facts, then working out adequate solutions to the many complex problems involved would become no more than a stimulating exercise of logistics and imagination. The new scheme of the "miracle strategy" might be imagined along the following lines:

• The only way to let increasing quantities of foreign crude oil, mainly from the Eastern Hemisphere, get into the U.S. market to fill the energy gap without endangering and depressing domestic prices, would be to raise overseas prices to such a level as to make delivered prices CIF the East Coast comparable to domestic prices. This was precisely the strategy adopted in 1948, when imports had to be admitted to the domestic market without administrative restrictions. Now, once again, imports had to be freed from administrative bottlenecks.

- The significance and consequences of this new look materialize in two respects: First, in the coming decades, the strategic interest of the United States and the security of its energy supply will demand high-level prices for international oil—that is to say, U.S. interests have markedly shifted to the side of producing countries in the face of consuming countries now engaging in decisive confrontations over prices. Secondly, the magnitude of the price increase should not be left to the uncertain bargaining balance of the parties directly involved since the new price level is predetermined by the threshold of competitiveness with U.S. domestic prices.

- Oil companies, mainly the majors, would be more than enthusiastic about the new strategy since it is the very nature of their high degree of integration to favor higher prices. Moreover, it would end the nightmare of seeing their profits trimmed off in the lucrative American market. They constantly opposed any price increase in the past decade because this was the imperative logic of the U.S. strategy. Now, they would react readily and favorably to any change in the strategy that would free the upward movement of prices. This would explain why they were so cooperative in enabling OPEC member countries to achieve their big victory.

- As part of the strategy, the political stability of producing countries should be reinforced, mainly in the friendly nations of the Gulf, where the really significant reserves are located. Therefore, the crisis in the Eastern Hemisphere should be dramatized to such a high level, compatible with the ability to keep events under control, as to make the final victory of the governments of producing countries a really big one. In this way, they would be credited with the only significant victory against imperialism and the monopolies that has ever been achieved in developing countries without social revolution. On the other hand, the substantial increase in their revenue would enhance their prestige and reinforce their popularity by providing the necessary funds for economic development and social welfare. What Lord Strathalmond of BP, one of the two top negotiators for the oil companies, had to say in Teheran is significant: "No wonder, for the settlement gave the six Gulf States even more than they had set as their minimum acceptable demands." Speaking of Iranian Minister of Finance Jamshid Amouzegar, the top negotiator for OPEC, he added, "We probably both knew how the negotiations were going to come out a good bit before we finished. He got a little more, not much, than we were thinking about when we started."[26]

- A logical corollary to the necessity of maintaining a peaceful friendly atmosphere in the Gulf would be to eliminate all sources of past conflicts and seeds of potential problems between oil companies and host governments. In particular, all policy-induced and non-professional price disparities in the Gulf should be eliminated, and the over-all pricing pattern should become smooth and harmonious so that no "jealousy" between one country and another could survive in the area, and there would never be any claim under the most-favored-nation clause. Accordingly, oil companies took the initiative to eliminate the long-standing "political" penalties that had been a source of persistent friction in the area, such as the investment discontinuity penalty (10¢/Bbl) affecting the heavy crude oils of 31° API gravity and below, the 6¢/Bbl penalty affecting Iraqi postings FOB Khor el Amaya, and the 10¢/Bbl underpricing of Bai/Hassan Jambur blend FOB Banias.

● A decisive adaptation of international oil economics to the new strategy dictated by the reevaluation of U.S. interests was to be achieved in the Teheran round. The vigilant White House sent the under-secretary of State to observe the bargaining, and the oil companies were made to understand that the round should be a rapid and comprehensive one. In the second round, the Tripoli revolutionaries should be handled with great care and tact so as to avoid any break that would make the whole system collapse. Negotiations should succeed, even at a higher price than the simple extrapolation of Teheran conditions, which was conceded as the price for accommodation with the intransigent revolutionaries. Even more, an overpricing of Libyan oil, the "leader" of African low-sulfur crudes, would serve the U.S. strategy by making them less competitive in Europe, thus diverting them "naturally" to the sulfur-pollution-conscious U.S. market, which could afford to pay the incremental price as a sulfur premium. Under these conditions, the United States (and, paradoxically, not the nearby European market) would become a "natural market" for African oil.

● The new strategy would make it possible to combat the inflationary trend of the American economy without touching U.S. domestic crude prices, since the heavy burden of Eastern Hemisphere price increases would be felt by European and Japanese economies and industries, which would then become less competitive. The acute dollar crisis that followed shortly afterwards (August 1971) clearly illustrates the real importance of this aspect.

● The main and ultimate objective of U.S. strategy is to restore self-sufficiency in energy, or at least to limit dependence on foreign imports to a minimum. This would be achieved by extensive domestic exploration and by the development of alternative sources that are uneconomic by prevailing standards. Both possibilities require increasing prices for U.S. domestic oil and therefore for imported foreign oil. These price increases would be automatically and "naturally" provided by the escalation clauses in the different price settlements—such as 2.5 percent plus 5¢/Bbl through 1975. By that time, the delivered cost of Eastern Hemisphere crude oil would exceed $4/Bbl, which was precisely the threshold of profitability for the production of hydrocarbon liquids from tar sands and shale oil. The U.S. reserves of shale oil are so enormous that their extensive development would fill a large part of the U.S. energy gap. (The 11 million acres of shale lands in the Green River Basin of Colorado, Utah, and Wyoming may contain as much as 1.8 trillion barrels of "known reserves," which is 60 times the present U.S. "proved reserves" of crude petroleum and which represents a potential that could meet U.S oil needs for hundreds of years.)[27] Under these conditions, the five-year period of stability, secured at a high cost, payable by the Europeans and the Japanese, appears as a reasonable transition period needed by the United States to review the situation, to investigate new sources of energy and develop necessary technologies, and to draw up a comprehensive strategy serving to ensure secure and adequate supplies of energy in the coming decades through the year 2000.

The scheme of price increase implemented through the Teheran settlement appears as a strategic move to meet U.S. needs in the 1980s. Nevertheless, in view of the above hypotheses, it would rather appear as a tactical move intended to make it possible to fill the gap throughout the transition period during which the United States will be actively preparing to achieve its far-reaching strategic objectives of energy self-sufficiency in the next century. The strategy for tomorrow is, in reality, a tactical step in the strategy for the day after tomorrow.

As indicated earlier, this analysis might seem to be no more than imaginative intellectual speculation. However, it would be very difficult to believe that the profound changes and upheavals that ravaged international oil business in the 1971 price settlement could have taken place without the benediction, if not the cooperative support, of the U.S. Government. In fact, the scheme we have imagined above is not all that speculative since its mechanism was schematically outlined by W. J. Levy in the early days of 1970, although his purpose was to denounce the task force report rather than suggest a new strategy.[28] "If the United States adopts the recommended tariff system, it would be virtually and indirectly inviting OPEC demands for a 50¢ or even $1.00 a barrel increase in the OPEC tax take, raising world crude costs by as much. This would be the immediate and nearly certain political response of the major oil exporting countries to the considered elimination of the quota system."

It should be clear, however, that the scheme suggested above does not signify, by any means, that the Eastern Hemisphere crisis was not a real and profound one and that the confrontation between OPEC member countries and the oil companies was a predetermined tricky comedy. The United States was not a direct partner in the conflict and was really anxious not to get involved in it. However, it so happened that the timing of the crisis, which was ignited by the Libyan revolution, coincided with the culmination of the U.S. energy crisis, which called for an urgent over-all reassessment of American strategy. The temptation was very great indeed to take advantage of the mounting conflict in the Eastern Hemisphere and to direct it to the benefit of the United States. The decisive action of the United States was to change the rules of the game and to allow the confrontation to develop on new and original grounds, the outer limits of which were determined by U.S. strategic objectives and limitations.

Within this new environment, the shock between the two parties was a violent and merciless one. Several oil companies feared drastic pressure and retaliation, or even nationalization. At times, Western European countries were seriously worried about the continuing flow of their oil supplies. The victory was achieved thanks to the astuteness, great skill, and refined sense of maneuvering of the top negotiators for producing countries. But the greatest credit should unquestionably be given to the Libyan revolutionaries, who had the courage to defy, alone, the almighty companies, and to pave the way for others through the precedents established in the price settlement of September 1970.

FUTURE OUTLOOK

In the oil business, the 1960s will be remembered as the era of the buyer's market. This was a period of sustained oversupply, with steadily rising crude oil reserves and persistent pressure on selling prices, despite the remarkably high average annual increase in consumption of 7 to 8 percent. The relative abundance of low-cost oil produced by independents and marketed at dumping prices outside integrated circuits introduced effective competition to a part of the market currently described as "sales to third parties," where the pressure on "realized prices" was enhanced by the gradual buildup of exports of crudes and procucts from the Soviet bloc. The downward trend of prices that resulted was welcomed and encouraged by European countries, and the governments of producing countries did not have the power or maturity to oppose it with significant effectiveness.

It is generally admitted that in the coming years, the international oil industry will be definitively characterized by a seller's market for crude oil and products. This fundamental change will be due to the fact that the huge U.S. domestic market will be open again, in one way or another, to imports of such large quantities as to absorb actual and potential overproductive capacity in the Eastern Hemisphere.[29]

On the other hand, a similar reversal of the situation will be experienced by the Soviet bloc, although to a lesser extent. The best estimates predict that the Eastern bloc as a whole may actually be importing something like 100 million tons a year by the end of the 1970s. Eastern European countries have been regularly importing increasing quantities of Middle Eastern oil, since the late 1960s, on a barter-deal basis with national oil corporations in Iran, Saudi Arabia, and other nations.

The demand for oil by the United States and the Soviet bloc will be actively challenged by energy-hungry Western Europe and Japan, which will be almost totally dependent on Middle Eastern and African oil for their supplies. Combined European and Japanese oil imports are expected to increase from some 18 million barrels per day in 1972 to 30 million in 1980.[30] To illustrate the magnitude of the fantastic demand for oil in the coming years, it is predicted that, in the 1970s, humanity will consume as much oil as its total consumption since the first oil discovery by Colonel Drake more than a century ago. In a public statement early in March 1971, an eminent oil expert predicted that "between now and 1990, total world consumption will be about 465 billion barrels, nearly equal to all proved reserves in the free world today. The United States alone is expected to consume 150 billion barrels of that amount, though present U.S. reserves, including Alaska, are estimated at only 50 billion barrels."[31]

The pricing issue has been a fashionable subject throughout the 1960s, and many schools of thought have challenged each other to foresee the prospective trends of price evolution. Those who contemplated carrying the exercise through the 1970s were frustrated by the Teheran and Tripoli Agreements. Crude oil prices will no longer be a debatable issue; they will remain high and stable and will continue to increase according to the escalation mechanism provided for in the price agreements and under the pressure of increasing demand and the relative shortage of supplies in an environment characterized by an oppressive psychosis of energy crisis. The competition between the United States, Western Europe, and Japan to have the priority of access to oil supplies largely contributes to the unprecedented escalation of realized prices in a so-called open market. Arab producing countries in the Gulf area havehad no difficulty disposing of their participation oil by tendering specified volumes on a competitive short-term basis. Prices escalated rapidly over the first half of 1973 from the prevailing spot market performances. Japan Line was the first to purchase participation oil from Abu Dhabi at $2.27 per barrel in February 1973; a few weeks later the Abu Dhabi Government was assailed by offers to purchase the same crude at prices averaging over $2.80 per barrel. In late June 1973, Qatar sold its participation oil to a U.S. independent at a price believed to be higher than the posted price.[32]

What, then, is the outlook of the international oil industry for the foreseeable future? On the professional side, there will be extensive exploration efforts to discover new sources of supply not only to ease dependence on the unpredictable Middle East but also to add the billion barrels of new crude reserves that will be badly needed by the end of the century to keep pace with demand. Efforts will concentrate on offshore exploration all around the world, and there are fair prospects, or at least hopes, that new major producing areas will be developed.

In correlation with the fantastic expansion of demand, the financial needs and commitments of the international oil industry will reach very high levels, which will be further aggravated by the irresistible inflation that is ravaging the world economy. Under these conditions, the major oil companies will lose one of the main advantages they derive from their large size and high degree of integration—their ability for self-financing by retained funds. They will be increasingly dependent on financial markets to raise the necessary funds, which will settle at new and unexpectedly high levels. A recent review by First National City Bank of financial analysis and capital requirements for the noncommunist world "seems to suggest that the figure could rise to as much as twice the 20 billion dollars annually previously estimated for the end of the decade."[33] In the opinion of Chase Manhattan Bank, the total financial needs of the industry for the 1970s may reach $500 billion,[34] while total needs for the period 1970-85 are put at a staggering figure of about $1 trillion.[35] Only about $600 billion can be financed by retained funds, and thus the industry will have to raise some $400 billion in the financial markets, which is about seven times the total funds raised by the oil companies from external sources in the period 1958-72.

In the face of the capital gap that will be encountered by the oil companies, the revenue of producing countries will pile up to several billion dollars each year, and a large part of this amount will be diverted back to the world's major financial markets and thus add further disorder and confusion to world monetary problems. It is not easy to evaluate accurately oil revenues in the coming years not only because of the uncertainties about price evolution and production prospects but also because of the troubled state of the international monetary system since dollar devaluations and currency fluctuations are to be added onto crude oil prices, according to the Geneva Agreements. One of the most authoritative estimates made by the State Department predicts that total oil revenues of Middle Eastern and North African producers might reach $21.1 billion in 1975 and $70.6 billion in 1980. Details per country are shown in Table 17.1. It should be noted that the estimated production figures for 1980 are higher than what is generally admitted by the industry, and especially by the producing countries themselves. Iran, for example, has said its production

TABLE 17.1

Estimated Production and Revenue
(production: thousands of barrels per day;
revenue: billions of dollars annually)

Exporting Country	1975 Production	Revenue	1980 Production	Revenue
Middle East				
Iran	7,300	5.2	10,000	14.3
Saudi Arabia	8,500	6.0	20,000	28.6
Kuwait	3,500	2.4	4,000	5.6
Iraq	1,900	1.3	5,000	7.1
Abu Dhabi	2,300	1.7	4,000	5.6
Others	1,800	1.1	2,000	3.5
Subtotal	25,300	17.7	45,000	64.7
North Africa				
Libya	2,200	2.2	2,000	3.4
Algeria	1,200	1.2	1,500	2.5
Subtotal	3,400	3.4	3,500	5.9
Total	28,700	21.1	48,500	70.6

Source: James E. Akins, U.S. Department of State, "The Oil Crisis: This Time the Wolf Is Here," Foreign Affairs, April 1973.

Note: The original revenue figures stated by Akins were increased by 11.9 percent to reflect the Geneva Agreement of June 1, 1973, intended to balance the effect of the dollar devaluation of February 1973.

will level off at 8 million barrels per day; Kuwait has said its will be kept at 3 million; and other countries are likely to be tempted to limit expansion rates. Yet the figures retained in the State Department estimate represent effective world needs unless there is a war or a major recession and there are no alternative sources of supply in the foreseeable future. The only alternative to a short-fall before 1980 will be Saudi Arabia, and its projected production of 20 million barrels per day, set by Minister of Petroleum Ahmad Zaki Yamani as a goal, seems almost too high to be feasible.[36]

The oil-producing countries will undoubtedly devote their increasing revenues to support their own development. However, except probably Iran, which will have no serious trouble in absorbing funds in its vast development projects, the Arab countries will be unable to find enough outlets for their funds even when adding inter-Arab development projects and assistance plans to the expenditure potential of the Arab exporting countries themselves. Cumulative Arab income from 1973 through 1980 will be probably over $210 billion, of which at least $100 billion will accumulate as surplus liquidities by 1980. At 8 percent, just the income from this enormous sum would be $8 billion, larger than the current expenditures of Kuwait, Saudi Arabia, and the Federation of Arab Emirates combined.

The Arab World will thus emerge as a major financial power, and the way its funds are utilized will be a matter of crucial importance to the world. In the coming decades, most of the oil-exporting countries will hold important positions in international financial markets, which will increase their bargaining power and political influence, not only over the oil industry but also over the industrial and economic development of Western nations, through the huge funds they will be able to manipulate and invest in key sectors.

The international community has no choice but to provide sound and profitable investment opportunities for Arab funds and to make them attractive enough to convince Arab leaders to make use of them. The task is unrewarding, since most Arab nationalists are reluctant to help finance Western development, especially as long as the West continues to provide its unconditional support to Israel. The dilemma is a real one; the alternative to such investments would be that the Arabs accumulate their money and simply float it from one currency to another, or from one country to another, depending on how each country reacts to Arab problems. This situation is most unlikely to materialize since the whole international monetary system would collapse well before and the Arab losses would be commensurate with the hugeness of their funds. The real alternative would then appear to be to limit production so that surplus revenue might be better safeguarded by keeping oil in the ground (financial conservation). This would cause a problem for the developed world far greater than control of key sectors by the Arabs of the floating billions.

Price increase was intended to pay for professional stability in the Middle East, so that the problems will mainly materialize on political grounds; they are most likely to be big problems. When one considers the politics of Eastern Hemisphere oil, it is inappropriate to talk about Gulf, Mediterranean, or African oil. It would be more appropriate to consider Arab oil and non-Arab oil. "Arab oil" would take in North Africa and the Middle East, excluding Iran. In 1972, Arab oil production accounted for about 816 million tons, corresponding to 31.5 percent of the world total. By comparison, domestic U.S. production for the same year was about 532 million tons.[37] But the real importance of Arab oil derives from the immensity of its reserves. Current conservative figures published in professional journals set Arab oil reserves at about 55 percent of the world total. In fact, this figure is much lower than reality. We have already indicated, for example, that Saudi Arabian reserves are stated to be about 87 billion barrels,[38] official figures disclosed by the Saudi Arabian Government approach 150 billion barrels, and the U.S. State Department estimates that the present proven reserves are over twice that figure and that the probable reserves could double the figure again.[39] Arab oil reserves are more likely to be, at least, two-thirds of the world's total. In connection with this abundance, the costs of production are generally very low, averaging 8-10¢/Bbl for the prolific Gulf producers.

More significantly, the supremacy of Arab oil will most likely strengthen in the coming years despite the exhaustive exploration in other areas. Arab lands have the richest oil-bearing potential of known sedimentary basins; large areas have not yet been fully explored; and the constant improvement of technology will undoubtedly enhance the value of proven areas. In this respect, the verdict of economics is the determining factor. It is generally estimated that, in the early 1970s, the cost of finding an extra barrel of crude oil in the Gulf is about 3 cents as against 17 cents in Venezuela and $1.10 in the United States, not to speak of the presumably even higher costs of offshore or Arctic exploration.

In an editorial in the <u>Oil and Gas Journal</u> of December 1969, an early warning was clearly formulated: "Our future standard of living may well be at stake. [The United States] and all the rest of the world for that matter, will one day need much more Middle East oil. The world's only really major surplus reserves lie in Arab countries."

Besides the importance of crude reserves, the necessity of distinguishing Arab and non-Arab oil derives essentially from the fact that security and continuity of Arab oil supplies are directly and intimately connected with the explosive political situation associated with the endless Arab-Israeli conflict, and with its uncontrollable impact on the irrationality and unpredictability of popular emotions and mass reactions. The December 1969 <u>Oil and Gas Journal</u> editorial is meaningful in this regard:

> Too few Americans are aware of the strong pro-Israel image we have created abroad. Even fewer are conscious of the fact that inherently friendly Arab leaders have almost literally been driven into the arms of the Russians by a succession of events such as these:

- The enthusiastic welcome given Israel's prime minister, contrasted with the relative backroom receptions extended to visiting Arab heads of state.
- The American release of a fleet of our newest jet fighters to Israel.
- U.S. inability, as recently publicized by the State Department, to prevent American citizens from serving in the Israeli military.

Oilmen returning from Arab lands report that anti-American feeling is escalating alarmingly. Arab radicals have seen their influence extended thanks to our official attitude. And the more conservative rulers are losing not only their faith in America but their influence to moderate an inflamed public.[40]

In the opinion of most Arab officials and nationalists, oil companies should contribute directly and actively to support the Arab cause and should mobilize their public relations departments and their influence over the mass media in order to enlighten the American public and especially the U.S. voter about the dangers of the conflict in the Middle East. Oil companies should begin to realize that it would be in their very interest to invest in worldwide campaigns to bring about a "peaceful and just" settlement in the Middle East. In fact, most Arab officials and observers are disappointed and angry with the attitude of the companies in this respect.

The increasing U.S. dependence on imports of Arab oil is seen as a determining factor in influencing future U.S. policy in the Middle East. More and more, alarming cries are voiced in the United States, both in the oil industry and official circles, urging the administration to adopt a less partisan attitude in the Arab-Israeli conflict and to show a minimum understanding of Arab views and demands.[41] Unfortunately, there are yet no signs of any change in the pro-Israel policy of the United States, and more effort must be devoted to making a change.

The second hot spot in the Eastern Hemisphere will be the political stability and the strategic security of the Gulf area. At the end of 1971, the British troops withdrew from the area because the shaky economy of the United Kingdom could no longer support the cost of military presence. Not only might rivalries among the Gulf states develop into occasional problems, but also Iranian ambitions and plans to dominate the area politically and militarily might be challenged by the Arabs. The great powers will be behind the scenes, and it is currently believed that one of the most important strategic objectives of the United States is to "protect" the Gulf area from Soviet penetration and Chinese influence. Some political observers have added a further strategic dimension to the interpretation of the Teheran

Agreement.42 According to their analysis, the Teheran Agreement was a predetermined arrangement intended to make the European consumer support the cost of the defense of this vital and highly stategic region, in order to avoid the direct involvement of the United States after the withdrawal of the British. Incremental oil revenues would contribute, to a great extent, to financing the buildup of an impressive military machine headed and supervised by the Iranians.

The evolution of crude oil prices beyond 1975 will depend, to a large extent, on U.S. assessment of its strategic interests and objectives by that time. However, it would be safe to predict that crude prices will unmistakably move upwards. The ultimate ceiling will be set by the competitive cost of producing hydrocarbon liquids from alternative sources such as coal, tar sands, and shale oil, which may one day enable the United States to recover a large part of its energy self-sufficiency. But the uncertainties of technology and inevitable inflation will contribute to push the economic ceiling higher and higher than present levels.

The prospects of potential instability in the Middle East in connection with political upheavals and subversive action would seem to be more relevant to intelligence activities than to prospective or strategic decisions of oil executives. The main preoccupation of the latter will be in the unpredictable field of their relations with the governments of producing countries. We have outlined, in the first chapters of this study, the psychological and emotional grounds for these relations and emphasized the generally hostile attitude of public opinion, especially in Arab countries. Oil companies, and especially the international majors, have always been privileged targets for nationalist attacks, more or less tinged with Marxist dialectic and leftist terminology. They were indicated because of their colonial and imperialist background in the area and of their "continuous collusion with imperialism." The governments of producing countries, although they could be sympathetic to the West on political grounds, had to be very cautious in their relations with oil companies so as to avoid any unwarranted provocation of nationalistic and highly sensitive public opinion. The oil companies themselves are to be blamed for not having taken any serious steps to improve their public image and to modify the profoundly hostile attitude of public opinion. They seemed to be preoccupied exclusively with commercial objectives aiming to push exploitation to its furthest limits and with devoting relentless efforts to the safeguarding of their privileges.

To varying degrees, the governments have been obliged to take actions that would appease the mounting hostility of public opinion. But they were very reluctant to take any drastic or revolutionary step that might have adverse political significance beyond the limits of the oil industry. The situation came very near to a general explosion in the aftermath of the six-day war of 1967 and has not returned to normal since then. The governments were generally in favor of natural and progressive evolution of their relations with the vast companies but could not dare to defy them openly.

Then there was the Libyan confrontation and victory in 1970 and the decisive rounds in Teheran and Tripoli in early 1971. To the surprise of the governments, the untouchable and powerful oil companies proved to be no more than a "paper tiger." No matter how the OPEC victory was made possible, the important thing is that the governments of OPEC members are convinced and believe that the victory—their victory—is to be credited to their united front and their consolidated political and professional forces, which, from now on, will put them in a strong bargaining position vis-à-vis oil companies, which are now forced into a defensive strategy. The temptation was very great indeed to reinforce their new advantages and to reiterate their victory by moving further towards trimming off company privileges and achieving larger government control over the industry. This united front proved to be quite effective, once again, in the confrontation over the revaluation of crude oil prices as a consequence of the dollar devaluation, which resulted in the Geneva Agreement of January 20, 1972, which was later amended on June 1, 1973.[43]

Since the "genie" came out of the bottle in the early 1970s, the international oil industry has been undergoing a profound and rapid mutation on all fronts. The background to all these changes was dominated by the emerging energy crisis in the United States, which induced a worldwide psychosis about tight supply and crude shortage. The long-term prospects of increasing dependence on crude oil imports from the Eastern Hemisphere were underscored by the actual shortage in products' supply due to insufficient expansion of U.S. refining capacity and to the difficulty of rapidly adding the new capacity needed because of the aggravation of pollution problems. Voluntary rationing was practiced by some distribution companies to avoid running out of stock. The situation reached such alarming level that, in early July 1973, Nixon urged Americans to reduce consumption of gasoline and electricity and to adopt a nationwide "energy conservation" ethic.[44] Consequently, markets for both crude oil and products became definitely seller's markets, and prices kept going up in major supply areas. Freight rates equally followed the upward trend, despite the available increasing capacity of tankers.

This environment contributed to strengthening the position of producing countries and encouraging nationalistic claims to become more aggressive and intransigent. However, a basic split appeared in the front of exporting countries concerning the nature and the evolution of their relations with the concession-holding companies. Divergence of opinion and action actually reflected ideological differences and political rivalries between the countries and regimes rather than professional arguments and considerations. Revolutionary regimes wanted to achieve prompt and full control of oil production within their countries and to put an end to the presence and "privileges" of Western companies. Moderated and so-called conservative regimes aimed to achieve the same nationalist objectives through progressive participation and without a political break with the West.

On the revolutionary side, the successful achievement of Algerian nationalizations of all producing assets in 1970 and 1971 and the failure of the companies' retaliatory attempts were viewed by Arab nationalists as the most important event in the recent history of the industry. The traumatic effects of the 1951-53 experience of Mossadegh in Iran were dissipated, and nationalists thus proved the maturity and the ability of Arab nations to take care directly of the development and exploitation of their natural resources. This precedent was reinforced by the nationalization of British Petroleum's producing assets in Libya in December 1971, presented as a punitive action against British "complicity" in the Iranian occupation of three strategic Arab islands in the Gulf area. On June 1, 1972, Iraq nationalized the Iraq Petroleum Company, thus imposing a final and brutal settlement on the remaining complex problems that had clouded the Iraqi scene since the 1961 expropriation of 99.5 percent of the IPC concession under Law 80 by General Kassem, which resulted in a general freezing of oil activities and expansion. After several months of uncertainty, a final agreement was reached on February 28, 1973, which was surprisingly favorable to the Iraqis since the companies practically dropped their claims regarding the 1961 expropriation and the huge North Rumaila oil field.[45]

On the moderate side, governmental participation in existing concessions was achieved after lengthy negotiations; 25 percent equity participation became effective January 1, 1973 in Abu Dhabi, Kuwait, Qatar, and Saudi Arabia,* and its implementation seems to be proceeding smoothly. The participation scheme will provide the political background and the professional environment for the oil industry in the Gulf area.[46] The mechanism of the participation deal is a very complex one. A brief survey of its main features is given in Appendix B.

Whatever the ideological option or the political approach adopted by one country or another, the dominating common factor in their relations with the oil companies is an atmosphere of emulation, competition, and a striving to achieve better performances and larger

---

*The participation agreement between Kuwait and the oil companies is still to be ratified by the Kuwait parliament.

victories. Any achievement by a producing country is regarded as a precedent-setting minimum ground for further claims by others, although such claims might materialize under different political conditions. The evolution of government-company relations, to the continuous detriment of the latter, has progressively escalated along with the aggravation of the energy crisis psychosis and with the successive "victories" of exporting countries. The credibility of BP nationalization by the Libyans was enhanced by the success of the Algerian precedent, and both certainly encouraged the Iraqis to undertake irremediable measures against the IPC group.

The campaign for participation conducted by Saudi Oil Minister Ahmad Zaki Yamani contributed to reinforcing the prestige and the leadership of Saudi Arabia in the moderate camp. This situation was a challenge to Iranian ambitions. Although the Saudi minister was designated by OPEC to negotiate collectively on behalf of the Gulf exporting countries, including Iran, the Shah announced on June 24, 1972 that Iran was not interested in participation as defined by OPEC; this was just three days before the OPEC conference that was to take decisive measures in connection with the stalemated participation negotiations. The Shah's announcement was made in London to emphasize his earlier assurances that Iran's oil would always be available to the West, in contrast to the supply uncertainties of Arab oil.

Instead of participation, the Shah announced a new deal with the international oil companies that forgoes formal participation in capital and management (and thereby avoids further compensation and future investment). Iran agreed to confirm the consortium's scheduled three five-year renewal periods after the initial agreement expires in 1979, in return for the consortium's undertaking to raise production and export capacity to 8 million barrels daily and to deliver certain quantities of the oil produced to the National Iranian Oil Company (NIOC) for marketing in its own downstream ventures, both inside and outside Iran.[47] The Iranians were convinced that the terms of this deal were much better than the contemplated advantages to be obtained through participation.

In fact, this agreement was never put into effect. Its ratification was withheld pending the finalization of participation negotiations with the Arab states of the Gulf. To their surprise, the Iranians found the terms of the participation agreement more favorable than their own deal; they could not sign an agreement that would leave them far behind the Arabs. The desire of achieving better performance induced a fundamental reversal of the Iranian policy. A new deal was announced by the Shah on March 16, 1973 and signed a few weeks later on May 24. The new agreement supersedes the 1954 consortium agreement and runs for a period of 20 years. It provides for NIOC to take over all the consortium's operations and facilities while entering into long-term sales contracts of crude oil at predetermined prices.[48]

The implementation of the participation agreement in the Arab Gulf states triggered another escalating campaign in Libya. Libya initially claimed that 51 percent participation was a minimum and later pushed demands to a full 100 percent participation, to be accepted "voluntarily" by the companies against compensation on the basis of net book value of assets involved. Oil companies operating in Libya were subject to periodic intimidation campaigns to force them to comply with governmental demands, and nationalization was presented as the only alternative to a negotiated arrangement. The nationalization of independent Bunker-Hunt, former partner in the nationalized assets of BP, was viewed as a warning measure. The companies seemed to be willing to accept the 51 percent participation concept but found themselves lagging behind what the Libyans were prepared to accept. In the midsummer of 1973, the situation is confused and confusing, and the future evolution of the oil industry in Libya seem quite unpredictable.

The continuous erosion of the dollar value and the severe inflation that is sweeping over Western economies add further trouble to the international oil scene. The atmosphere of energy crisis has helped revive the aggressiveness of some exporting countries and make some nationalists believe that defeating oil companies has become a routine matter, which requires only patience and skillful maneuvering. The OPEC success in obtaining full satisfaction in the Geneva Agreement of June 1, 1973, to compensate for the dollar devaluation and monetary fluctuations, was crucial in that it revealed the genuine weakness and defensive attitude of the companies. Temptation is growing to abuse this situation and escalate demands still further. In midsummer 1973, the oil companies are facing several potential problems on different fronts. Abu Dhabi is claiming a substantial premium for the low sulfur content of its crudes, retroactive to 1962.[49] Kuwait intends to renegotiate the terms of the participation agreement, which the Kuwaiti Parliament has been reluctant to ratify, and Saudi Arabia is reported considering the renegotiation of buy-back prices of participation oil.[50] Several oil companies in Libya are facing the serious prospects of being nationalized, whereas Algeria and Iraq are trying to push OPEC into renegotiating the entire range of 1971 arrangements made in Teheran and Tripoli.[51]

The future of the oil companies in the Middle East seems gloomy and charged with uneasiness and uncertainty. However, one should not overestimate the gravity of the situation when evaluated under the long-term strategic prospects of the international majors. The real power of the majors lies in their integrated and concerted control of world markets and their ability to dispose of the largest part of free-world production outside the United States to meet increasing demands. The majors' power is most likely to increase because of the aggravation of the supply crisis and because control of outlets cannot really be challenged on political or professional grounds. The absence of any concerted action on the part of consumers and the failure of European countries to agree on a common energy policy coincide to put the major oil companies in a privileged position and to designate them as the main channel for oil between export terminals and consumption centers. Whether oil would be produced under concession, participation, or nationalization terms, the majors would still be assured of disposing of the largest part of world production. The long-term sales contracts concluded between Algeria's Sonatrach and U.S. companies and between Iran and the consortium's member companies and the bridging and phase-in arrangements for participation oil are quite significant in this respect. The antagonism between exporting countries and the oil companies lies mainly on political and psychological grounds, and there are no basic conflicts of interest between them on professional grounds. Once satisfactory arrangements are achieved, in one way or another, to settle the political conflicts, then the professional complementarity of interests would emerge as a dominating factor. In other words, the oil companies would manage to adapt to any change in their relations with the exporting countries. Practically, the production function of the integrated chain of the industry would remain professionally under their control, although it might come under the political sovereignty of the exporting countries. Furthermore, price increases would be reflected onto products' proceeds, and the move would not be seriously opposed, because of the energy crisis psychosis and the inflationary trend in the Western economies.

In the coming years, there will certainly be a concordance of interests between the United States, the exporting countries, and the oil companies to keep prices high and stable and to favor their upward process of evolution. The pricing battle is definitively over, and the only loser is the consumer.

NOTES

1. G. Chandler (Royal Dutch/Shell), "Oil in the Seventies," Beirut, May 13, 1971. Paper reproduced in Middle East Economic Survey (MEES), August 6, 1971.
2. Oil and Gas Journal (OGJ), June 21, 1971, p. 71.
3. M. A. Adelman, March 28, 1971 at the Seminar of the Transportation Center, Northwestern University.
4. G. Chandler, op. cit.
5. Ibid.
6. In an address to the Iranian Parliament on December 8, 1970, as reproduced in MEES, December 18, 1970, Supplement.

7. Petroleum Intelligence Weekly (PIW) June 14, 1971.

8. According to reliable sources, the attitude of the Iranian delegate to the Caracas Conference was very moderate. He envisaged calling for a price increase of about 10 cents a barrel and discarded the idea of restricting production as a mean of pressure, should the oil companies fail to comply with this demand. The question of future price escalation was not raised.

9. J. C. Burrows and T. A. Domencich, An Analysis of the United States Oil Import Quota (Lexington, Mass.: D.C. Heath, 1970), p. 1.

10. OGJ, October 12, 1970, p. 62.

11. OGJ, July 19, 1971, pp. 25 and 40.

12. "Turmoil over U.S. Oil Imports," Petroleum Press Service, October 1968, p. 364.

13. OGJ, May 19, 1969, p. 74.

14. PIW, October 2, 1967, p. 6.

15. OGJ, May 6, 1968, p. 79.

16. OGJ, May 5, 1969, p. 79.

17. OGJ, November 23, 1970, p. 29.

18. OGJ, March 31, 1969, p. 38.

19. For details, see OGJ, February 10, 1969, p. 39, and April 28, 1969, p. 54.

20. OGJ, December 29, 1969, p. 73.

21. Petroleum Press Service, September 1969, p. 322.

22. The principal conclusions and recommendations of the 400-page report of the cabinet task force were published in the Oil and Gas Journal of March 2, 1970, as were the minority's statements. A brief analysis will be found in the same issue, p. 25.

23. James E. Akins, "The Oil Crisis: This Time The Wolf Is Here," Foreign Affairs, April 1973.

24. OGJ, August 2, 1971, p. 32.

25. OGJ, July 20, 1970, p. 20.

26. Fortune Magazine, "How the Arabs Changed the Oil Business," August 1971, p. 197.

27. Petroleum Press Service, "Search for an Oil Sale Policy," August 1968, p. 287.

28. PIW, January 19, 1970; and OGJ, January 19, 1970, p. 24.

29. Akins, op. cit.

30. Petroleum Press Service, April 1973, p. 127.

31. Wayne G. Glenn, of Continental Oil, OGJ, March 8, 1971, p. 34.

32. PIW, May 21 and July 2, 1973.

33. PIW, September 6, 1971.

34. OGJ, December 21, 1971, p. 24.

35. Petroleum Press Service, October 1972, p. 364.

36. Akins, op. cit.
37. Petroleum Press Service, January 1973.
38. OGJ, December 15, 1969, p. 25.
39. Akins, op. cit.
40. OGJ, December 15, 1969.
41. Testifying before a joint session of the Foreign Affairs, Near East, and Foreign Economic Subcommittees of the House of Representatives, William F. Penniman, a Middle East oil consultant, warned that "extreme anti-American reactions" would occur in the event of any outbreaks of hostilities in the Middle East. He said, "Whether or not the U.S. actually helps Israel in its struggles with Arab countries, the Arabs think the U.S. is helping Israel, and we have done nothing to correct that impression. . . . An even-handed policy is a step in the right direction but no longer sufficient in itself. . . . I can't understand why a pro-Arab policy is necessarily anti-Zionist. I'd rather we use the idea that it's pro-American." PIW, May 21, 1973.
42. See, in particular, the French periodical Le Spectacle du Monde, March 1971 and August 1971.
43. Arab Oil and Gas 5, 102 (June 16, 1973).
44. Time Magazine, June 25, 1973, p. 49; and PIW, July 9, 1973, p. 5.
45. Arab Oil and Gas 5, 105 (July 1, 1973).
46. For an over-all discussion of the participation issue, see T. Rifai, in Arab Oil and Gas 5, 100 (May 16, 1973).
47. PIW, July 3, 1972, p. 6.
48. MEES, May 25, 1973.
49. PIW, April 2, 1973.
50. MEES, June 15, 1973.
51. PIW, July 2, 1973.

CHAPTER 18

POSTSCRIPT: THE UNCERTAINTIES OF THE FUTURE

1973: THE VICISSITUDES OF THE CRISIS

Transitory Period of Mutation

Chapter 17 was written in mid-summer 1973 at a time when no one could foresee the major crisis that was about to shake the world's economics and politics and that led to the surprising development of the fourth Arab-Israeli war in October 1973. However, the seeds for such fundamental changes in balance of power and in energy economics were already maturing in favorable grounds. In fact these changes are not really surprising, and we have outlined their main features or motivations in our assessment of "Future Outlook," where strong emphasis was put on the direct link between the pricing issue and the Arab-Israeli conflict, on the inevitable further price escalation, and on the emergence of producing countries as a predominant power in the international oil industry. The real surprise lies in the magnitude of the price increase, triggered by the most dramatic war between Arab countries and Israel in which oil was largely used as an efficient weapon.

The foregoing analysis, as developed throughout this book, seems today quite outdated, though it is realistic enough insofar as it reasonably reflects the most advanced currents of thoughts that, until the end of 1972, had prevailed in the international petroleum industry. This emphatically underlines the nature and the dimensions of the fundamental changes that have affected the industry since then.

The most significant change has actually taken place on the psychological level. The producing countries' relatively easy successes on almost all fronts, the oil companies' seeming resignation before the producers' escalating demands, the acceleration of inflation and the monetary upheavals that have characterized Western economies, and the psychosis created by the alarmist energy crisis have all given the producing countries a feeling of infinite power and pride. The aforementioned atmosphere of emulation and competition played a major role in the precipitation of events in both the revolutionary and moderate regimes. All of this has worked toward placing the producing countries in a position of superiority and toward inducing them to eliminate the majors' hold of the oil companies.

On the professional level, the most spectacular change was undoubtedly the extraordinary boom of the so-called free crude oil market. The general prevailing feeling until the end of 1972 was that the producing countries would be incapable of marketing their production entitlement under the participation agreement without practicing a policy of "dumping," which would be disastrous for the stability of posted prices. A complex system of buy-back arrangements had been provided for recycling most of the government's share into the companies' integrated circuits. To general surprise, however, not only did the producing countries find it easy to sell their production shares, but demand was so high that market prices very soon exceeded postings and the oil companies, big and small, were found barging into one another in the Arab Oil ministers' waiting rooms to have the privilege of being designated as the lucky buyers. The supply-demand disequilibrium was aggravated by the threatening energy crisis, which generated these self-defense reactions:

1. The independent refiners were no longer sure of being provided with supplies from the majors; the Europeans and the Japanese feared the "unfair" competition of the United States, which was favoring a price increase; and everybody was rushing to buy today rather than tomorrow, so certain was it that prices were going to rise.

2. This unexpected evolution surprised the producing countries themselves and led them to wonder whether the participation agreement had not actually been a fool's deal for safeguarding the companies' interests at a cheap price. Without calling the latter to account, it rapidly became clear that the participation terms were no longer consistent with existing conditions. In Kuwait the Parliament refused to ratify the agreement concluded with the companies, prompting the government to call for new negotiations aimed at giving Kuwait an initial 60 percent participation.

3. Qatar was the first country to sell its production share at a price above posted prices. Abu Dhabi demanded an increase in its posted prices through the introduction of a special premium for the low sulfur content of its crude. The companies having rejected this demand, the Abu Dhabi Government decided to "freeze" its production

expansion programs. Furthermore, Abu Dhabi's Oil and Industry Minister Manaa al-Otaiba declared on several occasions that the participation agreement had to be renegotiated to increase the government's initial share as well as the buy-back price for its production share. Finally, Saudi Arabia, which had designed and defended the participation concept and provisions, demanded and obtained an increase in the price of bridging and phase-in crude of the government share sold back to the companies.[1]

On the political level, the prospect of the growing dependence of U.S. energy requirements on imports from the Arab countries gave new strength to the nationalists' claim that Arab oil could be an active weapon in the confrontation with Israel. Given the large size of its reserves, Saudi Arabia stood as the unchallenged leader of the new crusade and King Faisal threw his prestige into the balance in order to exert pressure on the United States to adopt a more evenhanded attitude in the Mideast conflict. The use of Arab oil as a political weapon encouraged the U.S. majors, particularly Aramco's shareholders, to launch pro-Arab information and propaganda campaigns in the United States.[2]

Immediate Consequences of the War

The oil weapon was dramatically imposed as of the outbreak of the fourth Arab-Israeli war, on October 6, 1973. However, despite the passions and military imperatives, this weapon required delicate handling, and could not exert immediate pressure on the United States, whose dependence on Arab oil was still marginal. Since the nationalization of U.S. interests in the Gulf and in Saudi Arabia was then impossible (for political and business reasons), reduction or interruption of exports mainly harmed the Japanese and the Europeans, who showed more sympathy to the Arab cause. It was also believed that a selective embargo on exports to the United States would be difficult to control. Finally, the Arab exporting countries' oil ministers decided on October 17 at a meeting in Kuwait to cut back Arab oil production by an immediate 5 percent and by an additional 5 percent every month until the Arab territories occupied during the six-day war were liberated. Actually, the effective cutback rates exceeded 10 percent and most Arab countries decreed an embargo on oil exports to the United States, accused by the Arabs of having given heavy military support to Israel during the war, and to Holland, which had publicly taken pro-Israeli stands during the first days of the conflict.

This policy of production cutback and selective embargo on certain exports was applied gradually during the second half of

October 1973 and the uncertainty that surrounded the attitude of some exporting countries contributed to further accentuating the climate of crisis and shortage. The credibility of the Arab determination to rely heavily on the "oil weapon" was demonstrated by the decision of Saudi Arabia to take a leading, militant role in this respect. By announcing an immediate cutback of 10 percent instead of the 5 percent decided in Kuwait, plus an embargo on all oil exports to the United States and the Netherlands, the largest exporting country, which the Americans had regarded as its safest source of supply, was now most determined to show its firmness. The real turning point in international oil politics resulted from the fact that Washington didn't believe such a dramatic change would happen.

This had deep political and economic repercussions in the consuming countries, particularly in Europe where the solidarity of the members of the European Economic Community was severely strained. On the whole, although the real effects of these measures were not immediately felt, the results obtained on the political level were sufficiently conclusive to break down the skepticism of those among the Arab leaders who held doubts on the timing or effectiveness of utilizing the oil weapon for the Arab cause. On November 4 and 5, 1973, the Arab oil ministers meeting in Kuwait decided to raise the production cutback rate to 25 percent compared to the September 1973 level, to formally confirm the embargo on exports to the United States and Holland, to further reduce production by 5 percent on December 1, and to reserve preferential treatment to the friendly countries (particularly France and Britain) that were to continue to be supplied with oil normally on the basis of their average imports during the first nine months of 1973. Moreover, the oil ministers of Algeria and Saudi Arabia were charged with undertaking a tour in the Western capitals in order to explain the Arab position and demands.

The "Oil Weapon"

The oil weapon was handled by the Arabs with skill and firmness. Most consuming countries were driven into a corner and were forced to push aside their own political difficulties or hesitations in taking a stand on the Palestinian problem in the hope of being listed by the Arabs among the privileged "friendly" countries. This was notably the case with Japan, on which strong pressure was exerted to make it break its diplomatic and economic relations with Israel. As for the Europeans, who were subjected to contradictory influences and were torn between national egoism and communal solidarity (vis-à-vis Holland), they got around the difficulty by making a declaration common to the European Community, relatively favorable to the Arab views.

In acknowledgment of this gesture the Arab oil ministers meeting in Vienna on November 18 decided to suspend their decision to cut back their production by 5 percent as of December 1 while pointing out that the European initiative was rather symbolic.

A summit meeting of the Arab heads of state was held in Algiers from November 26 to 28, 1973 during which the political utilization of the oil weapon was solemnly confirmed. The consuming countries were classified into three categories—friendly, neutral, and unfriendly—and the furture evolution of Arab production was subordinated to the political developments of the situation. Two essential conditions for the normalization of oil production were expressly mentioned, namely, the total evacuation by Israel of the occupied territories, including Jerusalem, and the recovery by the Palestinian people of their legitimate rights.

Used very skillfully, this "carrot and stick" policy was rewarding for the Arabs. It alerted world public opinion on what was really at stake in the Arab-Israeli conflict and attracted attention to the problem of the depletion of natural resources. The resulting shortage brought about a remarkable soaring of prices, which practically increased threefold, if not fourfold, in less than three months. These satisfactory results must have seemed sufficient for the Arab oil ministers meeting in Kuwait on December 25, 1973, inducing them to ease the production restriction measures and to establish a "truce" on the oil front. This move was welcomed with great relief in Europe and Japan where the fear of a serious economic recession was causing increasing misgivings. The easing of the Arab restrictions (no limits on exports to the friendly countries, whose list was extended to include Japan and Belgium) was equivalent to a de facto stoppage of the political oil war. However, this opened the way to a new economic war where the escalation of prices may upset the world economic and monetary balance, at least in the short run.

The serious economic preoccupations facing the world since the beginning of 1974 were indeed the result of a deep and gradual deterioration of the Western world's economic and monetary system. True, the problem has changed in scope and dimension with the considerable oil price increases, but it should not be forgotten that the producing countries' demands initially were only self-defense reactions against the continuous inflation exported from the industrialized countries. In fact, the genuine weakness of Western economic system was largely responsible for the prevailing situation and the acceleration of the mutations affecting international oil economics acquired new momentum with the gradual loss by the producing countries of their complexes as they grew aware that the Western world had become weak and vulnerable.

On the monetary level the continuous erosion of the value of the dollar and galloping inflation throughout the world had seriously aggravated the oil price situation, which had been precarious since the conclusion in 1971 of the Teheran and Tripoli Agreements. The aggressive behavior of certain progressionist countries vis-à-vis the companies imposed a new pace onto OPEC's action. The spectacular success achieved on June 1, 1973 in Geneva, offsetting the effects of the dollar devaluation, played a determining role by revealing the intrinsic weakness of the companies' defensive position. This was soon followed by calls for the revision, if not annulment, of the Teheran and Tripoli Agreements. The positions of Algeria, Iraq, and Libya were dominant. Finally, OPEC decided to invite the companies to renegotiate these agreements without disclosing the nature or scope of its demands.[3] The negotiations between the companies and the six Gulf exporting countries were scheduled in Vienna on October 8, and it was believed that they would mainly deal with price adjustment in terms of inflation.

First Wave of Increase, October 16, 1974

When the Middle East war was two days old, on October 8, 1973, the delegates of 20 oil companies met at Vienna to discuss oil prices with six members of OPEC. The timing, from the companies' point of view, was disastrous. The Arab representatives were excited and unified by the early Arab successes on the different Israeli fronts and were soon passing around newspaper photographs of massive American supplies to Israel. The war had added fervor and unity to their bargaining.

But with or without a war, the companies were in a very feeble bargaining position. As one oilman put it, the Teheran Agreement of 1971 was now clearly a "house of cards" that had already crumbled. The Iranians had already made it clear that they wanted to go up to $5 a barrel.

The grievances of OPEC were, in fact, very real; the inflation factor of 2.5 percent a year had not kept pace with actual inflation, and the price of other commodities, as the Shah never ceased to point out, was far outstripping oil. And OPEC was now in a position—which the companies took some time to realize—to enforce their own price. OPEC's 13 years of disunity and humiliation were over and the day of rekoning was near.

After four fruitless days in Vienna, with the war gathering force in the background, the six OPEC countries were still standing firm, and it was clear that they would not settle for less than $5 a barrel. The companies, however, were seemingly unaware of the deterioration of their bargaining position so that they continued to behave just as they used to in the "good old days." In order to gain time they insisted on adjourning the meeting for two weeks to consult with their home governments. Consequently, the Vienna Conference disbanded and was never recalled: the OPEC countries announced that they would meet on their own the next week in Kuwait. The news of this critical breakdown, which had such historic results, was swamped by the war news. What would have happened if the companies had agreed, there and then, on an increase up to $5 a barrel? It is a question that haunts at least some of the companies' negotiators.

During the following days, the oil lobby in Washington was trying desperately, and vainly, to bring about the slightest change in the unconditional U.S. political and military support of Israel. While the airlift of U.S. military supplies to Israel was going on, in Kuwait the OPEC member countries were beginning to fully realize their strength. The decisions concerning the price increase were announced on October 16, 1973. The next day the Arab oil ministers meeting in Kuwait decided to implement the production cutback and selective embargo measures as indicated earlier.

To general surprise, the committee negotiating on behalf of the six Gulf countries meeting in Kuwait on October 16, 1973 announced a historical decision that rang the knell not only of the Teheran and Tripoli Agreements but also of the whole concept of the relations between the producing countries' governments and the oil companies. Oil prices were henceforward to be set unilaterally by the exporting countries without any consultation with the companies. To illustrate the far-reaching effects of this decision, the French newspaper Le Monde wrote that "by crossing the Suez Canal, the Egyptians lost their complexes. Here in Kuwait another canal has been crossed with the same audacity and determination by countries until then considered to be moderate."[4]

On the technical level, the principle behind the decisions taken on October 16 is relatively simple. The posted price was from there on to be determined with reference to the market price. The difference between the two prices was to remain constant and equal to that prevailing before 1971, or about 40 percent. The market price of the reference crude (34° API Arabian Light) was set at $3.65/Bbl, or an increase of about 17 percent compared to certain prices reportedly obtained by Petromin in the sale of certain participation crude shipments. The combination of these various factors led to a new posting for the marker crude of $5.119/Bbl, or 70 percent more than the price of $3.011 prevailing until October 16. The government take (royalty plus taxes) increased by slightly over 70 percent, rising for example in the case of the reference crude from $1.770 to $3.048/Bbl.

The Abu Dhabi Government took the occasion to unilaterally satisfy its demand for a special increase in the prices of its crudes on account of their low sulfur content. Apart from the above-mentioned general increase, a special premium was added, varying with the different crudes as follows:

$0.75/Bbl for Murban crude (0.75 percent sulfur content)
$0.55/Bbl for Zakum crude (1.05 percent sulfur content)
$0.25/Bbl for Umm Shaif crude (1.20 percent sulfur content)

Furthermore, additional increases of 6.7 and 4¢/Bbl respectively were decided with retroactive effect to February 14, 1971, the effective date of the Teheran Agreement, to compensate for the loss arising from the nonrecognition then of the premium for the low sulfur content of the Abu Dhabi crudes.[5]

In the weeks that followed, the precedent set by Abu Dhabi was extended to all the low-sulfur crudes of the Gulf. Sulfur premiums were therefore added to the Gulf crude 40° API, 1.17 percent sulfur) and 5.3¢/Bbl for Oatar Marine crude (36° API and 1.4 percent sulfur) with retroactive effect to February 14, 1971, so that the general 70 percent increase of posted prices was applied to these premiums, which became 7.1 and 9¢/Bbl respectively. Moreover, new premiums applicable from October 16, 1973 were added on the basis of 42¢/Bbl for Dukhan crude and 25¢ for Qatar Marine.[6] A premium of 25¢/Bbl was also added to the Iranian Light posting (34° API and 1.35 percent sulfur) from October 16, 1973.

The introduction of these premiums strengthened the privileged role of Arabian Light crude (34° API and 1.63 percent sulfur) as marker crude for the Gulf area. Indeed, the sulfur premiums were set on the basis of the reference content (1.63 percent) as already indicated in Chapter 6. Account taken of the possible fluctuation in the sulfur content of the various crudes (high-lighted and illustrated in Figure 6.2, the sulfur differential approximates 8-8.5¢/Bbl for each 0.1 percent of sulfur content below the sulfur content of the reference crude (1.63 percent). The ensuing oil price structure in the Gulf is illustrated in Table 18.1.

Libya, which had not participated in the Vienna negotiations or the Kuwait meeting, could not stay outside this fundamental reversal of international oil economics, not to mention its determination to be in the vanguard of the "victories" gained by the producing countries. On Friday, October 19, 1973 a new posted price of $8.925/Bbl was announced for Libyan 40° API crude, or an increase of about 94 percent compared to the then existing price of $4.606. The Libyan official communiqué detailed the structure of the new price as follows:

TABLE 18.1

New Structure of Posted Prices in the Gulf
Decided in Kuwait on October 16, 1973
(U.S. $/Bbl)

Crude Oil	Gravity (°API)	October 1, 1973			October 16, 1973				
		PP	GT[a]	TCIT[b]	PP	GT[a]	TCIT[b]	BRP	PC
Abu Dhabi[c]									
Murban	39	3.144	1.824	1.974	6.045	3.582	3.732	4.52	(0.15)
Marine (Umm Shaif)	37	3.110	1.720	2.020	5.537	3.192	3.492	4.63	(0.30)
Zakum	40	3.185	1.766	2.066	5.964	3.451	3.751	4.42	(0.30)
Saudi Arabia									
Arabian Light	34	3.011	1.770	1.870	5.119	3.048	3.148	3.65	(0.10)
Arabian Medium	31	2.884	1.693	1.793	4.903	2.917	3.017	3.50	(0.10)
Arabian Heavy	27	2.725	1.597	1.697	4.632	2.753	2.853	3.31	(0.10)
Arabian Light (Sidon)	34	4.205	–	–	7.149	–	–	–	–
Iran									
Iranian Light[c]	34	2.995	1.750	1.870	5.341	3.172	3.292	3.64	(0.12)
Iranian Heavy	31	2.936	1.714	1.834	4.991	2.960	3.080	3.56	(0.12)
Iraq									
Basrah	35	2.977	1.739	1.859	5.061	3.002	3.122	3.61	(0.12)
Kuwait									
Kuwait	31	2.884	1.715	1.775	4.903	2.939	2.999	3.50	(0.06)
Qatar[c]									
Dukhan	40	3.214	1.855	2.025	5.834	3.443	3.613	3.82	(0.17)
Marine	36	3.127	1.775	1.995	5.503	3.215	3.435	3.69	(0.22)

PP = Posted Price; GT = Government Take; TCIT = Total Cost Including Taxes; BRP = Basic Realized Price; PC = Production Cost

[a] The estimation of Government Take (67) is made on the basis of a 55 percent income tax and of an expensed royalty of 12.5 percent. The above-indicated production costs are just indications.

[b] The Total Cost Including Taxes does not reflect the effects of participation, which as of October 16, 1973 would add at least 15¢ to costs.

[c] The crude oil prices of Abu Dhabi, Iranian Light, and Qatar take into account the low sulfur premiums and the retroactivity of part of these premiums as indicated in the text.

Sources: Middle East Economic Survey, October 19 and November 30, 1973, and Petroleum Intelligence Weekly, October 22, 1973.

	$/Bbl
Basic posted price	6.979
Sulfur premium	1.336
Suez Canal premium	0.152
Temporary freight premium	0.458
Total	8.952

A comparison between the new Libyan price and the increase decided in Kuwait on October 16, 1973 is only possible in the light of the following remarks:

• The prevailing basic posted price, as defined in the 1971 Tripoli Agreement, excluding the low sulfur premium, or $3.227/Bbl, has been subjected to all the increases provided for by the Geneva Agreements, or 27.22 percent, in addition to the increase of 70 percent, applied in the Gulf area.
• The above-mentioned low-sulfur premiums were replaced by a basic premium of $1.05/Bbl, in line with the premiums applied in the Gulf, considering that the average sulfur content of the Libyan crudes approximates 0.35 percent. To this basic premium were applied the increase provided by the two Geneva Agreements, or a total of 27.22 percent, raising its final value to $1.336/Bbl.
• The initial 12¢ Suez Canal premium was subjected to the same increase of 27.22 percent, thus rising to 15.2¢/Bbl.
• The temporary freight premium was determined on the basis of Worldscale 1973 rates instead of those prevailing in 1971 as retained in the Tripoli Agreement with the result that the freight premium was defined on the basis of 0.063¢/Bbl for each 0.1 percent of AFRA LRI rates (and no longer LR II) above the reference Worldscale 1972 value. The 27.22 percent increase is also applied to the freight premium.

The Libyan communiqué specified that posted prices in Libya would be revised on the 23rd of each month in terms of the arithmetic average of realized prices prevailing on the market during the preceding month. Prices will be modified whenever this average exceeds 0.5 percent upward or downward.[7]

These various decisions took everybody by surprise in view of the unexpected size of the price increases and, especially, of the solemn affirmation of the producing countries' sovereignty on the unilateral setting of prices. The somewhat arbitrary aspect of the size of the increases is indicative of the remarkable strong position in which the producing countries found themselves as a result of the latent situation of shortage and crisis, aggravated by the outbreak of the war and by the restriction and embargo measures decided by the producers. The oil companies, unable to oppose effectively the producing countries' sovereign decisions, gave in all the more willingly, since they were in a position to shift the rise in cost onto the consumers without any difficulty, while realizing some additional profit. The very strong position of the producing countries vis-à-vis the companies was clearly defined in the communiqué released by OPEC in Kuwait on October 16 specifying that "in case the oil companies refuse to lift crudes on the basis of these arrangements (new posted prices), the producing countries will make available to any buyer the Arabian Light at $3.65/barrel FOB Ras Tanura."[8]

OPEC's Redefinition of the Oil Price Concept

Thus to the strategic concept of stable posted prices, immune from market fluctuations, designed and implemented by the international major oil companies practically since the 1923 Achnacarry Agreement, the OPEC member countries substituted, in a conjuncture particularly favorable to them, a commercial concept according to which posted prices will be variable and tied in a more or less rigid way to market prices. The same goes for fiscal revenues. Thus through a political decision the setting of oil prices was left to the play of supply and demand. However, given the situation of shortage and the producing countries' position of cartel, it appeared as if these countries wanted to shoulder market risks when practically they no longer existed.

The impact of these measures, as important and grave as they may be, was attenuated following the developments in the use of the oil weapon to serve the Arab cause through production cutbacks and embargo measures already mentioned. The world was left with no other choice but to accept the fait accompli, but it did so all the more readily since the shortage had, at the beginning of winter, reached such a degree that the real prices practiced on the market had far exceeded the new postings decided in Kuwait and Tripoli. These posted prices were readjusted on November 1 in implementation of the Geneva Agreement, but all eyes were turned to the prices reached in certain crude oil auctions not subjected to the Arab embargo. At the beginning of November 1973, 80,000 tons of a Tunisian crude (a blend of 42° API El Borma-Zarzaitine crude with 0.07 percent sulfate) were sold at an auction at the price of $12.66/Bbl.[9] About one month later Iran auctioned some 475,000 Bbls deliverable during the first quarter of 1974. The purchase prices, wrenched out through hard struggle by about 15 U.S., Japanese, and European companies, topped at $17.34/Bbl for Rostam crude (36.7° API and 1.61 percent sulfur),[10] but the record seems to have been reached in Nigeria where Japanese and U.S. companies consented to pay $22.60/Bbl.[11]

At the OPEC October 16 Kuwait meeting it had also been decided to charge the OPEC Economic Commission Board with designing a new general oil pricing system and more particularly with defining what should be understood by the "market price" set as a basis for calculating posted prices. The Economic Commission Board met in Vienna from December 17 to 20, 1973, without being able to bring out specific recommendations on the complex problems raised. Using a technical and economic approach, the OPEC experts debated formulae, notably; (1) tying crude oil prices to the prices of refined products in the consumption markets; (2) with reference to the direct sales prices practiced by the producing countries; or (3) with comparison to the development cost of other energy sources.

The technical conclusions of the OPEC Economic Commission Board submitted to the Ministerial Committee of the six Gulf OPEC member countries, meeting in Teheran on December 22 and 23, 1973, obviously could not solve the fundamental price problem. The issue was settled by a political decision that profoundly upset the basic structures of world energy economics. Ignoring the principles adopted in Kuwait less than ten weeks earlier, the setting of posted prices no longer made any reference to realized prices. The new criterion adopted in Teheran was to fix a priori the government take per barrel and to deduct from it the posted price through the play of taxation and the technical production cost. Consequently, the government take per barrel of the marker crude (34° API Arabian Light) was sovereignly set at $7 giving a new posted price for this crude of $11.651 applicable from January 1, 1974, or an increase of about 131.5 percent over the posting of $5.035/Bbl prevailing in December 1973. The new posted prices were to be valid for a three-month period and an extraordinary OPEC conference was decided for January 7, 1974 "to discuss the bases of a long-term pricing policy and to review the possibility of establishing a dialogue between oil producing and consuming countries in order to avoid entering into a spiral increase in prices and to protect the real value of their oil"—so stated the official communiqué published by OPEC on December 23, 1973, in Teheran.[12]

New Price Structure, December 23, 1973

This decision of incalqulable consequences, which threw the world into confusion, was not easy to take. Iran, taking advantage of the situation of shortage created by the Arab cutback and embargo measures, and referring to the Iranian crude "market price" of $17.34 realized two weeks earlier, wanted to increase fourfold its posted prices. According to the <u>Petroleum Intelligence Weekly</u>,[13] the starting-point of the Teheran discussions was the definition of the government revenue per barrel rather than the setting of posted prices. The Iranians urged raising the revenue tied to the marker crude from about $3 in December to $12-14/Bbl, which would have entailed a posting of over $20. This maximalist position was fought by Saudi Arabia, which advocated increasing the government take to about $5/Bbl, which would have resulted in a posted price of about $8/Bbl. Saudi Oil Minister Ahmed Zaki Yamani declared:

> "We [the producing countries] should be reasonable and show that we are responsible as members of the international community. If we take the Iranian prices as a basis for the revision of posted prices in the Gulf, we would ruin the present economic structure of both the industrialized and the developing countries. Very rapidly, the totality of money supply available for international trade would no longer suffice to pay for oil.[14]"

It seems that the Teheran discussions turned into bargaining between the "villains" (Iranians) and the moderates (Saudis). Faced with Saudi Arabia's threat to withdraw from the meeting, the Iranians eased their position by lowering their minimum revenue demands to $8/Bbl, which would have entailed a posted price of about $14. The final compromise of $7 was adopted in spite of the Saudi protests, eliciting a disillusioned comment from Minister Yamani who declared: "As far as Saudi Arabia is concerned, in our opinion a lower posted price would have been more equitable and reasonable. However, we went along with the majority."[15]

On the technical level the new Gulf pricing system centers around the following points:

    1. The posted price of the "marker" Arabian Light crude (34° API and 1.63 percent sulfur) is officially adopted as a basis for the setting of prices.

    2. The gravity differential is set at 6¢ per degree API above 34° API and at 3¢ per degree below 34° API.

    3. The low-sulfur premiums are added to the basic prices of all crudes, the sulfur content of which is lower than that of the marker crude.

    4. A new system of freight differential is introduced to take into account the disparities in the geographic location of the export terminals in relation to Quain Island, taken as a reference point for the Gulf.

The new posted price structure in the Gulf is illustrated in Table 18.2.

TABLE 18.2

Posted Prices of Main Gulf Crudes as of January 1, 1974
(U.S. $/Bbl)

	Gravity (API°)	Gravity Differential (API°)	Sulfur Premium	Freight Premium	Posted Price
Saudi Arabia					
Arabian Light	34°	—	—	—	11.651
Arabian Medium	31°	(0.090)*	—	—	11.561
Arabian Heavy	27°	(0.210)	—	—	11.441
Iran					
Iranian Light	34°	—	0.250	(0.026)	11.875
Iranian Heavy	31°	(0.090)	0.100	(0.026)	11.635
Kuwait					
Kuwait	31°	(0.090)	—	(0.016)	11.545
Abu Dhabi					
Murban	39°	0.030	0.700	(0.015)	12.636
Marine (Umm Shaif)	37°	0.180	0.250	0.005	12.086
Zakum	40°	0.360	0.550	0.005	12.566
Iraq					
Basrah	35°	0.060	—	(0.039)	11.672
Qatar					
Dukhan	40°	0.360	0.420	(0.017)	12.414
Marine	36°	0.120	0.250	(0.008)	12.013

*Figures in parentheses are negative.
Source: Compiled by the author based on published data.

As certain Western observers pointed out, still under the shock of their "1973 Christmas present" from OPEC, there were no adjectives strong enough to describe the Teheran increases. It would be more eloquent to review the evolution of posted prices, the government take, and the tax-paid cost per barrel of crude from 1971 to 1974. This evolution is illustrated in Tables 18.3, 18.4 and 18.5.

As expected, the Teheran decisions signaled an avalanche of similar crude price increases in the other producing countries. Venezuela raised by 81.9 percent its reference posting, or from $7.73 to 14.08/Bbl. So did Indonesia with an 80 percent increase, raising the

TABLE 18.3

Evolution of Main Gulf Crude Postings, 1971-74
(U.S. $/Bbl)

	Pre-Feb. 15	1971 Feb. 15	1971 Jan. 1	1972 Jan. 20	1972 Jan. 1	1973 Apr. 1	1973 June 1	1973 July 1	1973 Aug. 1	1973 Oct. 1	1973 Oct. 16	1973 Nov. 1	1973 Dec. 1	1974 Jan. 1
Saudi Arabia														
Arabian Light	1.800	2.180	2.285	2.479	2.591	2.742	2.898	2.955	3.066	3.011	5.119	5.176	5.036	11.651
Arabian Medium	1.680	2.085	2.187	2.373	2.482	2.626	2.776	2.830	2.936	2.884	4.903	4.957	4.822	11.561
Arabian Heavy	1.560	1.960	2.064	2.239	2.345	2.484	2.623	2.674	2.755	2.725	4.633	4.634	4.557	11.441
Iran														
Iranian Light	1.790	2.170	2.274	2.467	2.579	2.729	2.884	2.940	3.050	2.995	5.341	5.401	5.254	11.875
Iranian Heavy	1.720	2.125	2.228	2.417	2.527	2.674	2.826	2.884	2.969	2.936	4.991	5.046	5.006	11.635
Kuwait														
Kuwait	1.680	2.085	2.187	2.373	2.482	2.626	2.776	2.830	2.936	2.884	4.903	4.957	4.822	11.545
Abu Dhabi														
Murban	1.880	2.235	2.341	2.540	2.654	2.808	2.968	3.026	3.140	3.084	6.045	6.113	5.944	12.636
Marine (Umm Shaif)	1.860	2.225	2.331	2.529	2.642	2.796	2.955	3.013	3.126	3.070	5.537	5.599	5.446	12.086
Zakum	–	–	–	–	–	–	–	–	–	–	5.964	6.031	5.865	12.566
Iraq														
Basrah	1.720	2.155	2.259	2.451	2.562	2.711	2.865	2.921	3.031	2.977	5.061	5.117	4.978	11.672
Qatar														
Dukhan	1.930	2.280	2.387	2.590	2.705	2.862	3.025	3.084	3.200	3.143	5.834	5.809	5.737	12.414
Marine	1.830	2.200	2.305	2.501	2.614	2.766	2.928	2.980	3.092	3.037	5.503	5.563	5.412	12.013

Source: Compiled by the author based on published data.

TABLE 18.4

Evolution of the Government Unit Income per Barrel for Main Gulf Crudes, 1971-74
(U.S. $/Bbl)

	Pre-Feb. 15	1971 Feb. 15	1971 June 1	1972 Jan. 20	1972 Jan. 1	1972 Apr. 1	1973 June 1	1973 July 1	1973 Aug. 1	1973 Oct. 1	1973 Oct. 16	1973 Nov. 1	1973 Dec. 1	1974 Jan. 1
Saudi Arabia														
Arabian Light	0.989	1.261	1.325	1.448	1.516	1.607	1.702	1.736	1.804	1.770	3.048	3.083	2.898	7.008
Arabian Medium	0.930	1.203	1.265	1.384	1.450	1.537	1.628	1.661	1.725	1.694	2.917	2.950	2.868	6.954
Arabian Heavy	0.843	1.106	1.169	1.302	1.367	1.449	1.535	1.566	1.627	1.597	2.754	2.785	2.708	6.881
Iran														
Iranian Light	0.983	1.250	1.313	1.430	1.497	1.588	1.683	1.717	1.783	1.750	3.172	3.208	3.119	7.133
Iranian Heavy	0.944	1.222	1.285	1.400	1.466	1.555	1.647	1.681	1.746	1.714	2.960	2.993	2.969	6.988
Kuwait														
Kuwait	0.958	1.231	1.293	1.406	1.472	1.559	1.650	1.683	1.747	1.716	2.939	2.972	2.890	6.966
Abu Dhabi														
Murban	1.005	1.272	1.337	1.458	1.527	1.620	1.717	1.752	1.821	1.787	3.582	3.623	3.521	7.578
Marine (Umm Shaif)	0.366	1.239	1.288	1.391	1.460	1.553	1.650	1.685	1.753	1.720	3.192	3.229	3.137	7.162
Zakum	–	–	–	–	–	–	–	–	–	–	3.451	3.492	3.391	7.453
Iraq														
Basrah	0.933	1.240	1.303	1.419	1.487	1.578	1.671	1.705	1.772	1.739	3.002	3.036	2.952	7.010
Qatar														
Dukhan	1.052	1.316	1.381	1.493	1.546	1.641	1.740	1.776	1.847	1.812	3.443	3.482	3.384	7.432
Marine	0.924	1.196	1.260	1.351	1.464	1.556	1.651	1.686	1.754	1.720	3.215	3.252	3.160	7.162

Note: The government take includes a royalty of 12.5 percent of the posted price considered as an expense and a 55 percent income tax applied to the "gross profit" obtained by deducting the royalty and the production cost from the posted price. The production costs retained here are those published by Middle East Economic Survey of December 28, 1973.

Source: Compiled by the author based on published data.

TABLE 18.5

Evolution of the "Total Cost Including Taxes" of the Main Gulf Crudes, 1971-74
(U.S. $/Bbl)

	Pre-Feb. 15	1971		1972				1973							1974
		Feb. 15	June 1	Jan. 20	Jan. 1	Apr. 1	June 1	July 1	Aug. 1	Oct. 1	Oct. 16	Nov. 1	Dec. 1		Jan. 1
Saudi Arabia															
Arabian Light	1.099	1.371	1.435	1.548	1.616	1.707	1.802	1.836	1.904	1.870	3.148	3.183	3.098		7.108
Arabian Medium	1.040	1.313	1.375	1.484	1.550	1.637	1.728	1.761	1.825	1.794	3.017	3.050	2.968		7.054
Arabian Heavy	0.993	1.256	1.319	1.402	1.467	1.549	1.635	1.666	1.727	1.697	2.854	2.885	2.808		6.981
Iran															
Iranian Light	1.103	1.370	1.433	1.550	1.617	1.708	1.803	1.837	1.903	1.870	3.292	3.328	3.239		7.253
Iranian Heavy	1.064	1.342	1.405	1.520	1.586	1.675	1.767	1.801	1.866	1.834	3.080	3.113	3.089		7.108
Kuwait															
Kuwait	1.018	1.291	1.353	1.466	1.532	1.619	1.710	1.743	1.807	1.766	2.999	3.032	2.950		7.026
Abu Dhabi															
Murban	1.555	1.422	1.487	1.608	1.677	1.770	1.867	1.902	1.971	1.937	3.732	3.773	3.671		7.728
Marine (Umm Shaf)	1.166	1.439	1.558	1.691	1.760	1.853	1.950	1.985	2.053	2.020	3.492	3.529	3.437		7.462
Zakum	–	–	–	–	–	–	–	–	–	–	3.751	3.792	3.691		7.753
Iraq															
Basrah	1.053	1.360	1.423	1.539	1.607	1.698	1.791	1.825	1.892	1.859	3.122	3.156	3.072		7.130
Qatar															
Dukhan	1.172	1.436	1.501	1.633	1.716	1.811	1.910	1.946	2.017	1.982	3.613	3.652	3.554		7.602
Marine	1.174	1.446	1.510	1.651	1.684	1.776	1.871	1.906	1.974	1.940	3.435	3.472	3.380		7.382

Note: The "Total Cost Including Taxes" includes the government take and the technical cost of production excluding company profits.

Source: Compiled by the author based on published data.

price to $10.8. In Nigeria the posted price of Nigerian Light 34° API was increased to $14.69, or by 77 percent. Bolivia was content with a more modest 63 percent increase, raising its price from $9.30 to $16, at which level the Bolivian crude became the most expensive in Latin America. Ecuador, by announcing a 37 percent rise only, or from $10 to $13.70/Bbl, seemed very moderate in its claims. Actually, however, this was the eighth increase in one year adding up to a total rise of 548 percent in 12 months from an initial price of $2.5.

However, the record was certainly set by Libya where the posted price as of January 1 of the 40° API crude was fixed at $15.768. Commenting on this decision, Libyan Premier Abdul Salam Jallud declared that the increase decided by Libya was very reasonable since the real market situation would have warranted a price of $20 if not $25. The reference price for Algerian crude was set at $16.20/Bbl.

The declaration of the Libyan prime minister reflected a new "paternalist" attitude on the part of the new "masters" of the international oil industry. Already in Teheran, at the press conference of December 23 in which he announced and commented on the decisions to increase prices, the Shah of Iran stressed the fact that the producing countries had been understanding and reasonable by not taking fully into account the realized price of $17 reached by Iranian crude in setting the new posted prices and by being "responsible" before the international community in setting the government take for the marker crude at only $7. This argument was taken up textually by the official communiqué released by OPEC on this subject.[16]

The shock caused by the oil shortage as a result of cutback and embargo measures was so acute that it was some time before people realized that supply difficulties were only part of the trouble. What was more important was that the OPEC countries were now united and could raise the oil price as high as they wished. The moderates in OPEC were now more convinced that the radicals were right: that for decades they had been diddled. The major oil companies by now realized that they were no longer negotiating: They were waiting for unilateral demands. By the end of 1973 the oilmen in New York had a sour joke with their returning negotiators: "Congratulations on not accepting more than the Arabs demanded."

What, in fact, was the chief cause of the huge leap in the oil price has been argued by oilmen ever since. There is little doubt that the Middle East war had accelerated and precipitated the crisis: It unified the Arab members of OPEC and brought the Saudis into line, and without it the price might have stayed around $5 for some time. "The Arabs" became a kind of synonym for the OPEC cartel, but it was a misleading description, for the most militant figure in the demands for high prices had not been an Arab at all, but the Shah of Iran, and the other most demanding country in OPEC was Venezuela, 8,000 miles away from the Gulf area.

Even without the war, the basic fact could not have eluded the members of OPEC for very long: the consuming countries were dependent on a small group of exporters, who were able by restricting production to control the price. It was the same basic fact that had been discovered by John D. Rockefeller a century before when the whole oil monopoly had been controlled by "a handful of men"—now it was a handful of countries. The oil companies had seen the weapon that they themselves had forged suddenly turned on themselves.

1974: THE SHAPING OF A NEW WORLD

The shock described above was so profound and so violent that oil economics and politics became the most debated popular issues throughout the world. It was surprising, and sometimes amusing, to see the efforts of an increasing number of nonprofessionals to penetrate the mysteries of oil pricing or to understand the mechanisms and to assess the implications of fiscal arrangements and provisions. This exercise proved indispensible to the average citizen trying to comprehend daily news propagated by the mass media; this kind of information was the privilege of a restricted number of specialists. The exercise grew in complexity and confusion along with the changes that continued to affect the international oil industry through 1974. In fact, the historical structure and the conventional setup of the industry came to an end after the October war and the price increase decided at Christmas 1973, and a new era started to stabilize and to emerge through a series of decisions and arrangements that took place at varying levels in 1974.

The present paragraph is being written in the early days of 1975 at a time when the metamorphosis is not completed. It is quite difficult to predict the image of the oil industry in the near future; some of the major issues are still in the heart of the confrontation between the exporting countries, the consuming countries, and the oil companies, and the potential danger of a new war in the Middle East adds further and unpredictable complications. However, the irreversible process of evolution can be visualized with enough accuracy to outline the main features of the basic components of the international oil industry in terms of future patterns of supply and demand, the price level and structure, and the political and legal background to oil movement and transactions. The evolution of these issues in 1974 shall be briefly reviewed below before we consider how they might affect the balance of power and the probable structure of international oil politics and economics.

The Evolution of Crude Oil Prices

One of the most important features of the new image of the international oil industry is that crude oil pricing has become an exclusive privilege of the exporting countries. The price evolution in 1974 actually results from negotiations between the OPEC member countries themselves. Without going into much detail, OPEC decisions were usually elaborate compromises between two main trends:

1. The hard line, represented mainly by Algeria and Iran, insisted on continuing price increase to cope with and compensate for worldwide inflation.
2. The moderate line, represented by Saudi Arabia, claimed that the level of oil prices was already too high and should rather be reduced, and that any further increase in the cost of supply could result in a worldwide depression with tragic consequences to the industrialized countries, the developing countries, and the exporting countries altogether.

OPEC member countries were almost unanimous on such other important items as:

- the necessity to reduce the exorbitant profits made by the oil companies on their equity crude entitlements under the participation agreements so as to reach a homogenous price structure;
- the necessity to reinforce the unity and the cohesion of OPEC, eventually through controlling production level if necessary (except for Saudi Arabia);
- the necessity to stabilize prices in the long run at acceptable levels compatible with the protection of the purchasing power of the oil barrel and with the price of alternate sources of energy.

The pricing issue has become inextricably complex because of the multitude of fiscal patterns used for the computation of tax-paid cost and government take under the prevailing participation agreements: buy-back prices for participation oil and changing royalty and tax rates for equity oil, not to mention realized prices in sales to third parties or in government-to-government deals, specific premium for some crude oils, etc. The price puzzle became so complex that several tables with abundant explanations and comments became necessary to present and illustrate any of the successive decisions that were taken by OPEC in 1974. Such exercise hopefully is expected to be useless in light of the latest decision taken in Vienna on December 13, 1973, which implicitly aims at the establishment of a unique price for crude oil, a concept that will effectively materialize when the 100 percent participation pattern will be enforced by the Gulf producing countries.

The main highlights of price evolution in 1974 were as follows:

The Geneva Conference, January 7-9, 1974

This extraordinary OPEC conference, as decided in Teheran on December 23, 1973, was to elaborate a long-term policy for the stabilization of crude oil prices. The Saudi position in favor of lowering prices was strongly defended by Minister Yamani who emphasized in particular the tragic situation of developing countries. However, the conference resulted in a relative strengthening of the hard line since it was decided

- to freeze prices for a three-month period,
- to instruct the OPEC Economic Commission to undertake the necessary studies "with a view to establishing a pricing system for crude oil in the long run,"
- to virtually abandon the proposed direct dialogue with consumers,
- to abandon the fixed ratio of 1.4 to 1.0 between posted prices and market prices by the Gulf countries,
- to devote special attention to helping developing countries that faced financial problems in connection with the soaring cost of oil supply,
- to call on industrialized countries to "adopt necessary measures to contain the high inflationary trend in their countries and to control the oil companies in the way they increase prices of oil products to end consumers."

The question of oil company profits was a major preoccupation in OPEC, although nothing was decided about this in the Geneva meeting. For example, the tax-paid cost to Aramco on its equity entitlement of the marker Light Arabian was about $7.12/Bbl, whereas market price could be tentatively set at about $10.83/Bbl (buy-back price set at 93 percent of posted price). This induced a substantial distortion in the supply cost of crude oil and in the pricing structure of refined products. Reducing the "extraordinary" profit margin on equity crude was viewed as a means to reduce the bill paid by the consuming countries.

The price freezing was carried on through the second quarter of 1974.

The Quito Conference, June 15-17, 1974

This conference was expected by most observers to be a battlefield between Saudi Arabia and the rest of OPEC members on the major issues that were dominating the international oil scene: prices, tax reform, and control of production levels. A number of recommendations were presented:

1. To compensate for worldwide inflation in the first part of 1974 it was recommended that posted prices should be raised so as to yield a 9 percent increase in government take. On the Arabian Light marker crude, this would have involved a posted price increase of around $1.02/Bbl (from $11.65 to $12.67) and a rise of some $0.63/Bbl (from $7.00 to $7.63) in government take.

2. As a measure designed to curb excess profits by the oil companies on their equity crude (now 40 percent of production under the 60-40 participation model), it was recommended that the tax rate should be raised from the current 55 percent to the equivalent of 87 percent. This would have entailed an increase, on the basis of the current level of postings, of about $3.23/Bbl in government take on Arabian Light crude from $7.00 to $10.23. The rationale behind this move assumed the realized market price to be 93 percent of the posted price ($10.83/Bbl in the case of Arabian Light), giving the companies a notional profit margin of $3.73/Bbl if this is defined as the differential between the current tax-paid cost of $7.10/Bbl (government take plus 10 cents cost) and the realized price of $10.83/Bbl. Under the 87 percent tax formula this notional margin would have been cut down to 50 cents a barrel ($10.83 minus $10.33). If recommendations 1 and 2 were to have been implemented together, the effect would have been to raise government take/tax-paid cost by around $4.00/Bbl.

3. Considering the current state of slack demand in the major consumer markets and the consequent potential pressure on prices, it was recommended that member countries should consider instituting a production program.

The Saudis counterproposed a reduction of $2/Bbl in posted prices after which appropriate measures would be implemented to maintain the purchasing power of oil revenues. In regard to the second recommendation, Saudi Arabia emphatically outlined that it was not prepared to associate itself with any move to change the fiscal system in order to tax the companies' excess profits on equity crude since it was in the process of negotiating a 100 percent take-over of Aramco, in which case the whole concept of posted prices would become obsolete. Finally, the Saudis reconfirmed they would not recognize any restriction on the level of production until such time as price levels acceptable to Saudi Arabia be agreed upon with the other OPEC members.

Saudi Arabia having threatened to go its own way by reducing prices and raising production unilaterally if the others insisted on adopting the above-mentioned recommendations, the Quito meeting ended with a compromise stating:

    1. No change in posted prices for the third quarter of 1974. These prices will be reviewed in respect to the fourth quarter at an extraordinary OPEC conference scheduled to convene in Vienna on September 12, 1974.

    2. To increase, with effect from July 1, 1974, the rate of royalty payment by 2 percent or its equivalent in government take (entailing in most cases—notably in the Middle East and Africa—the raising of the royalty from 12.5 percent of the posted price value of crude oil produced to 14.5 percent). This would result in a rise of around 10-11¢/Bbl in government take on the main Gulf crudes; government take on Iranian Light, for example, would increase from $7.133/Bbl to $7.240/Bbl. However, pending the conclusion of its projected new arrangement with the foreign owners of Aramco, Saudi Arabia will not for the time being associate itself with this decision.

    3. The other significant development, which was not announced in the communiqué, was a "gentlemen's agreement" between all the member countries except Saudi Arabia to freeze (that is, not to increase) their production levels until the next conference in September.

The Vienna Conference, September 12-13, 1974

During the three-month period between the two OPEC conferences, the position of the different parties remained unchanged, although calls for limiting oil companies' profits on equity oil became unanimously stronger. Once again OPEC decisions represented a compromise between the position of Saudi Arabia and other member countries. It was decided to compensate for inflation by raising tax and royalty rates in order to bring about an increase of 33¢/Bbl in average government take. However, the posted prices on which taxes and royalties are calculated would remain frozen at their previous levels, which have been in force since January 1, 1974.

An OPEC communiqué emphasized that this increase "should not be passed to consumers, taking into consideration the excessive margin of profits still being made by the international oil majors on their upstream operations."

The main decisions of the Vienna Conference were as follows:

    1. Posted prices will be frozen for the fourth quarter of 1974. The posting for the marker crude, 34° API Arabian Light, therefore remains at $11.651/Bbl.

2. With effect from October 1, 1974, tax and royalty rates on the companies' equity crude (40 percent of production) will be raised in such a way as to bring about an increase of 33¢/Bbl, or 3.5 percent, from $9.41/Bbl to $9.74/Bbl on weighted average of government take (that is, 40 percent equity oil at tax plus royalty and 60 percent government crude at the market price of 93-94.8 percent of postings) to compensate for inflation in the industrialized countries. This will entail an increase in the royalty rate—currently 14.5 percent in most OPEC countries—to 16.67 percent, and a rise in the tax rate in the majority of member states from the prevailing 55 percent up to 65.75 percent.

3. In actual price terms this will result in the following increases in price and cost levels for Arabian Light market crude: government take on equity crude by $1.147/Bbl, from $7.113/Bbl to $8.260/Bbl; tax-paid cost on equity crude by $1.147/Bbl, from $7.213/Bbl to $8.360/Bbl; average government take on all crude production by 33¢/Bbl from $9.41/Bbl to $9.74/Bbl; and average cost to the oil companies on all crude production by 33¢/Bbl, from $9.51/Bbl to $9.84/Bbl.

4. The conference decided to establish a working committee to convene on October 23 in Vienna "to study and recommend a new system for long-term oil pricing." What OPEC had in mind, with effect from January 1, 1974, was to set prices for a period of one year subject to quarterly adjustments to compensate for inflation in the industrialized countries. (It should be noted that the increase of 3.5 percent for inflation for the last quarter of 1974 is the equivalent of an annual rate of 14 percent.) The price to be thus set, on which the inflation adjustment will be added, will be the average government take (which will serve as a sort of floor to actual market prices) rather than the posted price. This latter will continue to be frozen at the prevailing level, and will in fact for all practical purposes disappear in favor of the market-related price. Also in prospect is a further thoroughgoing overhaul of the fiscal system with a view to unifying price levels and further narrowing the differential between the cost of equity crude and the government market price—which was only partially achieved by the present measures—while still leaving an appropriate margin for the companies.

5. As regards production control, the conference decided to request the OPEC secretary-general to carry out a study on the subject of supply and demand. This decision follows strong opposition from Saudi Arabia and others to a proposal by the OPEC Economic Commission for across-the-board production cutbacks by all countries for a certain period of time to help mop up the surplus that had appeared during the summer months. Meanwhile, a number of countries have announced their decision to make a voluntary cut-

back in their production levels. In addition to countries like Kuwait, Iraq, Venezuela, Libya, and Algeria, which had already cut back output by leaving their unsold production in the ground, these included Abu Dhabi, which after the conference announced a cut in production of 300,000 barrels daily for the month of September.

The OPEC decisions came against a background of various proposals put before the conference from various quarters. These included:

1. a proposal by the majority of the members of the OPEC Economic Commission to raise the average cost of crude to the companies by 14 percent to around \$10.68/Bbl. This would have involved a surcharge of roughly \$3.25/Bbl on equity crude, raising tax-paid cost from, say, \$7.21/Bbl to \$10.46/Bbl;

2. an Iranian proposal to set a price of around \$9.40/Bbl for the companies' crude (roughly the same as at present) and \$9.90/Bbl as a sort of minimum open market price, leaving the companies with a 50¢ margin and effectively doing away with the posted price;

3. a Saudi proposal (which would have had much the same practical effect as that of Iran) to combine the increases in government take on the companies' crude with a reduction in posted prices to give some relief to the consumers—that is, to raise government take on equity crude while lowering the government market price with a view to having the two meet at some acceptable level.

The Abu Dhabi Meeting, November 9-10, 1974

The above-mentioned Saudi proposal was in fact accepted in principle by Algeria, Kuwait, and Iran, but it was strongly opposed by a number of other countries, notably Iraq and Libya, who rightly feared that the lowering of actual cost of supply to consumers would induce a similar move in the price of direct sales made by these countries outside the posted price system. In order to get around this hostility and to facilitate the adoption of a single-price system, Saudi Arabia elected to have such measures adopted and implemented by the main Gulf exporting countries so as to put other OPEC members face to the fait accompli when they were to meet on December 12 as scheduled. To this effect a surprise meeting was convened in Abu Dhabi; its context and results are well stated in the following official communiqué released at the conclusion of the meetings:

> At the invitation of the United Arab Emirates, the Gulf oil producing countries held a meeting in Abu Dhabi during the 9th and 10th November 1974 in order to consider the Saudi Arabian proposal that had been presented

to the 41st conference of the Organization of Petroleum Exporting Countries held in Vienna on 12th and 13th September 1974 and which envisages the lowering of crude oil posted prices and the raising of the royalty and income tax rates applicable to the companies. The meeting unanimously elected His Excellency Mana Said al'Otaiba, the Minister of Petroleum and Mineral Resources and head of the UAE delegation, as president of the meeting. The Kingdom of Saudi Arabia, the United Arab Emirates, and the state of Qatar agreed at the meeting to implement the Saudi proposal while the state of Kuwait, recognizing the right of one or more of the Gulf countries to implement the Saudi Arabian proposal, preferred that as far as Kuwait is concerned, such a decision should be taken at the next ordinary conference of the Organization of Petroleum Exporting Countries to be held in Vienna on 12th December next or at an extraordinary conference to be convened prior to the date mentioned. Iran, while keeping its position, believes that the matter relating to determination of crude oil prices should be dealt with at the OPEC Ministerial Conference to be held in Vienna on 12th December 1974. The Iraqi delegation believes that the oil pricing issue is a matter of vital importance to all member countries of the Organization of Petroleum Exporting Countries and stresses that any decision on this issue should be taken at the OPEC Conference and not in the meeting. The above-mentioned countries who intend to implement this proposal have decided to send a mission to the OPEC member countries to explain their position prior to the next OPEC Conference. The three countries that approved the Saudi proposal decided to implement it with effect from 1st November 1974 through the end of July 1975.

The Kingdom of Saudi Arabia, the United Arab Emirates and the State of Qatar agreed to implement the Saudi proposal with effect from 1st November 1974 through July 1975 in view of their belief that the oil companies operating within their territories realized excessive profits on the export of their crude and because these profits accrue in consequence of the big difference between actual market prices and tax paid cost under the present taxation system. The countries in question considered a portion of these profits should be returned to the consumer and have therefore decided as follows:

1. to lower the crude oil posted price by 40 U.S. cents per barrel with effect from 1st November 1974 (with no allowances for differences in quality, gravity or location),
2. to raise the rate of royalty to 20 percent (from 16.66 percent in effect since 1 October),
3. to raise the rate of income tax applicable to oil companies to 85 percent (from 65.7 percent in effect since 1 October).

Another decision not stated in the communiqué was to price buy-back oil sold to the concessionary companies at 94.85 percent (instead of 93 percent) of the new lower posted price.

The combination of these four elements drastically raised overall crude oil costs for the major oil companies. Exact figures would vary from one case to another depending on such factors as participation terms, relative repartition of crude exports as equity oil, buy-back oil and direct sales by governments to third parties, production costs, etc. Table 18.6 gives a theoretical exercise applicable to the marker Light Arabian crude assuming a production cost of 12¢/Bbl and a 40/60 participation pattern under which all of the 60 percent participation oil would be bought back by the companies; different figures would be obtained if direct sales are considered at actual prices other than the companies' buy-back prices (93 percent or 94.85 percent of postings).

As compared to the situation prevailing since October 1, it appears that the Abu Dhabi decisions resulted in: (1) an increase of the tax-paid cost of $1.552/Bbl to $9.92; (2) a decrease of the buy-back-price by $0.164/Bbl to $10.671; (3) the combination of the two factors, which results in increasing the weighted average cost to companies by $0.522/Bbl to $10.37.

The Vienna Conference, December 12-13, 1974

Two main proposals were presented to the OPEC members in regard to the formulation of a new pricing system. One was the Iranian proposal for the abolition of the system of taxes and royalties calculated on notional posted/tax reference prices in favor of a single unified price with a built-in margin for the operating oil companies (set at 50¢/Bbl). The second stemmed from the November 11 decision taken in Abu Dhabi in favor of the Saudi proposal as indicated above. As usual, the conference decision represented a compromise between the different views and proposals. The Iranian original scheme was adapted to accomodate the financial effects of the Abu Dhabi decision. As indicated in the OPEC communiqué,

TABLE 18.6

Changes in Prices and Costs of Light Arabian Crude Oil
(U.S. $/Bbl)

	Pre-October 1	October 1	October 1
1. Posted price	11.651	11.651	11.251
2. Production cost	0.120	0.120	0.120
3. Royalty, percent rate	12.5	16.67	20.00
$/Bbl	1.456	1.942	2.250
4. Tax, percent rate	55.00	67.75	85.00
$/Bbl	5.541	6.305	7.549
5. Tax-paid cost, 2 + 3 + 4	7.117	8.367	9.919
6. Buy-back price, 93 percent	10.835	10.835	10.463
94.85 percent	—	—	—
7. Government take, 3 + 4	6.997	8.247	9.799
8. Companies' average cost			
Equity oil, 40 percent	7.117	8.367	9.919
Participation oil, 60 percent	10.835	10.835	10.671
9. Average cost	9.348	9.848	10.370

Source: Compiled by the author.

the Conference . . . decided to adopt a new pricing system based on the financial effect of the decision taken on 10 and 11 November in Abu Dhabi. In accordance with this decision, the average government take from the operating companies will be $10.12 a barrel for the marker crude. This decision is effective from 1 January 1975 through 30 September 1975.[18]

The new pricing system will incidentally entail the scrapping of the posted price for all practical purposes. A very significant feature of this scheme is that OPEC has abandoned its earlier plan—formulated in the last Vienna Conference in September—to index oil prices as from January 1975. Instead, OPEC has opted for a nine-month period of price stability that, if one estimates the current world inflation rate at 14 percent or more, would entail an erosion of at least 10 percent or over $1/Bbl in the price of oil in real terms.

The communiqué goes on to add in respect to the new pricing system that "for the time being it should be noted that Saudi Arabia, joined by Qatar and the UAB, are negotiating with the foreign oil companies a new arrangement which could be made retroactive."

What this means in effect is that the new pricing system decided upon by OPEC has itself been overshadowed by the prospect of the forthcoming 100 percent take-over of Aramco by the Saudi Arabian Government—an example that will no doubt be speedily emulated in the other Gulf states where the companies retain an equity interest in oil operations. And, as explained below, Saudi Arabia expects that the new unified market price for both majors and independents entailed by its projected 100 percent take-over will settle at a level even somewhat lower than the $10.12/Bbl referred to in the OPEC decision.

Paradoxically, in view of OPEC's manifest desire to simplify rather than complicate the oil price puzzle, the new system outlined in the communiqué is already giving rise to varying interpretations among the member states as to its practical implementation. There are two major problems here: first, what is to be the market price for direct government sales to third parties? and second, how does one interpret the phrase "average government take from the operating oil companies?"

Regarding the first point, the fact is that—though the buy-back to the operating companies is calculated at 93 percent of the lower posted prices decided on at Abu Dhabi—no agreement was reached between the member states on the level of the market price for direct government sales—this of course being a crucial question in determining the effective margin per barrel of the major companies, that is, the differential between the average cost of crude to the majors and the open market price. Most of the member countries consider that direct sales should also be made at 93 percent of the postings decided on at Abu Dhabi, that is, $10.46/Bbl for the marker crude. Saudi Arabia, however, declined to enter into any commitment along these lines, having stated that it would not sell below the equity crude government take level of $9.80/Bbl plus production cost.

Concerning the second point, a majority of the OPEC countries construe the phrase "average government take from the operating oil companies" to mean that $10.12/Bbl—regardless of how the figure was arrived at in the first place—should be the government take for each barrel of crude oil lifted by the companies irrespective of any variation in the mix between equity and buy-back crude. Under this interpretation, OPEC member states will enjoy considerable flexibility on means of achieving the $10.12 average revenue level—whether by tax, royalty rises, make-up payments, or however. On this basis, the cost to the major oil companies would work out at a flat irreducible $10.24/Bbl (that is $10.12 plus 12¢ production cost) on the marker crude.

Saudi Arabia, on the other hand, feels that the figure of $10.12/Bbl should be construed as a guideline for establishing the average government take resulting from 40 percent equity crude and 60 percent sold at 93 percent of postings, and that therefore the operating companies should be allowed to lower the average cost of crude to them by taking less buy-back oil in relation to equity crude. In this connection Saudi Oil Minister Ahmad Zaki Yamani indicated that the average government take from oil lifted by the operating companies in Saudi Arabia will definitely be less than $10.12/Bbl, since they will be taking less than 100 percent of production.[19]

In light of the above it can be seen that what might be described as the majority interpretation rests on two fixed elements: an open market price for direct government sales of $10.46/Bbl, and an average cost to the operating companies of $10.24/Bbl. And this of course entails a fixed margin of 22¢/Bbl for the companies.

On such a basis, according to this interpretation, the government take on each of the various Gulf crudes under the new scheme would equal the residue of the following: 93 percent of the Abu Dhabi posting (that is, October 1974 postings minus 40¢ on each), minus the company margin applicable to the Arabian Light marker crude (22¢/Bbl), minus the cost of production. The differentials for quality and location between the various Gulf crudes would therefore remain as they were under the previous posted price system.

Under the Saudi Arabian interpretation, should the concept of average government take be linked to actual liftings by the oil companies in comparison with the base participation scheme (40/60), the companies' margin could be as high as a ceiling of 54¢/Bbl if they elect to lift only their 40 percent equity crude oil at a tax-paid cost of $9.92/Bbl as compared to the government sale price of $10.46.

The future will tell how the situation will develop. The issue will become less complex when the governments of the producing countries complete the 100 percent take-over of the operating companies—as is expected in the next few months. Meanwhile, both Kuwait and Iraq promptly announced effective measures in view of implementing the OPEC decision. As from January 1, 1975, British Petroleum and Gulf Oil will have to pay a "single acquisition cost" of $10.147/Bbl, or a 71¢ hike from the companies' average cost of $9.44 effective October 1, 1974. The new $10.147 cost (including an 8¢ production cost) is exactly the amount Kuwait needs to achieve a government take of $10.07, which is equivalent to OPEC's theoretical "take" of $10.12 on the marker crude.[20] Iraq equally announced a reduction in its direct sale price for Basrah crude to $10.50/Bbl for 750,000 tons to be lifted by a Japanese group in the first quarter of 1975. This price, close to the $10.48 indicated by the theoretical OPEC formula, is down 36¢ from the $10.86 that the Japanese group was currently paying.[21]

According to the Saudi interpretation, it would appear that the companies will be intentionally straitjacketed in several countries in a situation where the more they lift, the higher their average overall per-barrel costs, and the lower their margin. And if they raise their third-party sales price substantially above $10.46, independent buyers will be able to get the oil cheaper from the OPEC state oil companies. This way, it is expected, major oil firms will be "encouraged" to take only what they need for their own integrated networks and leave an increased market share to state firms.

The volume of liftings by the oil companies thus promises to be the next key negotiating point. The OPEC conference took no formal action fixing these volumes, and the matter was left to individual negotiation. And the international industry may see wholesale lopping-off of even more third-party sales than has already occurred and a corresponding movement of independent buyers toward the OPEC state firms. Market shares of the majors might well become dependent on the decisions of the producing countries regarding allocation of crude oil volumes.

Regarding Mediterranean crude, the freight premium, previously set at around $1.00/Bbl, is likely to be reduced to 60¢ or less to reflect the continued sharp erosion between AFRA LR3 on the Gulf-Rotterdam voyage and LR1 on Mediterranean-Rotterdam run. However, there were objections that AFRA scale is inadequate for this purpose since it does not accord sufficient weight to the spot market. The formula is therefore likely to be adjusted to give more weight to spot rates, and an expert committee was scheduled to discuss this matter at the Algiers conference of OPEC oil and foreign ministers on January 24, 1975.

Traditionally, largely for historical reasons of tax convenience, the major part of oil company profits has been generated in the upstream production stage, and these profits have provided the bulk of the necessary investment capital for all phases of the oil industry. Clearly this can no longer be the case, either under the new OPEC rule or under the prospective new Saudi Arabian type of 100 percent participation deal. From now on it looks as though a substantial part of the profit-making function will have to be shifted downstream. In other words, each stage of the integrated operation will in the future have to yield an adequate return on its own, and this will naturally require a radical realignment in the structure of the industry.

The Evolution of Supply and Demand Patterns

The Participation Issue

As outlined above, the provisions and mechanisms of participation agreements in force in the Gulf producing countries became an essential part of the evolving crude oil pricing concept and structure. It has been recalled earlier in this chapter that the political and psychological evolution in the Middle East lead the producing countries to consider the initial participation agreements implemented in 1973 as a fool's deal intended to safeguard the companies' interest at a cheap price. After the multiple "victories" achieved as a result of the October war, and thanks to the skillful utilization of the "oil weapon," the Gulf producing countries progressively became conscious of the fact that the balance of power in their favor was so great that they could afford to achieve full control of oil production while having the oil companies continue their technical services as operators and to lift oil and market it worldwide under terms and conditions unilaterally dictated by the producing countries.

This hard-line nationalistic approach was actually adopted well before the October war by the Kuwait National Assembly when it refused to ratify the initial 25 percent participation agreement and called on the government to achieve a minimum 60 percent participation at better terms, mainly in respect to buy-back entitlements and prices. Such an agreement was actually signed on January 29, 1974, but the question of buy-back prices remained unresolved for a long time.

A similar agreement was signed in Qatar on February 20, 1974. Buy-back provisions were agreed upon about two months later. The oil companies were to purchase at least 60 percent of the government's 60 percent share of production (that is, 36 percent of total production) at an average price corresponding to 93 percent of posted prices. Buy-back prices were to be reviewed on a quarterly basis after a six-month period.[22]

The leading role in the participation crusade, however, was readily recovered by Saudi Arabia. In June 1974 a provisional agreement was concluded with Aramco, retroactive to January 1, 1974, raising government participation to 60 percent and stipulating that buy-back volumes were to be determined at government discretion; buy-back prices were set at 93 percent of postings. This agreement took place amid rumors suggesting that Saudi Arabia was considering already negotiating a 100 percent take-over of Aramco. This probably explains why no compensation was actually mentioned or paid for the increase of the government equity share (Saudi Arabia had paid $500 million in 1973 to acquire 25 equity shares in Aramco on the basis of "updated book value") since compensation was one of the main controversial issues in the negotiations for the new deal.

It is to be recalled that Saudi Arabia provisionally refrained from implementing increases in taxes and royalties decided upon by OPEC in June and September pending the outcome of negotiations on the projected new arrangement with Aramco. At the same time, however, the Saudis agreed that such increases were justified in view of the excessive level of oil company profits—though they maintained that the rise in tax and royalty rates should be accompanied by a reduction in posted prices to give some relief to the consumers.

However, as time went by, Saudi Arabia's patience began to wear thin in view of Aramco's resistance to the Saudi concept of 100 percent participation. In the main the resistance seemed to stem from the companies' discontent with the financial implications of the Saudi proposal, particularly the requirement to purchase their crude liftings at open market prices rather than at a preferential price. The deadlock was further intensified when the Saudis, exasperated by the lack of progress and by what they considered to be inadmissible tactics on the part of the companies, demanded a written acceptance in principle of the broad lines of the Saudi proposal before substantive talks could proceed.

By mid-October 1974, with still no breakthrough in sight, the Saudis seemingly decided that if the 100 percent plan was not to be implemented owing to the companies' opposition, then at least the fiscal aspects of the current 60-40 arrangement should be adjusted to conform to the required standard. And this line of reasoning led directly to the Abu Dhabi decision.

The "dynamite" of the Abu Dhabi scheme—with its direct threat to the whole financial and market position of the companies—evidently did its work in unblocking the road toward a settlement satisfactory to Saudi Arabia. In fact, the companies obviously had little choice but to yield—though they might well have gotten a better deal from their point of view had they agreed to come to terms earlier.[23]

In the early days of December 1974, it became clear that the Saudi tactics had paid off. Informal indications from authoritative Saudi sources suggested that the government received an offer in writing from the four foreign owners of Aramco (Exxon, Socal, Texaco, and Mobil) that would, in principle, satisfy Saudi Arabia's requirements. Some major points of detail remain to be finally settled, but it is anticipated that these will be dealt with speedily in the near future.

In general terms, the shape of the new deal between Saudi Arabia and the oil companies is expected to be somewhat as follows:

1. The Saudi Arabian Government will take over the Aramco operation 100 percent. The cost of acquiring the venture will be settled along the same lines as those applicable in the Gulf 60-40 participation agreements, that is to say more or less on the basis of net book value (currently estimated at $2 billion).

2. The former owners of Aramco will be guaranteed access to a certain volume of crude oil—the actual quantitiy has not yet been

fixed—at the prevailing market prices. That is to say that the government will be selling its crude oil to the former Aramco owners and independent third parties at the same price, without discrimination.

    3. The former owners of Aramco will continue to provide whatever management, operational, and technical services may be required by the government. In compensation of these services, the companies will receive a fee, the amount of which will be determined in the light of a proper appraisal of the value of the services in question. However, it should also be noted that the role of the new "Saudi Aramco" is likely to extend beyond the confines of oil production and export operations to other infrastructural and industrial ventures in the kingdom. This will obviously broaden the scope of the technological services to be provided by the oil companies, as well as adding a new dimension to the projected Saudi-Aramco deal, which is unprecedented in any other oil agreement elsewhere.

    4. Separate arrangements will be made for continued exploration by the former Aramco owners in the Aramco area to provide for an appropriate return on the exploration investment that the government would still require the companies to make.

Such a radical new deal in the world's largest oil exporting country will obviously have far-reaching implications, notably:

    1. Following the Abu Dhabi decision to raise tax and royalty rates, Saudi Arabia was planning to offer 40 percent of Aramco production (that is, two-thirds of the government's 60 percent share)—amounting to some 3.4 million Bbls/day on the basis of the current output of 8.5 million—for direct sale on the world market to independent third parties, specifically excluding the Aramco owners and the other international majors. The price fixed for such direct sales would have been significantly lower than the previously mooted level of 93 percent of the new posting. The result of such an overwhelming flow of government crude onto the open market at a price substantially below the price of buy-back crude to the Aramco owners (94.8 percent of postings)—and taking into account the steep rise in tax-paid cost for equity crude consequent upon the Abu Dhabi measures—would have been nothing less than the destruction of the majors' traditional domination of the international oil market: first by depriving them at a stroke of a very large slice, in terms of volume, of the world oil trade; and second by reducing their profit on crude oil production to a negative margin. Following the breakthrough on the Aramco front, this plan is no longer needed. Saudi Arabia will in due course be making additional direct sales of crude oil on a government-to-government basis or to independent companies, but not in such a spectacular fashion as was called for in the post-Abu Dhabi plan.

2. The 100 percent deal in Saudi Arabia will naturally extend to the existing participation deals in the Gulf and elsewhere. Already Abu Dhabi, Kuwait, and Qatar have publicly announced their intention to start negotiating similar deals with the concessionary companies concerned.

3. It is obvious that the new 100 percent participation deal will render obsolete the prevailing fiscal structure based on the government's tax and royalty take on equity crude oil calculated from national posted/tax reference prices. This actually explains the fact that Saudi Arabia, joined by Qatar and the United Arab Emirates (Abu Dhabi), have practically and provisionally dissociated themselves from the new pricing system decided by OPEC in Vienna on December 13, 1974, pending the formal conclusions of the 100 percent participations agreements, the pricing previsions of which would be implemented with a retroactive effect to January 1, 1975.

Further negotiations are needed to agree on a number of important points of detail before a complete picture of the new arrangement in Saudi Arabia can emerge, namely the size of the fee to be paid to the former owner companies for services to the new "Saudi-owned Aramco," the volume of guaranteed crude oil deliveries to the companies, the determination of the market price at which the companies will purchase the oil, and the separate arrangements to be made vis-à-vis exploration.

One of the most interesting and distinctive features of the projected deal in Saudi Arabia is that the new Saudi Aramco is slated to play a major role in the kingdom's development program beyond the purely oil sector. Aramco, it should be noted, is unique among the operating consortia in the Middle East in that it is a fully developed corporate structure in its own right—not just a skeleton service company for its owners. This being so, it obviously makes sense for the Saudis to utilize to the full this existing structure—with its access to the technology of four of the world's major energy companies—while developing Saudi personnel to take over.

Thus the Saudi Aramco could become a technological powerhouse for the development of the kingdom in a variety of fields. For example, it is already possible that Aramco may be entrusted by the government with the establishment and management of a comprehensive electric power network for the whole of the Eastern Province and Riyadh.

One of the main aims of the Saudi strategists who drew up the 100 percent take-over scheme was to establish a durable arrangement that, owing to its distinctive nature and particular advantages to Saudi Arabia, could not be overtaken or outflanked by any new oil agreements in other producing countries. It now looks as though the Saudis are well on the way to achieving that objective.

Bilateral Direct Supply Deals

The successful implementation of embargo and production cutback measures by the Arab producing countries created a sort of panic psychosis in the consuming industrialized countries who feared that the major oil companies would no longer be able to meet satisfactorily all of their requirements. In particular, those countries who have adopted pro-Arab attitudes since the early days of the conflict were anxious to get the dividend of such political investment by enjoying a privileged treatment from the Arab producers, especially since the reallocation of resources by the companies in order to keep supplying the countries subject to Arab embargo would deprive them of a part of their normal entitlements.

Under such conditions of acute shortage, securing direct sources of supply became the main preoccupation regardless of the eventual cost. Independent refiners, which were practically no longer supplied by the majors, rushed to the only source left available to them, namely, occasional auction sales by producing countries, thus contributing to the skyroketing of price levels. France was the first country to benefit from its good relations with the Arab countries by signing in December 1973 a direct government-to-government deal with Saudi Arabia for the supply of about 27 million tons of crude oil over three years. This deal was the precursor of a much broader agreement negotiated at top political levels that was then believed to stretch over 20 years and to involve as much as 800 million tons of Saudi crude. This agreement, which has not yet been finalized, is likely to be concluded after the 100 percent take-over of Aramco by Saudi Arabia.

On January 17, 1974, Japan's Ministry of International Trade and Industry (MITI) signed an agreement in Baghdad under which a $1 billion loan was granted to Iraq to be used in the implementation of a number of oil and industrial projects. The loan is to be reimbursed by the supply of some 160 million tons of crude oil and petroleum products over 10 years. Other European countries know to be anxious to secure their crude oil supplies via direct deals— mainly Britain, West Germany, and Italy—were reported to have attempted such deals in Iran, Saudi Arabia, Kuwait, and Libya without any apparent success.

These moves were strongly opposed by the United States, which was actively engated in a worldwide campaign to solve the current energy crisis and to oppose the mounting hegemony of exporting countries. At a press conference held in Washington on January 10, 1974, U.S. Secretary of State Henry Kissinger emphasized the determined U.S. opposition to negotiations by individual countries seeking bilateral agreements with oil-producing countries. Although the

United States, with its substantial resources, is in a better position than most to engage in such negotiations, it agreed that "unrestricted bilateral competition would be ruinous in terms of world stability and economy."[24]

Once the crisis was virtually over and the supply conditions came back to normal, the attractiveness of such direct deals became less evident. Because of the overwhelming control of downstream operations by the majors, few European countries were able to impose a certain "type" of oil to the international companies' subsidiaries operating on their territories as is the case in France. Furthermore, direct deals do not provide any price advantage or guarantees, and the terms of trade on a barter basis cover limited domains that reduce proportionally the expected balance of payments improvement. In fact, direct government-to-government sales proved to be more costly than traditional supplies by the majors since they were generally invoiced at 93 percent of posted prices, which was substantially higher than the companies' average cost as outlined earlier when reviewing the price evaluation in 1974.

This situation is likely to change in the future subsequent to the implementation of the new single-price system that will drastically reduce the companies' margin and will induce them to limit their liftings to their own integrated requirements. More and more oil will become available to the national oil companies of the exporting countries to be sold through direct deals that will be concluded essentially on commercial basis.

Production Programming

One of the means contemplated by the consuming countries to curb crude oil prices was to restrict consumption with a view to having slackening demand exert a downward pressure on prices. Candidly or tactically it was presumed that by both reducing consumption and avoiding waste a significant cutback in oil demand in the industrialized countries—such as the 15 percent reportedly suggested by the United States at the meeting of the finance ministers of the United States, Britain, France, West Germany, and Japan held in Washington at the end of September 1974—would create such a massive surplus-producing and export capacity in the OPEC countries that sales competition would inevitably ensue among them, thereby splitting the group's solidarity and bringing about a collapse of prices.

This was the official U.S. doctrine regarding ways and means to bring about a price reduction. Calling for the constitution of a consumer's front, President Gerald Ford put his personal prestige in the debate and set a target cut of 1 million Bbls/day in oil imports by the end of 1975. The U.S. Federal Energy Administration estimated production capacity of OPEC at about 37 million Bbls/day as compared to an average production of about 31 million Bbls/day. With a 15 percent cutback in demand it was reckoned that this 6 million Bbls/day idle capacity would be roughly doubled. The basic reasoning behind this doctrine is that where significant idle capacity exists there is always a tendency toward its utilization and henceforth competitive forces would set into action.

Even supposing that any such coordinated demands cutback scheme were to be endorsed and implemented by all the industrial nations, which is quite unlikely, the prospects of success in forcing the producers to lower their prices would seem to be very dim indeed. This scenario tends to minimize the importance of the psychological, not to say mythical, motivations that support the action of exporting countries against any kind of pressure or domination now that they have not only recovered all their rights as sovereign nations but also realized the dominating force of their union. Any attempt to intimidate them would activate their ardor at a time when they are utterly convinced of their strength. From a practical view point, considering that oil revenues far exceed what exporting countries can ever spend or invest, reducing demand would be welcomed as a happy opportunity to conserve their depletable petroleum reserves and to relieve them from the pressure of cumbersome money they do not need. The desire to consolidate oil prices at their high levels would be so strong as to prompt them to agree on a coordinated plan to reduce production accordingly. This was clearly formulated by most OPEC members in the different conferences that took place in 1974 as recalled above, both to counter the threat of concerted action from the consumers as well as to oppose any eventual Saudi action aiming at increasing production unilaterally with a view to lowering prices. Nonetheless, Saudi Arabia proved to be as determined as other OPEC member countries in opposing the U.S. plan.

A few days after President Ford's speech, Saudi Oil Minister Yamani bluntly stated in a panel discussion in Washington on October 4, that "conservation alone will never put real pressure on OPEC," pointing out that there is no conceivable level of consumption cutback that could cause the producers financial hardship, since Saudi Arabia and Iran alone could reduce about 5.5 million and 2 million Bbls/day respectively and still have sufficient revenues.[25]

TOWARD A NEW INTERNATIONAL EQUILIBRIUM

It is reasonable to predict that crude oil prices are most unlikely to undergo any significant decrease in the near future. The "oil revolution" that took place in October 1973 after a three-year period of gestation is a fact of life that cannot be reversed even if the factors that contributed to its build-up were to revert to normal. The economic rationale used by OPEC during the 1960s to call for and justify the necessity of price increase has been finally accepted a posteriori as a valid argument by both oil companies and consuming countries. Inflation, oil companies' huge profits, compensation for many years of previous underpricing and the sovereign rights of nations to dispose of their natural resources at best conditions to finance their development are generally considered by producing countries as sufficient justification for the increase of oil prices that have been kept almost unchanged for about two decades. It might be argued, however, that the precipitous quadrupling of oil prices in October 1973 and the ten-fold increase of oil revenues since the Teheran-Tripoli Agreements in early 1971 have gone far beyond any reasonable compensation for past injustice and do represent a real danger to world economic and monetary stability. The October war and subsequent Arab embargo are seen by some observers as the direct cause of the "oil revolution" in that it has created the proper environment for OPEC to impose its own pricing criteria and level. In fact, the Arab-Israeli conflict acted as a catalyst triggering an explosion that otherwise would have occurred more smoothly. It has actually caused oil prices to rise much higher in a shorter period of time, thus adding several dollars per barrel to what would have been a more acceptable price level.

The question is whether Arab oil prices would be reduced if the Arab-Israeli conflict is satisfactorily settled. If they do, prices of non-Arab oil will have to follow, the move and the world would be relieved from a major financial problem.

The leaders of the Arab oil countries have never officially promised a specific price reduction in return for settlement of the Arab-Israeli dispute on their stated terms. But they have indicated that after such a settlement they would be far more amenable to it than now. This is not surprising. Nothing has higher priority in the eyes of the Arabs than regaining all the occupied territories and repatriating the Palestinian refugees. These goals would be well worth a few dollars a barrel, particularly since of the seven Arab OPEC members only two—Algeria and Iraq—can actually absorb all the oil revenues they currently receive. The problem is that neither the United States nor the other major Western industrial nations are willing, or able, to guarantee to the Arabs their maximum demands in return for a lower oil price.

Hence, if there is to be a true settlement, rather than the imposition of terms by the stronger party on the weaker one, it will have to result from give and take on both sides with neither side satisfied with the outcome and the settlement accepted, at least initially, reluctantly, provisionally, and with many external and internal reservations. Hopefully, both sides will eventually learn to live with such a settlement. But when it is signed or otherwise agreed on, the Arabs will hardly feel a moral obligation to lower the oil prices in gratitude for the help received from the West.

An equally if not more important consideration in not lowering the price of oil in the wake of an Arab-Israeli settlement is that such an action could bring about the demise of OPEC, the most successful international cartel in modern times. This instrument enabled its members to move from relative poverty to real riches within a matter of years. The Arab members with surplus oil revenues—Saudi Arabia, Kuwait, Abu Dhabi, Qatar, and Libya—would be opposed by the eight other members of the organization who want oil prices to stay at least where they are now but preferably to move up in line with world inflationary trends. Six of these members—including Iran and Venezuela, the second and third largest exporters—are not Arabs and have, therefore, nothing to gain by accepting any price cut in return for a settlement of the Arab-Israeli dispute. Algeria and Iraq would oppose the price cut for economic reasons. The fact that Iraq refused to join the Arab production curtailments in October 1973 and actually increased its exports during the embargo period is a clear indication of that country's priorities.

If OPEC breaks up because of Arab insistence on lowering prices for political reasons, the price decline could become uncontrollable, subject only to market forces. No Arab oil producer would want to bring this about. The notion that Saudi Arabia, the most likely proponent of a price reduction related to an Arab-Israeli settlement, could single-handedly bring down the OPEC price because of its vast production potential is somewhat exaggerated. An increase in export capacity by 2.5-3.5 million Bbls/day, which is all the country would be physically capable of in the next three years, could be largely offset by corresponding reductions in countries with surplus oil revenues.[26]

A sustained greater increase over and above commercial requirements would take a long time to develop and would also be opposed by most of Saudi Arabia's upper-echelon technical and financial administrators, some of whom consider even current production levels excessive. They would certainly find it difficult to justify the enormous cost of creating this capacity for the sole purpose of forcing prices down.

Predictions of a substantial and durable reduction in Arab oil prices following a hypothetical Arab-Israeli settle ment are most unrealistic. This would have been possible had such settlement occurred in the few months following the price increase. It is now already too late for such a deal since the world has progressively, although reluctantly, adjusted to the new price levels, and the so-called unbearable increase in energy cost has already been overshadowed and "digested" by the two-digit inflation currently sweeping the world. World-wide wages as well as prices of services and commodities have substantially risen in such a way that a decrease in oil prices would appear anachronistic.

Another question is whether oil prices are likely to decrease as a result of intimidation campaigns launched by the consuming nations, especially the United States.

From the beginning the United States has adopted a hard-line policy against OPEC pricing initiatives, and especially against the Arab exporting countries who have imposed total embargo on oil exportation to the United States, which openly played an active role in supporting Israel against the Arabs politically as well as by arranging a massive airlift of military supplies during the war. The official U.S. doctrine was to convene the consuming nations to form a unified front that would be strong enough as to force the "OPEC cartel" to an acceptable compromise. Both President Nixon and Secretary Kissinger put their prestige in the battle by taking personal initiatives to launch the crusade against the exporting countries. In a letter sent on January 9, 1974, to the heads of government of eight industrialized countries, President Nixon invited them to attend a meeting at the foreign minister level on February 11 to work out a "consumer action program." "A concerted effort of this kind is but a first and essential step toward the establishment of new arrangements for international energy and related economic matters. To this end, a meeting of consumer and producer representatives would be held within 90 days."

At the same time, President Nixon sent a letter on January 9 to the governments of the OPEC member countries expressing the hope that "the results of the forthcoming meeting will lead to an early joint conference of consumer and producer nations."

These initiatives were strongly opposed by most of the producing countries who warned against any attempt to create a confrontation between producer and consumer blocs. President Boumediene of Algeria declared that the U.S. proposals "have a precise meaning. They aim to establish a new form of protectorate or condominium over part of our raw materials and production. We are determined to fight all attempts of this nature."[27] The Shah of Iran bluntly threatened a cutback in Iranian oil production (not subject to any embargo) if any attempt were made by the consumers to bring pressure on Iran.[28]

The Washington conference took place on February 11, 1974, in which the United States announced a seven-point program to solve the world energy crisis, providing in particular for the sharing of available sources in emergency cases. This conference actually led nowhere, and the proposed consumer-producer meeting never took place. Failing to attain this objective, the United States called for the institution of a formal body grouping the consuming nations with the aim of coordinating their policies and action. The International Energy Agency was created in September 1974 within the OECD in Paris, grouping 16 member countries under the leadership of the United States.

France was absent from this Paris-based organization. It has actually opposed the U.S. views all the way and took direct initiatives to help elaborating long-term solutions acceptable a priori to the exporting countries. Opposing the hard confrontation line sponsored by the United States, President Giscard d'Estaing, called for an international conference grouping consuming, exporting, and developing nations. This initiative was welcomed by most OPEC member countries, and the French policy of accommodation stands a better chance than the aggressive attitude of the United States.

Political confrontation and economic retaliation cannot force OPEC members to accept an oil price cutback.

* * *

A full chapter would be needed to review briefly the financial implications and consequences of the "oil revolution" and to analyze the problem of recycling the so-called petrodollars. The initial forecasts made in the early days of 1974 concluded that OPEC oil revenue in 1974 should range between $105 and $115 billion according to different assumptions regarding participation oil. By comparison, OPEC oil revenue in 1972 amounted to about $15 billion, ten times less. By the end of 1974 these forecasts proved to be realistic. Out of this a minimum of $60 billion should be considered as surplus funds exceeding all possibilities of expenditure by the countries concerned.

OPEC revenues are likely to be higher than the above-mentioned figures as a result of the new price system adopted in Vienna on December 13, 1974 setting average government take at $10.12/Bbl and of the prospective 100 percent take-over of oil operating companies. This will certainly more than offset any decrease in world consumption that cannot exceed a certain limit without endangering the Western economies, keeping in mind that extensive development of other sources of energy will be a lengthy and costly process of evolution. Assuming OPEC sales would grow at an annual average rate of 5 percent, it is estimated that accumulated OPEC surplus funds could reach more than $1,000 billion by the early 1980s. These surplus funds will actually be in the hands of a few underpopulated countries, mainly Saudi Arabia ($30 billion revenue per year), Kuwait ($10 billion), Libya ($9 billion), Abu Dhabi ($6 billion), and Qatar ($2.5 billion).

Despite its population and its ambitious development plans, Iran will be able to spare a large part of its $22 billion annual revenue. Some early conservative estimates put the accumulated surplus by 1980 at $400 billion. Quite clearly, any such accumulation of surplus petrofunds (whatever estimate one accepts) is unsustainable. This means that no such movement of funds on this magnitude will actually take place. Something will inevitably go wrong.

The very size of the surplus spread among so few countries is unprecedented and has imposed intolerable strains on international financial mechanisms. The interest payments on the anticipated accumulation of the minimum figure of $400 billion surplus would alone put the system as its exists today under the severest pressure.

However, one important fact is immediately discernible: If OPEC is to run a surplus that will not, by general agreement, go away, then the non-OPEC countries must by definition run a deficit of equal size. How to tackle an imbalance of trade of such a magnitude is beyond imagination. Consumers cannot dispense with OPEC oil but will shortly be unable to pay for it. The recycling of petrofunds has rapidly become one of the most important issues in world affairs.

It is clear, or it should be clear, that no return to the old order is possible. The OPEC victory has become possible because of the cohesiveness of their action at a time when the balance of power shifted into their favor. Their force is tremendously increasing not only because of the consolidation of their early victories, but also because of the rapid build-up of an impressive financial empire that would eventually prove a much more efficient and persuasive weapon than oil itself. The transfer of power has taken place and must be accepted as permanent within the inherent permanence of all institutions. The only danger the oil exporting countries might face in the future is to be their own victims should they ignore the necessity of coming to an acceptable compromise with the consumers. History tells that most revolutions have been victims of extremism.

NOTES

1. MEES, September 28, 1973.
2. Under the active pressure of the Saudi government, the parent companies of Arabian American Oil Company (Aramco) launched a few advertising campaigns in the United States to sensitize American public opinion to the Arab cause in the Middle East conflict. Thus, Mobil Oil published an announcement in the New York Times of June 21, 1973, and a former Exxon executive clearly sided with the Arabs in a public conference (MEES, July 6, 1973). On July 26, 1973, the chairman of Standard Oil of California addressed a letter to the com-

pany's 300,000 or so shareholders and employees stressing the necessity for the United States to revise its policy toward the Arab world and to adopt a more even-handed attitude in the Mid-East conflict (Arab Oil & Gas, August 16, 1973). This initiative raised very strong reactions from the Jewish community of California (Financial Times, August 29, 1973).

3. During the 35th Conference, held in Vienna on September 15 and 16, 1973, Resolution XXXV.160 was adopted providing that "The Conference resolves:

> 1. that the Member Countries concerned shall negotiate, individually or collectively, with the oil companies with a view to revising the Tehran, Tripoli and Lagos Agreements in the light of the prevailing conditions and expected future trends in the crude oil and oil product markets, as well as the world inflation;
> 2. to this end a Ministerial Committee, composed of the Heads of Delegations of the Member Countries bordering the Gulf, be established in order to negotiate collectively the revision of the terms of the Teheran Agreement with the representatives of the companies, on the 8th of October, 1973, in Vienna; and
> 3. to empower the said Committee to call for an Extraordinary Meeting of the Conference if it is deemed necessary.

4. Le Monde, October 18, 1973.
5. MEES, October 19, 1973.
6. MEES, November 16, 1973.
7. MEES, November 2, 1973.
8. The text of the communiqué published in Kuwait on October 16, 1973, in the name of the six Gulf producing countries, specifies that

> in case the oil companies refuse to lift crudes on the basis of these arrangements, the producing countries will make available to any buyer the Arabian Light at $3.65/barrel FOB Ras Tanura.
>
> The new posted prices are based on actual market prices in the Gulf as well as in other areas, corrected for gravity differentials and geographic location.
>
> From this day on, actual market prices will determine the level of the corresponding posted prices, maintaining the same relationship between the two prices as existed in 1971 before the Teheran agreement.

The correction for changing posted prices upwards or downwards will be made when the actual market prices of crude oil exceed or drop below the corresponding level of the new announced prices by one percent

The Geneva agreements (on the compensation for monetary movements, notably for the dollar devaluation) shall continue to be in force.

9. PIW, November 12, 1973.

10. According to Petroleum Intelligence Weekly of December 17, 1973, the sales prices of Iranian crudes were the following:

Crude Oil	Gravity (API°)	Sulfur (%)	Price ($1,661)
Rostam	36.7	1.61	17.34
Iranian Light	34.0	1.35	17.04
Iranian Heavy	31.0	1.58	16.04
Sassal	34.3	1.90	16.20
Darius	33.2	2.50	16.00
Bahrengansan	32.0	–	14.00
Cyrus	19.0	–	9.00

11. PIW, December 31, 1973, p. 5
12. MEES, January 11, 1974, Supplement.
13. PIW, December 31, 1973.
14. Le Monde, December 23-24, 1973.
15. MEES, December 28, 1973.
16. Official OPEC communiqué published on December 23, 1973 in Teheran.
17. MEES, November 18, 1974.
18. MEES, December 20, 1974.
19. MEES, December 13, 1974.
20. PIW, December 23, 1974.
21. MEES, December 20, 1974.
22. MEES, April 19, 1974.
23. MEES, December 6, 1974.
24. MEES, January 4, 1974.
25. MEES, October 11, 1974.
26. These figures are estimates by John Lichtblau of the Petroleum Industry Research Foundation, as indicated in PIW, October 21, 1974.
27. Le Monde, February 5, 1974.
28. The Daily Telegraph, February 6, 1974.

APPENDIXES

APPENDIX A

E S S O

Esso International Inc.
60 West 49th Street
New York 20, N.Y.

Cargo Sales Department
I. C. Anderson—Vice President

February 26, 1962

Dear Sir,

    Attached are two copies of a memorandum embodying most of the points we discussed recently in London regarding the posted price of Zelten crude oil.

    In the attached memorandum we have stated that it is normal to post a price for any new crude oil that may become available. We believe there are sound reasons for so doing. For example, it tends to establish a wide recognition of that particular crude oil's value compared with crude oils from other sources. In the long run, it also promotes more stable and orderly marketing. It is recognized that there will be periods when industry surpluses will exist, such as today, and competition (including Iron Curtain countries) will necessitate discounts off published posted prices. Such situations do not void the desirability of maintaining a published posted price, but in practice, may make it easier to cope competitively with discounts, as discounts will tend to conform to some uniformity.

    We realise that a great deal more could be written on the subject of posting prices, but we desired to keep the attached memorandum to a reasonable length. We hope you will find it of value.

Very truly yours,

I. C. Anderson

ICA: com
Att. 2

ZELTEN POSTED PRICES

When a new source of crude oil is discovered and this crude oil is ready to be marketed, it is customary to establish a posted price for this particular oil.

There are several bases on which to approach the matter of establishing a new posting. To any given refiner, for example, the value of crude oil depends upon his refining equipment, efficiency, etc., and his realization from the sale of refined products. Obviously the results vary significantly from one refiner to another. No one posting, therefore, will be ideal for all refiners.

In earlier days the U.S. was the major crude producer as well as consumer. Over the course of time the posted price for crude developed some sort of relationship with product prices even though the relationship did not stay constant. Although the U.S. has long ceased to be an important petroleum exporting country, a good deal of crude and product pricing in other areas of the world reflect the early influence of U.S. prices. Geography, over-all supply versus demand, and greatly changed transportations rates have also influenced today's price levels.

As Esso Libya approached the time they would have crude oil for export, the matter of a posting for such crude received a great deal of attention. Early samples of the crude were analysed in the laboratory and in small scale pilot units. This was done to obtain the best possible feel of what the crude would be like and what a refiner could expect to get out of it. It will be appreciated that such results only approximate later results in large scale commercial units which will be run under various conditions.

Even though it would have been possible to try to estimate the probable value of the new crude to widely different refiners in various geographical locations and attempt to strike some sort of an average, this was not the approach taken. For various reasons it was decided to develop a posting for Zelten crude which reflected, to the best of our ability, its proper relationship to the postings of those crude oils with which it had to compete for market outlet. In so doing, the character of the crude and its geographical locations were taken into account.

The location of Zelten is a distinct advantage. The laboratory work previously referred to, however, soon told us that Zelten was quite different from competing Middle East crudes, even those approximating it in gravity. It was obvious that it contained more gasoline and less fuel oil than the average European demand pattern required. This is a market where gasoline already sells at distress prices and is a problem to refiners. The one striking feature of Zelten is its low sulphur content. The problem then is to be able to use this characteristic effectively in view of the high wax content which particularly effects the handling of fuel oil. The problem is highlighted when it is considered that Zelten fuel oil is solid—will not flow through pipelines until melted by heat—in most European countries, even in the summertime!

It was felt that in setting the initial posting for Zelten, we could adjust for geography with a great deal more assurance than for quality. To get a feel of the real net effect of the positive and negative features requires considerable commercial experience. We felt the volume of production for a while could be handled by shipping to a number of our affiliates, thus minimising the high wax content problem. Accordingly, it was felt that the established price could bear a direct relationship, without further quality adjustment, with those crude oils from the Middle East which Zelten would seek to replace. In this comparison, crudes such as Kuwait, Iraq, Arabian and Iranian most certainly have to be taken into account.

The oil industry today is aware that the prices posted at Eastern Mediterranean reflect obsolete ocean transportation rates and should be adjusted downward to be in proper relationship with the Persian Gulf posting. Nevertheless, since the Sidon and Banias postings have not as yet been so adjusted we used them in our study. The following table shows the parity posted price of Zelten related to the present postings for the crudes mentioned above, corrected for differences in geography. Actually all the crudes, other than Kuwait, give excess gasoline yields compared with demand in Europe and so no gravity adjustments have been made; the price of the Middle East crudes used is exactly as published today.

	31° API Kuwait	34° API Arabian Ras Tanura-Sidon	34° API Iranian	36° API Iraq	41 API Qatar
Posted Price FOB Loading Port, $/B	1.59	1.80	2.17	1.73	2.21	1.95
Frt. to Rotterdam	.60	.58	.26	.69	.26	.57
C & F Rotterdam	2.19	2.38	2.43	2.42	2.47	2.52
Frt. Marsa el Brega/ Rotterdam	.22	.22	.22	.22	.22	.22
Parity Posting FOB Marsa el Brega	1.97	2.16	2.21	2.20	2.25	2.30

If the various parity postings in the above table are averaged the number would indicate about $2.18 for Zelten at Brega. It was further considered, however, that Qatar is less well received than the other crudes mentioned and that Kuwait has a built-in quality adjustment. Eliminating these two crudes results in an average of $2.21 and it was decided to use this figure for the initial posting for Zelten. It should be mentioned that this figure also relates to the best present indication of term charter rates for tankers. This rate appears to be about Scale III—50 percent as evidenced by many charters made in the tanker market.

In the absence of commercial experiences, we have made limited attempts to sell Zelten to non-affiliated companies and then only at full posted price. To date, no interest by customers has been shown on such basis. Late data, from our affiliated refineries show about the same economic results when buying Zelten at $2.21 FOB Brega as when buying Iraq at $2.21 FOB Banias. The results on both Zelten and Iraq in our refining picture are poorer than those obtained when processing Kuwait, Arabian or Iranian at their respective posted prices, the disadvantage in some cases being nearly 20¢/barrel.

Crude Quality Differences

The quality differences between Zelten and Middle East crudes have been referred to above. It may be desirable to illustrate in more detail how these differences exert an influence on the setting of a posted price.

Zelten is higher gravity, lower sulphur, and contains more wax than the principal Middle East crudes. For example, the gas oil product from Zelten contains about twice the amount of wax as a similar product from Arabian or Kuwait crudes, giving both cloud and pour difficulties. To satisfy market requirements, therefore, it may be necessary to further refine the product or to blend the Zelten distillate off with lower pour material. The wax is also heavily concentrated in the fuel oil as shown in the following table expressed in terms of the temperature at which the fuel oil will flow.

Crude	Pour Point 650-plus° Bottoms
Arabian	45° F
Kuwait	65
Light Iranian	65
Hassi Messaoud	65
Iraq	85 Est.
Zelten	105 (90)
Feul Oil Specification	65-70

The high wax content of Zelten also introduces another very interesting effect. The density of paraffinic compounds is less than naphthenic or aromatic compounds. The result is that the API Gravity of Zelten is higher than would be deduced from the product yields obtained from the crude. From a yield standpoint, Zelten performs more like a 35 to 36° API crude than like 39 to 40° API. This . . . was one of the main reasons for giving less effect to gravity in pricing the crude than would be expected from the usual conventional treatment. . . . It will be noted that 39 to 40° API Zelten contains only about as much gasoline as a 35.5° API average Middle East crude would yield under general European type refining. Its fuel oil yield, on the other hand, is greater than would be expected from its API gravity and approximates the yield from a 36.5° API average Middle East crude.

The very low sulphur content of Zelten crude is of interest to any prospective customer. He will try to gain advantage from this feature in at least two ways. Firstly to the extent that the difficulty in handling the high pour fuel can be overcome, particularly in winter, the fuel may command a few cents per barrel premium over regular heavy fuel. Unfortunately, such outlets are limited and will take care of only a small amount of the available fuel from processing Zelten. The second use is in blending off some of the higher sulphur fuel oils derived from certain Middle East crude oils to make a more desirable over-all fuel oil product. The value of this in terms of a price penalty for high sulphur is not great but it does facilitate selling the fuel oil in competition with others who may not have such blending stock available.

Another feature which has a very practical meaning to a purchaser is the fact that not as much Zelten can be processed in the same equipment that may now be used for Arabian or Kuwait crudes. For example, our experience has indicated nearly a 10 percent loss in distillation capacity when processing Zelten rather than Arabian. This increases investment and operating costs and represents a real debit for the Zelten crude. This effect may vary from refiner to refiner depending upon the exact nature of their facilities.

Overall Conclusion

Any particular quality feature of a given crude oil has a different economic impact on different refiners and no single debit or credit, in terms of cents per barrel, is correct for all purchasers. In the case of Zelten, it is felt that its debits compared with Middle East crudes such as Kuwait, Arabian, Iraq and Iranian tend to outweigh its main advantage of low sulphur. Because of the difficulty of pinpointing this difference in terms of price, it has only been reflected in effect by the emphasis given to its API relationship to the other crudes. It is felt that if anything, this has been the fairest approach that can be taken in the setting of the Zelten posting. Geographical differences, of course were fully reflected.

ICANDERSON: com
2-23-62

APPENDIX B

MAIN FEATURES OF
THE PARTICIPATION AGREEMENT

Following is the text of an explanatory memorandum released by the international oil companies to present and explain the complex "general agreement" for government participation. The full text of the agreement itself was published as a supplement to Middle East Economic Survey dated December 22, 1972 and to Petroleum Intelligence Weekly dated December 25, 1972.

1. Status of Participation Negotiations

As conceived by the producing-country governments, participation means a share in the ownership of the operations, providing an active role in the management and the right to export their equity share of the production.

In order to achieve this objective, five Gulf producing states (Abu Dhabi, Iraq, Kuwait, Qatar, and Saudi Arabia) designated Sheikh Ahmed Zaki Yamani as their negotiator, who insisted that the negotiations be conducted on an industry basis. The industry designated G. T. Piercy of Exxon, A. C. DeCrane, Jr. of Texaco, and C. C. Pocock of Shell as their negotiating team.

The "General Agreement" which has just been signed by Saudi Arabia and Abu Dhabi, the respective concession holders, and their shareholders sets out the principles and conditions under which the governments may buy their interests and specifies that participation will become effective January 1, 1973. Detailed implementing agreements must now be agreed to by each concession holder and its host government.

The agreement provides that these Gulf states will acquire an interest, within their borders, in the crude oil and natural gas producing operations and facilities. There is no provision in this agreement for participation in other operations such as transportation or refineries, and this has been left for future individual country negotiations.

2. Participation Schedule

The companies agreed to increase the initial level from the 20 percent they had previously proposed to 25 percent provided that that level would be maintained until 1978. This plateau at the initial level facilitated the establishment of a system for disposition of the governments' crude which would assure continued supplies from the producing countries involved.

The host governments may elect to acquire additional increments of participation as set out in the schedule below:

Effective Date	Percentage Increment	Total Percentage Level of Participation
Initial Increment	25	25
Jan. 1, 1978	5	30
Jan. 1, 1979	5	35
Jan. 1, 1980	5	40
Jan. 1, 1981	5	45
Jan. 1, 1982	6	51

3. Compensation

Establishment of the principles to be used in determining the amount of compensation for the participation increments was the result of hard bargaining on the part of both sides.

The agreement provides that all of the companies' unrecovered investments in the portion of the business that the countries are purchasing would be returned to them, making due allowance for inflation that has occurred since the investments were made. Thus, for each year, the total expenditures for exploration, development, and other capital investments will be determined. From this figure an amount will be deducted which is equal to the taxes saved by virtue of the governments having recognized depreciation and amortization of these expenditures. The net expenditures (total expenditures less taxes saved) are then to be adjusted to the value of today's dollar by using an agreed Middle East construction price index.

The compensation to be paid by each of the Gulf states will be certified by an outside auditor and the amount set out in the respective implementing agreements. Payment will be made either in U.S. dollars or sterling as specified in each implementing agreement. Compensation for future increments of participation will be determined using the principles described above with appropriate updating of the figures as required.

4. Disposition of Oil

Principles: The agreement contains a complex set of interrelated provisions which are designed to:

(a) enable each of the parties to have access to the quantities of crude needed to meet their current and forecasted requirements.
(b) provide a market for sizeable volumes of the government's oil during a transition period.
(c) give the operating company adequate notice to enable it to install additional capacity as may be required to meet the forecasted needs of all the parties.

How the System Works: A detailed notification, scheduling and crude purchase/sale system has been conceived in order to accomplish these objectives. For convenience, various portions of the oil under consideration have been designated as described below. The volumes of each of these classifications of oil are interrelated, and are stipulated so that, in total, the needs of the parties are described above will be met. While there is considerable flexibility as to the quantities of oil the governments may elect to export, there are provisions which prevent sudden and drastic changes in the volumes which the companies will be called upon to handle.

Basic Right: Both the companies and the government will have a right to their equity shares of the total availability of each grade of crude oil.

Bridging Oil: During the first three years of the agreement, each company will purchase certain quantities of "bridging oil" from its respective government. This purchase is intended to assist the company in meeting its current supply obligations. The volumes are a declining percentage of the government's basic right to the crude oil.

Phase-in Oil: In addition, the companies are required to accept certain maximum quantities of "phase-in" oil from the respective countries, at the countries' option. This oil will be marketed by the companies in order to provide the producer countries with an outlet for their oil, while the countries are developing their own markets. Each increment of government participation carries with it an obligation on the part of the companies to accept additional quantities of phase-in oil over a period of 10 years. The countries have elected to require the companies to purchase the maximum quantities for the first four years of the agreement.

Forward Avails: During the planning period any party may submit a forecast of his own requirements and the operating company will construct capacity equal to the sum of all these requirements. To the extent that any party's requirements exceed his entitlements under other provisions of the agreement, he will purchase appropriate volumes of "forward avails" oil at an agreed price. Here again, though, once a "forward avails" quantity has been committed for sale, there is an appropriate scaling down period to prevent sudden disruption of a given supply.

Overlift Oil: If, at the end of a calendar year, one party has chosen to lift more than its entitlement, an adjustment is made by transferring an appropriate volume of "overlift oil" at the agreed price.

Thus, insofar as the concession's reserves will allow, the provisions of the agreement dealing with disposition of oil will serve the needs of all the parties.

5. Costs

The companies will have available to them crude oil under a number of different provisions of the agreements, and each of these categories of crude bear a different cost. The acquisition of crude under some provisions is fixed, under others it is at the option of the countries, and still others operate at the option of the companies.

6. Other Provisions

In addition, the general agreement provides:
(a) that each participating producing country shall have the right to take an active part in management. The extent to which major management decisions will require the agreement of all or the majority of the parties, is left to the implementing agreements. Examples of these major management matters include those dealing with:
—exploration and development programs, capital and operating expenditures.
—sale or disposition of major assets.
—selection of key personnel and employee compensation.
(b) that each party shall bear its share of all expenditures including those required to maintain and expand capacity.
(c) that all existing agreements between the parties will remain in effect in accordance with their terms. These include:
—the concession agreements with all amendments made to this date including: all rights and obligations of the companies; current expiration date.
—the 1971 Teheran and 1972 Geneva posted price agreements.

SELECTED BIBLIOGRAPHY

Adelman, M. A. "Efficiency of Resource Use in Crude Petroleum." Southern Economic Journal, October 1964.

_____. "Oil Prices in the Long Run (1963-75)." Journal of Business of the University of Chicago, April 1964.

_____. The World Oil Outlook. Natural Resources and International Development. Baltimore: John Hopkins Press, 1964.

Ashford, W. H. "The Economics of Large Tankers." Journal of the Institute of Petroleum 53 (September 1967): 290.

Bell, H. S. Petroleum Transportation Handbook. New York: McGraw-Hill, 1963.

Bermudez, A. J. The Mexican National Petroleum Industry: A Case Study in Nationalization. Stanford, Cal.: Stanford University Press, 1963.

Bradley, P. G. The Economics of Crude Petroleum Production. Amsterdam: North-Holland Publishing Company, 1967.

Burrows, J. C., and Domencich, T. A. An Analysis of the United States Oil Import Quota. Lexington, Mass.: D. C. Heath, 1970.

Cassady, Ralph, Jr. Price Making and Price Behavior in the Petroleum Industry. Petroleum Monograph Series 1. New Haven, Conn.: Yale University Press, 1954.

Cattan, Henry. The Evolution of Oil Concessions in the Middle East and North Africa. New York: Oceana Publications, 1967.

Ching Chin Chen. "Crude Oil Prices and the Postwar Japanese Oil Industry." Ph.D. Dissertation, Massachusetts Institute of Technology, 1967.

CPDP, Comité Professionnel du Pétrole. Periodical Publications, Paris.

De Chazeau, M. G., and Kahn A. E. Integration and Competition in the Petroleum Industry. New Haven, Conn.: Yale University Press, 1959.

El-Sayed, Mustafa. L'Organisation des Pays Exportateurs de Pétrole. Paris: Libraire générale de Droit et de Jurisprudence, 1967.

Fontaine, Pierre. Le pétrole du Moyen-Orient et les trusts. Paris: Les Sept Couleurs, 1960.

Frank, Helmut J. Crude Oil Prices in the Middle East: A Study in Oligopolistic Price Behavior. New York: Praeger Publishers, 1966.

Frankel, P. H. Oil: the Facts of Life. London: Weidenfeld and Nicolson, 1962.

─────── . "Structure of World Oil Industry." 2d Management Conference on Economics of Petroleum Distribution, Northwestern University, March 1966.

─────── , and Newton, W. L. "Comparative Evaluation of Crude Oils." Journal of the Institute of Petroleum 56, 547 (January 1970).

─────── . "Economics of Petroleum Refining—Present State and Future Prospects." Journal of the Institute of Petroleum 54, 530 (February 1968).

Granier de Lilliac, R. Evolution des prix du pétrole. Paris: Annales des Mines, September 1966.

Grant, E. L., and Ireson, W. G. Principles of Engineering Economy. New York: Ronald Press, 1964.

Hamilton, Daniel. Competition in Oil. Cambridge: Harvard University Press, 1958.

Hartshorn, J. E. Oil Companies and Governments. 2d ed. London: Faber and Faber, 1967.

Hirst, David. Oil and Public Opinion in the Middle East. London: Faber and Faber, 1966.

Hubbard, Michael. The Economics of Transportation Oil to and Within Europe. London: MacLaren and Sons, 1967.

Institut Français du Pétrole. Coûts de transport et approvisionnements en pétrole brut à long terme. Paris: Ed. Technip, 1970.

Issawi, Charles, and Yeganeh, Mohamed. The Economics of Middle Eastern Oil. New York: Praeger Publishers, 1962.

Jacobs, John I. and Co. World Tanker Fleet Review (London).

Julien, Claude. L'Empire Américain. Paris: Ed. Grasset, 1968.

Kubbah, A. A. Q. Libya: Its Oil Industry and Economic System. Baghdad: Arab Petro-Economic Research Center, 1964.

Laudrain, Michel. Le Prix du pétrole brut. Paris: Ed. Genin, 1958.

Leeman, Wayne. The Price of Middle East Oil. Ithaca, N.Y.: Cornell University Press, 1962.

Lenczowski, George. Oil and State in the Middle East. Ithaca, N.Y.: Cornell University Press, 1960.

Lipton, M. "Government and Future of World Oil." 2d Management Conference on Economics of Petroleum Distribution, Northwestern University, March 1966.

Longrigg, S. H. Oil in the Middle East. 3d ed. London: Royal Institute of International Affairs, Oxford University Press, 1968.

Lovejoy, W. F., and Homan, P. T. Economic Aspects of Oil Conservation Regulation. Baltimore: John Hopkins Press, 1967.

Lubell, Harold. Middle East Oil Crisis and Western Europe's Energy Supplies. Baltimore: John Hopkins Press, 1963.

Lufti, Ashraf. "Nationalization of Oil: The Problems Involved." Middle East Economic Survey 12, 5 (November 29, 1968).

_____. OPEC Oil. Beirut: Middle East Research and Publishing Center, 1968.

McLean, J. G. "The Importance of the New Comers in the International Oil Business." Middle East Economic Survey 11, 24 (April 12, 1968).

_____, and Haigh, R. W. The Growth of Integrated Oil Companies. Cambridge: Harvard University Press, 1954.

Martinez, Anibal R. Our Gift, Our Oil. Vienna, 1966.

Masseron, J. L'économie des hydrocarbures. Paris: Ed. Technip, 1969.

Mauss, F. Les combustibles liquides. Paris: Ed. Technip, 1963.

Maxwell, J. B. Data Book on Hydrocarbons. Princeton, N. J.: D. Van Nostrand, 1960.

Mendershausen, H. Dollar Shortage and Oil Surplus in 1949-50. Essays in International Finance no. 11. Princeton, N.J.: International Finance Section, Department of Economics, Princeton University, 1950.

Mikdashi, Zuhayr. A Financial Analysis of Middle Eastern Oil Concessions, 1901-65. New York: Praeger Publishers, 1966.

Nelson, W. L. Petroleum Refinery Engineering. 4th ed. New York: McGraw-Hill, 1958.

──────. "Questions on Technology," "Process Costimating," "Productivity Costimating," and other series in Oil and Gas Journal.

O'Connor, Harvey. The Empire of Oil. New York. Monthly Review Press, 1955.

──────. World Oil Crisis. New York: Monthly Review Press, 1962.

Penrose, E. T. "Government Partnership in the Major Concession of the Middle East: The Nature of the Problem." Middle East Economic Survey 11, 44 (August 30, 1968).

──────. "The International Oil Industry in the Middle East." Middle East Economic Survey 11, 40 (August 2, 1968).

──────. The Large International Firm in Developing Countries. London: George Allen and Unwin, 1968.

Perry, J. H. Chemical Business Handbook. New York: McGraw-Hill, 1973.

──────. Chemical Engineers' Handbook. New York: McGraw-Hill, 1963.

Platt's Oil Price Handbook and Oilmanac. New York: McGraw-Hill, yearly.

Richman, P. B. Taxation of Foreign Investment Income. Baltimore: John Hopkins Press, 1963.

Rifaï, T. "Economics of Sulphur Pollution and Its Impact on the International Petroleum Industry." Vienna, OPEC, Seminar, June-July 1969.

Schulman, James. "Transfer Pricing in Multinational Business." Ph.D. Dissertation, Harvard University, 1960.

Schwadran, Benjamin. The Middle East, Oil, and the Great Powers, 1959. New York: Council for Middle Eastern Affairs Press, 1959.

Shell International Petroleum Company Ltd. Current International Oil Pricing Problems. London, 1963.

Stauffer, Thomas R. "The ERAP Agreement: A Study in Marginal Taxation Pricing." 6th Arab Petroleum Congress, Baghdad, 1967.

Stocking, George W. Middle East Oil. Nashville, Tenn.: Vanderbilt University Press, 1970.

Swensrud, Sydney A. "The Relation Between Crude Oil and Product Prices." Bulletin of American Association of Petroleum Geologists 23 (1939): 765-788.

Tanzer, Michael. The Political Economy of International Oil and the Underdeveloped Countries. Boston: Beacon Press, 1969.

Turner, F. C. "Methods of Crude Oil Valuation." OPEC Seminar, Vienna, June-July 1969.

United Nations, Economic Commission for Europe. The Price of Oil in Western Europe. Geneva, March 1955.

U.S., Federal Trade Commission. "The International Petroleum Cartel," Staff Report, 82d Congress Committee Print no. 6, August 22, 1952.

Votaw, Dow. The Six-Legged Dog. Berkeley and Los Angeles: University of California Press, 1964.

Wuithier, P. Raffinage et génie chimique. Paris: Ed. Technip, 1965.

Yamani, A. Z. "Aspects of Oil Policy for the Arab Countries and the Relation of Arab Policy to that of OPEC." Middle East Economic Survey 11, 32 (June 7, 1968).

_____. "Participation Versus Nationalization—A Better Means to Survive." Middle East Economic Survey 12, 33 (June 13, 1969) and 34 (June 20, 1969).

ABOUT THE AUTHOR

TAKI RIFAÏ has worked for over eight years with the Institut Français Du Pétrole (IFP) in various fields that include research, process development and evaluation, industrial economics, feasibility studies, and overall development planning. He has specialized in international petroleum economics with special reference to price structure and pricing policy.

The author was responsible for Middle Eastern affairs in the International Relations Department of the IFP for over five years. He has served occasionally as a UNESCO consultant to a major oil producing country. He held the position of Economic Advisor to the Libyan Ministry of Petroleum during which time he witnessed the Libyan revolution of September 1, 1969.

Presently, Dr. Rifaï is manager at the Banque Arabe et Internationale d'Investissement, a newly created financial institution that is equally subscribed to by Arab and Western interests. The BAII intends to promote business opportunities and to develop economic cooperation between Arab and Western countries. It plans to deal with the complex problems that are likely to arise along with the rapid accumulation of surplus liquidities in the oil producing countries.

Dr. Rifaï was born in Syria and educated in Paris where he graduated from the Ecole Polytechnique and the Ecole Nationale Superiéure Des Mines. He obtained his doctorate in applied physical chemistry from the University of Paris.

RELATED TITLES
Published by
Praeger Special Studies

THE ENERGY CRISIS AND U.S. FOREIGN
POLICY
      Edited by Joseph S. Szyliowicz
      and Bard E. O'Neill*

ISRAEL AND IRAN: Bilateral Relationships and
Effect on the Indian Ocean Basin
      Robert B. Reppa, Sr.

MIDDLE EAST OIL AND U.S. FOREIGN
POLICY: With Special Reference to the U.S.
Energy Crisis
      Shoshana Klebanoff

THE SAUDI ARABIAN ECONOMY
      Ramon Knauerhase

THE SOVIET ENERGY BALANCE: Natural Gas,
Other Fossil Fuels, and Alternative Power
Sources
      Iain F. Elliot

THE UNITED STATES AND INTERNATIONAL
OIL: A Report for the Federal Energy Administration on U.S. Firms and Government Policy
      Robert B. Krueger*

*Also available in paperback as a PSS Student Edition.